ULTRASENSITIVE
LASER SPECTROSCOPY

QUANTUM ELECTRONICS — PRINCIPLES AND APPLICATIONS

A Series of Monographs

EDITED BY

PAUL F. LIAO
Bell Telephone Laboratories
Murray Hill, New Jersey

PAUL KELLEY
Lincoln Laboratory
Massachusetts Institute of Technology
Lexington, Massachusetts

YOH-HAN PAO
Founding Editor

ULTRASENSITIVE LASER SPECTROSCOPY

EDITED BY

DAVID S. KLIGER

Division of Natural Sciences
University of California
Santa Cruz, California

 1983

ACADEMIC PRESS

A Subsidiary of Harcourt Brace Jovanovich, Publishers

New York   London
Paris   San Diego   San Francisco   São Paulo   Sydney   Tokyo   Toronto

ACADEMIC PRESS, INC.
111 Fifth Avenue, New York, New York 10003

United Kingdom Edition published by
ACADEMIC PRESS, INC. (LONDON) LTD.
24/28 Oval Road, London NW1 7DX

Library of Congress Cataloging in Publication Data
Main entry under title:

Ultrasensitive laser spectroscopy.

(Quantum electronics--principles and applica-
tions)
  Includes index.
  1. Laser spectroscopy. I. Kliger, David S.
QC454.L3U44 1983   535.5'8   82-18417
ISBN 0-12-414980-4

PRINTED IN THE UNITED STATES OF AMERICA

83 84 85 86   9 8 7 6 5 4 3 2 1

CONTENTS

v

LIST OF CONTRIBUTORS

Numbers in parentheses indicate the pages on which the authors' contributions begin.

ROBERT R. BIRGE (109)   Department of Chemistry, University of California, Riverside, California 92521

HOWARD L. FANG (175)   The Standard Oil Company (Ohio), Corporate Research Center, Warrensville Heights, Ohio 44128

DONALD M. FRIEDRICH (311)   Department of Chemistry, Hope College, Holland, Michigan 49423

T. D. HARRIS (343, 369)   Bell Laboratories, Murray Hill, New Jersey 07974

F. E. LYTLE (369)   Department of Chemistry, Purdue University, West Lafayette, Indiana 42907

DAVID H. PARKER (233)   Department of Chemistry, University of California, Santa Cruz, California 95064

ROBERT L. SWOFFORD (175)   The Standard Oil Company (Ohio), Corporate Research Center, Warrensville Heights, Ohio 44128

ANDREW C. TAM (1)   IBM Research Laboratories, San Jose, California 95193

PREFACE

The availability of a wide variety of lasers has opened new vistas to the field of spectroscopy over the past decade. Whole new areas such as multiphoton absorption spectroscopy and coherent spectroscopy have been developed which would not be possible with conventional light sources. Much effort has gone into developing techniques which yield spectra with ultrahigh resolution or which study transient processes occurring extremely rapidly. While not a simple or commonplace event, it is well within the present state of the art to study spectra with kilohertz energy resolution or transient spectral events on a dekafemtosecond time scale.

In this volume we will examine a number of spectroscopic techniques which have high sensitivity in another sense. While methods of transient spectroscopy and high-resolution spectroscopy are extremely valuable, this book has focused only on techniques which are particularly sensitive for detecting very weak spectral transitions. These techniques are valuable for studying "forbidden" transitions which have very small transition probabilities or for detecting substances which might be present in very low concentration.

We have concentrated in this volume on those sensitive techniques which have become most popular in recent years. There are undoubtedly other sensitive techniques not covered in this text. Indeed, new techniques are being developed at a remarkable rate. By concentrating on those techniques which have been most widely used, however, we hope to give the reader a sense of the types of methods to which lasers have been applied for the study of weak transitions. Furthermore, this volume emphasizes experimental methods. It does not merely present literature reviews of the different fields discussed but is meant to give the reader a sense of how the different experiments are done and what are the strengths and weaknesses of each method.

The approach most often used to detect very weak absorption signals is to observe phenomena resulting from the absorption process rather than to measure absorption directly. Thus, as pointed out in the chapter discussing analytical applications of these methods by Fred Lytle and Tim Harris, one measures a signal from a small or zero baseline rather than subtracting two large intensity signals to yield a small difference.

In some cases one measures heat produced in a sample as a result of light absorption. Temperature increases as a result of light absorption can be measured directly but it is more common to take advantage of phenomena related to the temperature increase. Thus, Andy Tam explains in his chapter on optoacoustic spectroscopy how one can take advantage of the volume, or pressure, change associated with a temperature increase to measure weak absorption processes. Similarly, advantage can be taken of the change in refractive index with temperature to measure absorptions. This is described, by Howard Fang and Bob Swofford, in terms of the thermal lens effect and, by Don Friedrich, in terms of interferometric techniques.

While the above-mentioned photothermal methods yield very sensitive measures of absorption, it is often easier, for molecules with reasonable emission quantum yields, to use light emission to monitor absorption. This is described by Bob Birge in his chapter on excitation techniques. Another alternative, discussed by Dave Parker, is to detect ions produced as a result of light absorption. At high light intensities available with modern lasers, photoionization can become so sensitive that single-molecule events can be detected. Finally, intracavity techniques are described by Tim Harris. Here one measures absorption itself rather than indirectly related phenomena. This technique gains its great sensitivity by taking advantage of the nonlinear nature of gain and loss effects in laser cavities.

It should be clear by now that the techniques described in this volume vary tremendously. They vary not only in their procedural aspects but also in their applications and general utility. This is reflected in the chapters themselves. Some emphasize a variety of applications while others emphasize theoretical aspects of the techniques. These differences are a reflection of the different states the various fields are in at the moment. The overall aims of each chapter are, however, the same. Each technique is described in sufficient detail to give the reader a feeling for what it involves, the present state of the field, and advantages and disadvantages of the technique. Thus, it is hoped that the reader will be exposed to a variety of modern laser spectroscopic techniques and will see when it might be appropriate to apply each one.

1 PHOTOACOUSTICS: SPECTROSCOPY AND OTHER APPLICATIONS

Andrew C. Tam

IBM Research Laboratories
San Jose, California

Copyright © 1983 by Academic Press, Inc.
All rights of reproduction in any form reserved.
ISBN 0-12-414980-4

I. INTRODUCTION

The photoacoustic (PA) or optoacoustic (OA) effect, i.e., the generation of acoustic waves due to the absorption of modulated electromagnetic waves, is an old effect, discovered by Bell a hundred years ago (Bell, 1880). Besides being an old technique, the effect is also weak in that usually only a very small fraction (much less than one part per million) of the absorbed optical energy is converted into acoustic energy. Despite the fact that the PA effect is old and weak, there has been a great resurgence of interest in it in the past several years. Many theoretical investigations of the various cases of PA generations in gases, liquids, solids, and at interfaces between these media have been made, and new experimental applications of PA effects in various media are reported almost every week in the literature. This great uprise in interest in PA effects is due to the following reasons: (i) Intense light sources (various lasers and arc lamps) have become more readily available in recent years. (ii) Highly sensitive sound detectors (microphones, hydrophones, thin-film piezoelectric detectors, fiber detectors, etc.) have been developed. (iii) The PA technique has been shown to be one of the most sensitive spectroscopic techniques for gas samples as well as for condensed samples. (iv) The PA technique has many unique applications, for example, spectroscopic studies of opaque or powdered materials, studies of energy conversion processes, and novel nondestructive subsurface imaging.

The fact that the field of PA research and applications is rapidly expanding can be appreciated by noting that several review articles have been published recently for example, Colles *et al.* (1979), Dewey (1974), Harshbarger and Robin (1973), Hordvik (1977), Kanstad and Nordal (1980a,b), Kirkbright and Castleden (1980), Patel (1978a), Patel and Tam (1981), Robin (1976), Rosencwaig (1978), and Somoano (1978), and three books have appeared, Pao (1977), Rosencwaig (1980a), and Coufal *et al.* (1982a). The aim of the present chapter is to describe PA generation and applications, with emphasis on the most recent developments [like a noncontact PA detection technique developed by Tam *et al.* (1982)] and on studies not covered in detail in the review articles mentioned above.

Furthermore, recent works on other effects related to the PA effect (e.g., beam-acoustic effect, photorefractive effect, and photothermal radiometric effect) are discussed. These other effects are related to the PA effect because all of them are due to the thermal energy deposited in a sample after the absorption of an incident beam. The author has attempted to make this review reasonably up to date as of Spring, 1982. The literature covered is indicated at the end of this review, in Table IV, which also contains some references not discussed in the body of this article because of space limitations.

The name "optoacoustics" (OA) has also been frequently used for the effect of light-generated sound, especially for the experiments in the detection of small absorptions in gases, and in the use of pulsed light beams for gated piezoelectric detection of weak absorptions in condensed matter. Hence, the name "OA" is also used in the present chapter.

II. PA SPECTROSCOPY IN GASES

A. Theory

A great deal of interest in the OA effect in gases was generated by Kreuzer (1971), who reported that an ultralow gas concentration can be detected by OA monitoring using an infrared laser beam as a light source. A sensitivity limit of a concentration of 10^{-8} of methane in nitrogen was demonstrated, and a limit as low as 10^{-13} could be expected with an improved light source.

The theory of OA generation and detection in gases includes the considerations shown schematically in Fig. 1. Optical absorption is the first step, which results in the production of excited states. For a simple case of a two-level system involving the ground state and the excited state of densities N and N', respectively, we can calculate N' using the following rate equation:

$$dN'/dt = (N - N')R - N'(R + A_r + A_n). \tag{1}$$

Here A_r is the radiative decay rate of the excited state, A_n is the nonradiative decay rate due to collisions of the excited state, and R is the excitation rate due to the light beam of flux Φ photons $cm^{-2} sec^{-1}$ with an absorption cross section σ cm^2:

$$R = \Phi\sigma. \tag{2}$$

In many cases of OA work, the modulation frequency of the light is slow (e.g., ~kHz or less) compared to the excited-state decay rate. Fur-

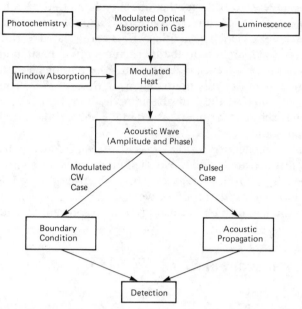

Fig. 1. Block diagram of some factors to be considered in the theory of OA detection.

thermore, the light intensity is usually weak enough so that $N \gg N'$ and the stimulated emission from the excited state can be neglected. Under these approximations of slow modulation and weak light, Eq. (1) reduces to

$$N' = NR/(A_r + A_n) \qquad (3)$$

or

$$N' = N\Phi\sigma\tau, \qquad (4)$$

since $\tau = (A_r + A_n)^{-1}$ is the lifetime of the excited state. The corresponding heat production rate H due to the excited-state density N' (which depends on position $\mathbf{r}$ and time t, because Φ is a function of $\mathbf{r}$ and t) is given by

$$H(\mathbf{r}, t) = N'(\mathbf{r}, t)A_n E', \qquad (5)$$

where E' is the average thermal energy released due to a nonradiative deexcitation collision of the excited state. If the deexcitation collision results in converting the excited state to the ground state, then the deexcitation energy E' is simply the energy of the excited state with respect to the ground state. Equation (5) contains the essence of the heat production term in most OA experiments using slowly chopped light beams in the kilohertz range or less. Similar results have been obtained by Kreuzer (1977). Equation (5) states the intuitively appealing fact that the heat-source term for the OA signal is proportional to the product of molecular density N, photon absorption rate $\Phi\sigma$, probability for nonradiative relaxation of the optically excited state τA_n, and the heat energy released per deexcitation E'.

We stated earlier that Eq. (5) is only applicable for the case in which the sinusoidal modulation (at angular frequency ω) of the optical beam is much slower than the total decay rate τ^{-1} of the excited state. If this condition is not satisfied, then we cannot put $dN'/dt \approx 0$, as is done to obtain Eq. (3), but we may rewrite Eq. (1) as

$$(d/dt + \tau^{-1})N' = NR, \tag{6}$$

where we have again assumed the absence of optical saturation, i.e., we have assumed $N' \ll N$ or $R \ll \tau^{-1}$. The incident light flux is assumed to be sinusoidally modulated, i.e.,

$$\Phi = \Phi_0[1 + \exp(i\omega t)] \tag{7}$$

where only the real part has physical meaning. We may drop the constant part in Eq. (7) since we are only interested in the modulated heat source which generates a corresponding OA signal. The solution of Eqs. (6) and (7) is

$$N' = [N\Phi_0\sigma\tau/(1 + \omega^2\tau^2)^{1/2}]\exp[i(\omega t - \psi)], \tag{8}$$

where

$$\psi = \tan^{-1}(\omega\tau) \tag{9}$$

is the phase lag of the modulation of the excited-state density compared to the optical excitation, and is large when the excited state decays more slowly than the modulation rate of the light intensity. Note that Eq. (8) reduces to Eq. (4) in the limit when $\omega\tau \to 0$. The heat production term H corresponding to Eq. (8) is again given by Eq. (5).

As seen in Fig. 1, the next step in the theory is the generation of acoustic waves by the heat source $H(\mathbf{r}, t)$ of Eq. (5). Morse and Ingard (1968) have given the inhomogeneous wave equation relating the acoustic pressure p and the heat source H:

$$\nabla^2 p - \frac{1}{c^2}\frac{\partial^2 p}{\partial t^2} = -\frac{\gamma - 1}{c^2}\frac{\partial H}{\partial t}, \tag{10}$$

where c is the velocity of sound and γ is the ratio of specific heats of the gas. All dissipative terms (due to heat diffusion and dynamic viscosity) have been neglected in Eq. (10).

Equation (10) is usually solved for the sinusoidal modulation case by expressing the Fourier transform of p in terms of "normal acoustic modes" p_j which satisfy the appropriate boundary conditions. Thus

$$p(\mathbf{r}, \omega) = \sum_j A_j(\omega)p_j(\mathbf{r}) \tag{11}$$

with the normal mode p_j being solutions of the homogeneous wave equation, i.e.,

$$(\nabla^2 + \omega_j^2/c^2)p_j(\mathbf{r}) = 0, \tag{12}$$

and p_j must be chosen to satisfy the boundary condition that the gradient of p normal to the cell wall vanish at the wall, since acoustic velocity is proportional to the gradient of p and must vanish at the wall. The resultant orthonormal modes in the cylindrical geometry are given by Morse and Ingard (1968) as

$$p_j(r, \phi, z) = g_j \cos(m\phi)\cos(k\pi z/L)J_m(\alpha_{mn}\pi r/R_0), \tag{13}$$

with a corresponding angular frequency ω_j given by

$$\omega_j = \pi c[(k/L)^2 + (\alpha_{mn}/R_0)^2]^{1/2}. \tag{14}$$

Here g_j is a normalization constant; L is the length and R_0 the radius of the gas cell; (r, ϕ, z) are the cylindrical coordinates of a spatial point; k, m, and n are the longitudinal, azimuthal, and radial mode numbers; J_m is a Bessel function; and α_{mn} is the nth solution of the equation $dJ_m/dr = 0$ at $r = R_0$.

From the above discussions, we see that the condition of vanishing pressure gradient at the cell wall requires that the acoustic pressure $p(\mathbf{r}, \omega)$ be expressed as linear combinations of eigenmodes p_j of the form of Eq. (13) for a cylindrical geometry. For the specific case of the excitation beam being along the axis of the cylindrical OA cell and in the weak-absorption limit, only the radial normal modes can be excited by the heat source $H(\mathbf{r}, \omega)$, i.e., we need only consider the radial normal modes

$$p_j(r) = g_j J_0(\pi\alpha_{0j}r/R_0) \tag{15}$$

with an eigenfrequency ω_j given by

$$\omega_j = \pi c\alpha_{0j}/R. \tag{16}$$

Having delineated the expansion basis p_j, let us now return to Eq. (11) to discuss how to solve for the expansion coefficients $A_j(\omega)$. We first note that the Fourier transform of Eq. (10) is

$$(\nabla^2 + \omega^2/c^2)p(\mathbf{r}, \omega) = [(\gamma - 1)/c^2]i\omega H(\mathbf{r}, \omega). \tag{17}$$

Substituting Eq. (11) into Eq. (17) and using the orthonormal conditions for the eigenfunctions p_j, we may solve for A_j as

$$A_j(\omega) = \frac{-i\omega[(\gamma - 1)/V_0] \int p_j^* H \, dV}{\omega_j^2[1 - (\omega^2/\omega_j^2) - (i\omega/\omega_j Q_j)]}.$$ (18)

Here V_0 is the cell volume, Q_j is the quality factor for the acoustic mode p_j (p_j^* is the complex conjugate of p_j), and the integral is over the volume of the cell. The term involving Q_j has been included in Eq. (18) to account for the mode damping and avoid the physically unreasonable situation of $A_j \to \infty$ as $\omega \to \omega_j$.

Equation (18) may be further simplified for the case of H being given by Eqs. (4) and (5) with Φ being given by Eq. (7). In this case

$$H(\mathbf{r}, \omega) = N\sigma\tau A_n E' \Phi_0(\mathbf{r})$$

$$= q\Phi_0(\mathbf{r})$$ (19)

where we have lumped the space- and time-independent coefficients of $\Phi_0(\mathbf{r})$ together as the coefficient q. We also assume that the light beam is Gaussian, i.e.,

$$\Phi_0(\mathbf{r}) = (\Phi_0/\pi a^2)\exp(-r^2/a^2)$$ (20)

where a is the beam radius, and that the beam propagates along the axis of the cell so that only eigenmodes of the form of Eq. (15) need to be considered. These assumptions [H given by Eq. (19) and the excitation beam being Gaussian and axial] are approximately satisfied in many actual OA experiments. The amplitude of the lowest-order radial pressure mode ($j = 1$) is then given by Eq. (18) as

$$A_1(\omega) = \frac{-i\omega(\gamma - 1)g_1 q\Phi_0 L \exp(-\mu_1)}{\omega_1^2[1 - (\omega^2/\omega_1^2) - (i\omega/\omega_1 Q_1)]V_0},$$ (21)

where

$$\mu_1 = a^2/(2\pi\alpha_{01}R)^2,$$ (22)

$$g_1 = \text{normalization factor for } p_1 = [J_0(\pi\alpha_{01})]^{-1},$$ (23)

$$L = \text{cell length},$$

and we have used

$$V_0^{-1} \int J_0 \frac{\pi\alpha_{01}r}{R_0} \exp\left(-\frac{r^2}{a^2}\right) dV = \exp(-\mu_1).$$ (24)

Close to resonance [i.e., $\omega = \omega_1(1 + \delta)$ with δ being small], Eq. (21) reduces to

$$A_1(\omega) = \frac{Q_1(\gamma - 1)g_1 q \Phi_0 L \exp(-\mu_1)}{\omega_1(1 - 2iQ_1\delta)V_0}. \tag{25}$$

Equation (25) contains the basic physics of the operation of a resonant OA cell. Resonant enhancement of the amplitude of the radial pressure $j = 1$ is obtained when the fractional detuning from resonance δ is less than $(2Q_1)^{-1}$. Also, in general, larger acoustic amplitude is obtained for larger specific heat ratio γ, larger light power absorbed $q\phi_0 L$, smaller beam excitation radius a, and smaller cell volume V_0.

Equation (25) is valid for the case of near resonance to the lowest radial mode. For the opposite case of far off-resonance (i.e., nonresonant OA cell), then Eq. (18) may be approximated as

$$A_j(\omega) \approx \frac{-i\omega(\gamma - 1)}{\omega_j^2 V_0} \int p_j^* H \, dV \tag{26}$$

for $\omega \ll \omega_1$, i.e., the light-beam modulation frequency being much less than the lowest-radial-mode resonance frequency. In this nonresonant mode operation (which is rather common in OA studies), the acoustic amplitude [Eq. (26)] lags behind the beam modulation by 90° (see Fig. 2a),

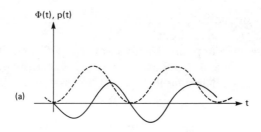

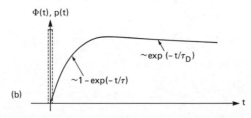

Fig. 2. Schematic of the OA signal $p(t)$ (full line) and the incident photon flux $\Phi(t)$ (dotted line) for (a) modulated cw excitation (nonresonant) case, and (b) pulsed excitation case. [Part (b) is drawn after Rosengren (1975).]

as is well known in OA studies where no slow heat release mechanisms are involved. Again, as in the resonant case considered earlier, a larger OA signal is obtained for larger γ, larger light absorption, or smaller cell volume.

Returning to Figure 1, we see that the final step of the theory of OA signal is the detection, which is frequently done with a microphone. If the microphone has a known frequency response, then all the various components A_j in Eq. (18) with frequencies ω_j within the microphone bandwidth will be detected, and suitable frequency analysis of the microphone signal should give the various A_j's.

Up to now, we have only treated OA signal generation for the case of a modulated cw beam so that boundary conditions are important because the cw beam allows "boundary reflections" to set up and interference of the original and the reflected acoustic waves occurs. However, for the case of pulsed OA excitation, boundary conditions are frequently unimportant when short-duration light pulses are used; this is because the time needed for the generated acoustic wave to reach the cell wall (assuming the distance traveled to be 1 cm) is roughly 30 μsec, which is typically much longer than the light-pulse duration (shorter than ~1 μsec for most pulsed lasers or pulsed Xe arc lamps, for example) and also much longer than decay times of excited states in most gases. Thus, interference of the generated acoustic wave and the reflected acoustic waves generally do not occur in contrast to the cw modulated case. However, pulsed OA generation does produce a "ringing" acoustic signal due to multiple reflections in the gas cell (see, e.g., Leo et al., 1980) as is already well known for pulsed OA studies in condensed matter (see, e.g., Patel and Tam, 1979a). Furthermore, if the frequency of the pulsed OA generation is a suitable subharmonic of an acoustic resonance frequency of the OA cell, resonant OA oscillation can be produced (Dewey and Flint, 1979).

For pulsed OA detection in gases, the net heat released up to time t can be written [based on Eqs. (4) and (5)] as

$$H_{\text{pulsed}}(t) = WA_n\tau E'[1 - \exp(-t/\tau)], \qquad (27)$$

where the incident light pulse is assumed to be of infinitesimal time duration occurring at time $t = 0$ with the total number of photons absorbed being W, and the other terms have the same meaning as defined for the modulated cw excitation case discussed earlier. Heat diffusion has been neglected in Eq. (27). Note that $H_{\text{pulsed}}(t)$ has the unit of energy while $H(\mathbf{r}, t)$ in Eq. (5) has the unit of energy per unit time. The pressure increase $p(t)$ of the irradiated column of gas of volume V can be estimated from Eq. (27) by using the ideal gas law:

$$p(t) = H_{\text{pulsed}}(t)R/MC_VV, \qquad (28)$$

where R is the universal gas constant, M is the gas molecular weight, and C_V is the specific heat per unit mass at constant volume.

The time dependence of $p(t)$ for the pulsed OA signal is indicated in Fig. 2b for the case of short optical pulse duration and long thermal diffusion time τ_D, given by

$$\tau_D \approx a^2/D, \qquad (29)$$

where a is the beam radius and D is the thermal diffusivity of the gas. We see that the initial rise of the OA signal $p(t)$ depends on the lifetime τ of the excited state involved, while the final slow decrease of $p(t)$ back to zero depends on the thermal decay time constant τ_D. Detailed discussion of the time-dependent OA signal after pulsed excitation, including the effects of excited-state lifetimes, has been given by Wrobel and Vala (1978).

B. Instrumentation for OA Studies of Gases

The general types of instruments used in OA studies are indicated in Fig. 3. The instruments needed are (i) light source, (ii) OA cell with transducer, and (iii) a means of modulating the light source (e.g., pulsing a laser or using a chopper), or modulating the sample absorption (e.g., using a modulated electric field for Stark modulation of the absorption). These items are considered in more detail below. The other instruments indicated in Fig. 3 include a scanning drive for the light source (e.g., a scanning spectrometer used in conjunction with a Xe arc lamp as the light source), an optical detector (preferably of flat optical response in the spectral region of interest, so that fluctuations in the incident optical

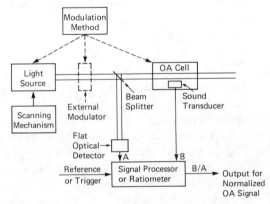

Fig. 3. Schematic of a general OA detection experiment.

power can be accounted for), and suitable means for signal processing and recording (e.g., lock-in amplifier for the modulated cw case, or boxcar integrator or microcomputer for the pulsed OA case). More details on the design considerations for OA systems for gases have been given, for example, by Kreuzer (1971, 1977), Dewey (1977), and Patel (1978a).

Two general classes of light sources are used for OA study. The first class includes arc lamps, filament lamps, and glow bars. They are used because they are inexpensive, usually compact and reliable, and cover broad spectral ranges from the UV to the far IR. However, their drawbacks include low spectral brightness, incapability of fast modulation or switching, and necessity of an external spectral selection element like a monochromator. The problem of low spectral brightness has been made much less serious with the recent introduction of Fourier transform photoacoustic spectroscopy techniques developed by Farrow et al. (1978b), Rockley (1979), Royce et al. (1980), etc., as described in Section III.J. The second class of light sources are lasers, which have high spectral brightness and collimation, can be readily modulated by extracavity or by intracavity means, and are of narrow spectral linewidth without the necessity of using low-throughput spectrometers. Their drawbacks include expense and limited tuning range, although some lasers are tunable over an appreciable spectral range (e.g., dye lasers can be tuned from the near UV to the near IR with the use of different dyes). The use of laser sources for OA detection has been discussed by many authors, e.g., Gelbwachs (1977) and Kelly (1977). Various direct laser sources such as gas lasers, solid-state lasers, and dye lasers have been used, and also such indirect laser sources as spin–flip Raman lasers (Patel, 1978a) have been used.

Many versions of OA cells are in use and some of them are schematically shown in Fig. 4. Discussions of the operational principles of several different types of OA cells, including resonant and nonresonant operations, can be found in Rosengren (1975) and Dewey (1977). Examples of simple OA cells are given in Fig. 4a and 4b. In the earlier work of Kreuzer (1971), a round metal tube with perforations is used, and an electret foil wrapped round the tube is used as the sound transducer. With the development of high-sensitivity microphones (driven by such commercial motivations as miniaturized hearing aids and cassette recorders), electret foils are not commonly used nowadays, but instead, some types of commercial microphones are preferred for higher sensitivity, higher reproducibility, or lower electrical pickup, as shown in Fig. 4b. The windows used are usually highly polished quartz or sapphire for near-visible operation, or polished NaCl crystals for IR operation. Although the windows may be polished and clean, light scattering from the window seems to be an important cause of sensitivity limitations in most OA studies, especially

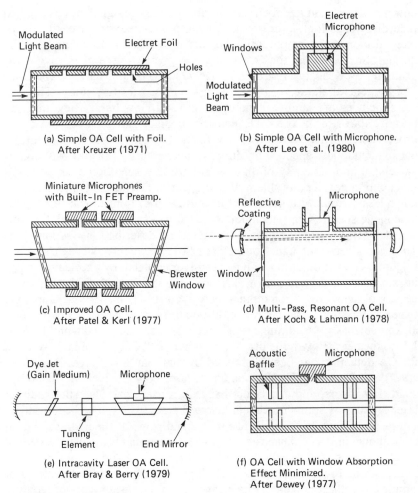

(a) Simple OA Cell with Foil.
After Kreuzer (1971)

(b) Simple OA Cell with Microphone.
After Leo et al. (1980)

(c) Improved OA Cell.
After Patel & Kerl (1977)

(d) Multi-Pass, Resonant OA Cell.
After Koch & Lahmann (1978)

(e) Intracavity Laser OA Cell.
After Bray & Berry (1979)

(f) OA Cell with Window Absorption
Effect Minimized.
After Dewey (1977)

Fig. 4. Schematic drawings of several types of OA cells for measurements in gases, used in the literature.

when the OA cell has been in operation for some time and a "film" of some absorbing or scattering material has built up on the window (Parker, 1973). Light scattering at the windows causes spurious OA signals (due to absorptions at cell walls or at the microphone) especially for "multipass" OA cells. To minimize light scattering, Brewster windows can be used, as indicated in Fig. 4c for the OA cell of Patel and Kerl (1977). In their improved OA cell, the cell body is made with a section of rectangular waveguide (U band), and several miniaturized electret microphones with built-in FET preamplifiers are attached to the outside of the flat surface of the waveguide with suitable holes.

To enhance sensitivities in an OA detection, laser sources can be used, multiple passes of the light beam through the cell can be designed, and mechanical resonance enhancement of the OA signal can be sought. All these factors to enhance sensitivity are ingeniously demonstrated in the OA apparatus of Koch and Lahmann (1978) as shown in Fig. 4d. Here, the laser beam enters through a small uncoated orifice of a lens, and the beam is multiply reflected 65 times into the cell before exiting through the same orifice. Furthermore, a resonant mode (e.g., an azimuthal resonance) can be preferentially excited by a suitable modulation frequency of the laser beam and by positioning the laser beam near the "antinode" of the resonant mode to maximize the overlap of the acoustic mode with the heating source H [see Eq. (18)]. Using these designs for optimizing the sensitivity, Koch and Lahmann (1978) achieved a detection sensitivity of 0.1 ppb of SO_2, using a UV laser beam of 1 mW power.

One way to achieve multiple passes of a light beam is to perform OA spectroscopy inside a laser cavity as indicated in Fig. 4e. Such intracavity OA spectroscopy has been performed by several researchers, for example, Bray and Berry (1979), Smith and Gelfand (1980), and Barrett and Berry (1979). Such an intracavity technique eliminates the necessity of an external multipass optical arrangement as in Fig. 4d, but the tradeoff is that careful alignment of the OA cell in the laser cavity is now necessary.

We have mentioned earlier that window absorption frequently poses serious limitations to the OA detection sensitivity. To maximize the window effect in cw modulation cases, acoustic baffles can be used as shown in Fig. 4f. Alternately pulsed OA methods can be used, as acoustic signals caused by window absorption can be discriminated against because of the longer propagation time from the window to the microphone compared to the time from the bulk of the gas. An even better solution to the window problem is a "windowless" OA cell, used by Gerlach and Amer (1980) for continuous monitoring of ambient air. For certain types of resonant OA cells, the effect of window heating can be reduced by positioning the entrance and exit of the light beam at nodes of the mode being excited. Also, if the "windowless" entrance and exit positions are at the nodes, then the amplitude of the acoustic mode excited is not significantly reduced because of the absence of windows.

Resonance enhancement of the OA signal is a useful technique to increase sensitivity with the tradeoffs that stable modulation frequency must be used, the cell dimensions and temperature must be kept constant, and no modulation-frequency-dependence studies can be performed. However, if sensitivity is the main concern, as in trace detections, resonant cells are generally preferred. Several types of acoustic resonances are possible as illustrated in Fig. 5. The longitudinal, azimuthal, and radial resonant modes have been discussed earlier in Section II.A. The Helmholtz resonator (Fig. 5d) has a resonance frequency given by

$$\omega_h = c(A_0/l_0 V')^{1/2}, \tag{30}$$

where c is the velocity of sound, A_0 and l_0 are the cross-sectional areas and length of the connecting tube, and V' is the "reduced" volume of the cell given by

$$1/V' = 1/V_1 + 1/V_2. \tag{31}$$

Helmholtz OA cells with large (e.g., factor of 100) signal enhancements are used, for example, in pollution-control studies (McClenny *et al.,* 1981).

When OA cells are used to study trace gases, the nature of the buffer gas (mixed with the trace gases) is very important, and careful selection of the buffer gases can significantly increase the sensitivity and flexibility of the OA detection as pointed out by Thomas *et al.* (1978) for a resonant OA cell and by Wake and Amer (1979) for the nonresonant OA cell. OA signal magnitude is increased if a buffer gas of low thermal conductivity and specific heat is used and an optimal buffer gas pressure exists. Also, the resonant frequency of the OA cell depends on the type of buffer gas because c changes with the buffer gas.

The sound transducer used in the OA cell (Fig. 3) is usually a commercial microphone. Some examples of favorite microphones are Knowles

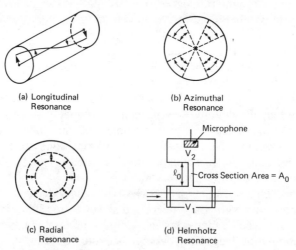

(a) Longitudinal Resonance

(b) Azimuthal Resonance

(c) Radial Resonance

(d) Helmholtz Resonance

Fig. 5. Various types of resonances in OA cells.

model BT-1759, Bruel and Kjaer model 4145, and General Radio model GR1961. Exotic "microphones" are also possible; for example, Miles *et al.* (1980) have described a thin-filament configuration which follows the motion of a gas and can be used instead of a conventional microphone for measuring small pressure fluctuations in OA spectroscopy. For highly corrosive gases where no microphones or other transducers can be directly inserted into the gas sample, totally remote sensing methods are preferred for the OA detection; an example is the use of a probe laser beam which is deflected by an acoustic pulse in a pulsed OA experiment as described by Tam *et al.* (1982). Alternatively, protected transducers can be used for corrosive gases; for example, Marinero and Stuke (1979b) have designed an OA cell with the transducer protected by a fused quartz membrane to study corrosive gases like ICl.

Various modulation methods can be used in OA experiments. Modulation methods for the light source include Q-switching, mode-locking, pulsing flashlamps, wavelength switching, and the use of mechanical modulators (choppers), electro-optic modulators, acousto-optic modulators or wavelength modulating devices. Modulation of the absorption characteristics of the sample is made possible by, for example, applying modulated magnetic or electric fields to the sample; thus, Kavaya *et al.* (1979) have demonstrated that Stark modulation of a gas sample by a modulated electric field is excellent for OA detection because the background signal in the Stark modulation mode is about 500 times smaller than that in the same OA cell in the conventional chopped-light-beam mode. This is because the background absorption (e.g., that caused by the cell windows or by the buffer gases) has little dependence on the electric field. In a sense, the Stark modulation method is equivalent to a wavelength modulation method, as used by Lahmann *et al.* (1977) for OA studies of liquids. Moses and Tang (1977) have already pointed out the advantage of background suppression in wavelength modulation methods. Castleden *et al.* (1981) have described a simple wavelength-modulated PA spectrometer which generates differentiated PA spectra: this provides an apparent enhancement in the resolution, and results in an increased precision of locating absorption features in the sample.

C. Measuring Weak Absorption Lines

The sensitivity of laser OA detection in gases has advanced to an absorption measurement capability of $\sim 10^{-10}$ cm^{-1} with a cell length of ~ 10 cm (see Patel and Kerl, 1977). This high sensitivity cannot be matched by other conventional absorption techniques such as extinction measurement, which measures absorption plus all scattering losses, and cannot be readily used to monitor absorption coefficients less than $\sim 10^{-3}$ cm^{-1}. There are a few other ultrasensitive techniques for spectroscopic detection that may have sensitivities better than the laser OA technique, notably the single-atom detection method of Hurst *et al.* (1979) using multistep laser excitation and ionization (see also Chapter 4 in this book); also there is the high-sensitivity technique of monitoring luminescence at a different wavelength from the excitation wavelength, and detection of a single atom is again possible with pulsed laser excitation and gated luminescence detection. Although the laser OA method lags behind the multiphoton ionization spectroscopy or the luminescence monitoring spectroscopy method in sensitivity, it does offer some advantages over these other methods, such as simplicity and generality. The simplicity of laser OA detection means that only a sound transducer is required. The generality means that nonradiative thermal relaxation in a gas system near atmospheric pressure occurs very generally (it may only be partial in certain highly fluorescent or chemically active systems), while ion attachment, diffusion, recombinations, and quenching phenomena may pose serious limits to the other two detection techniques.

The high detection sensitivity of the laser OA method has the obvious use for detecting very weak absorptions, such as those that are forbidden by the dipole selection rule or by spin conservation. An example is an overtone vibrational transition. If the vibrational potential of the molecule can be represented by a simple harmonic oscillator potential, then only the fundamental vibrational excitation is allowed for optical excitation, and all higher harmonics are forbidden because the dipole matrix elements vanish for the higher transitions. However, in a real molecule, higher vibrational transitions are still possible but weak, since the small anharmonicity permits an overtone absorption, which weakens rapidly for increasingly higher overtones.

The use of dye-laser OA spectroscopy to examine overtone absorption in gases seems to have been first demonstrated by Stella *et al.* (1976). In their experiment, the 6190-Å overtone absorption band of CH_4 and the 6450-Å overtone absorption band of NH_3 was measured by placing the OA cell in the cavity of a dye laser and scanning the laser across the absorption band. They were able to resolve rotational features. Bray and Berry (1979) and Reddy and Berry (1981) have studied high C–H stretch overtones by intracavity cw dye-laser OA spectroscopy, and determined the spectral line shapes that bear important information on the dynamics of high vibrational states in gas-phase polyatomic systems; they have measured high harmonics (up to the 9th) in benzene and other organic molecules, showing that the overtone excitation must correspond to a highly localized vibration of one single CH bond (i.e., the oscillation is a local mode instead of normal mode; see, for example, Henry and Siebrand, 1968; Henry, 1977). Smith and Gelfand (1980) have performed intracavity OA spectroscopy of HD in the 5–0 vibrational transition with the R(0), R(1) and R(2) rotational lines measured; although such overtone lines have previously been observed by long pathlength measurements, the high-sensitivity intracavity OA method of Smith and Gelfand (1980) permits the measurement at a pressure–pathlength product of only 10^{-6} of that previously used.

D. High-Sensitivity Trace Detection

The high sensitivity of laser OA detection also permits measurements of weak absorptions due to a strong absorption line of a trace constituent. This application of OA detection for trace analysis was realized in the 1930s, but the application has become particularly attractive with the availability of laser sources. The idea of trace detection by "spectrophone" detection in conjunction with laser excitations was first demonstrated by Kerr and Atwood (1968). Later Kreuzer (1971) reported that, using a 15-mW He–Ne laser at 3.39 μm, he could achieve a detection sensitivity adequate to measure 10 ppb of CH_4 in N_2. Kreuzer (1971) further pointed out that, with more intense sources of tunable IR radiation, concentrations of impurities as low as 10^{-13} could be detected.

Patel (1978a) reported some pioneering work of laser OA monitoring of pollutants. An apparatus used by Patel (1978a) is shown in Fig. 6. The

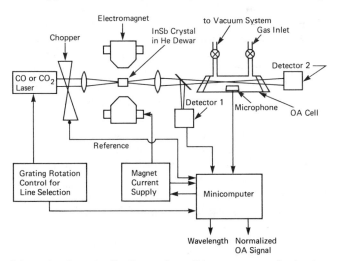

Fig. 6. Schematic of a spin–flip Raman laser OA spectrometer for in situ pollution detection. [After Patel (1978a).]

light source used is a spin–flip Raman laser (SFRL) with InSb as the active medium that is pumped by a CO or CO_2 laser. The IR output from the SFRL is tunable by a magnetic field, which is typically a superconducting magnet situated in the same liquid-helium Dewar as the InSb crystal. This detection technique turns out to be extremely useful for in situ pollutant detection, both in terrestrial stations and in the upper atmosphere.

Many other workers have recently demonstrated outstanding successes in detecting trace pollutants in nitrogen or in atmospheric air. Koch and Lahmann (1978) have used a cw frequency-doubled dye laser of 1 W power to detect SO_2 of concentrations as low as 0.1 ppb. Vansteenkiste *et al.* (1981) have reported the use of a $PbS_{1-x}Se_x$ diode laser of 96 μW power at 4.8 μm to detect the absorption of CO in a small nonresonant OA cell. A concentration of 50 ppm of CO in N_2 by volume can be detected by operating the cell in a double-pass mode. Such an OA detector, which can be highly miniaturized because diode lasers are used as the light source, may be very important for future motor vehicle emission controls. Of course, if higher-power laser sources are used, higher detectability can be achieved; for example, Gerlach and Amer (1978) have reported a detection of 0.15 ppm of CO. Claspy *et al.* (1976) reported a study of laser OA detection of explosive vapors, including nitroglycerine, ethyleneglycol dinitrate, and dinitrotoluene. A CO_2 laser source tuned to suitable lines in the 9–11 μm spectral range is used to minimize background absorption due to other normal constituents in air. Claspy *et al.* (1977) reported OA detection of NO_2 in air using a flash-lamp-pumped dye laser as the excitation source, and sensitivity of several ppb has been achieved. Earlier, Angus *et al.* (1975) had reported OA detection of 10 ppb of NO_2 using a modulated cw dye laser for excitation.

Atmospheric pollutants need not be only gaseous, but can be particulate as well. McClenny and Rohl (1981) have used an OA method to detect particulate carbon in the atmosphere, and elemental carbon concentrations of 0.3 μg/m^3 in air can be estimated by photoacoustic analysis of particles collected on Teflon filters. Further discussions of photoacoustic aerosol detections will be given in Section III.I.

Table I summarizes some of the OA detection sensitivities demonstrated for atmospheric pollutants by various researchers; these sensitivities are frequently limited by the intensity or tunability of the light source used, and should not represent fundamental limits.

E. Absorption by Excited States

Since OA detection can be used for trace detection, and since trace amounts of excited states can be produced in a gaseous system by suitable

TABLE I

Examples of Experimental Sensitivities Achieved for OA Detection of Atmospheric Pollutants[a]

Author	Pollutant Gas	Sensitivity (ppb)	Light Source Used for the OA Detection
Kreuzer et al. (1972)	NH_3	0.4	Use selected IR lines from a grat-
	C_6H_6	3	ing-tunable CO_2 or CO laser; out-
	NO	0.4	put power ~1 W
	NO_2	0.1	
Claspy et al. (1977)	NO_2	15	Pulsed dye laser
Angus et al. (1975)	NO_2	10	Modulated cw dye laser
Koch and Lahmann (1978)	SO_2	0.1	Frequency-doubled dye laser at UV spectral region
Gerlach and Amer (1978)	CO	150	CO laser oscillating in the (1,0) vibrational transition near 4.8 μm
Vansteenkiste et al. (1981)	CO	5×10^4	Lead–salt diode laser

[a] Results are obtained for a pollutant in a dry inert gas like N_2.

excitation mechanisms (optical, discharge, chemical, etc.), it is logical to expect the OA detection method to be well suited for measuring excited-state spectra and collision dynamics. This was first demonstrated by Patel et al. (1977) for a gaseous system of NO (Fig. 7). ^{14}NO is excited by a fixed-frequency CO laser line at 1917.8611 cm^{-1}, which produces $v = 1$ excited states. These excited molecules are examined spectroscopically with a tunable infrared beam obtained from an InSb SFRL pumped with another CO laser line at 1891.6545 cm^{-1}. The two collinear beams are focused into an OA cell of length about 1 cm, containing ^{14}NO of pressure ~1 Torr. As the SFRL is scanned, high-resolution OA spectra of $v = 1 \rightarrow v = 2$ vibrational–rotational transition of the $^2\Pi_{1/2}$ and $^2\Pi_{3/2}$ states of ^{14}NO

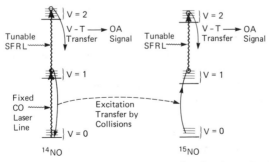

Fig. 7. Principle of excited-state spectroscopy of molecules using OA detection as demonstrated by Patel et al. (1977) and Patel (1978b).

are obtained. From the observed spectra, accurate frequencies of the $1 \to 2$ band centers for both $^2\Pi_{1/2}$ and $^2\Pi_{3/2}$ states are derived; also the Zeeman effect on the transitions can be studied by putting the OA cell in a magnetic field.

In the more recent work of Patel (1978b), excited-state OA spectroscopy is extended to cases where no fortuitous excitation by a fixed-frequency laser for the first excitation step is possible. In this case, molecular vibrational energy transfer may be used, as demonstrated (see Fig. 7) for the case of $^{14}NO(v = 1)$ to $^{15}NO(v = 1)$ transfer. In this experiment, a $1:1$ mixture of the two isotopic NO molecules is excited by a fixed-frequency CO laser, which excites only the ^{14}NO system. The vibrational energy transfer allows OA probing of the $^{15}NO(v = 1)$ state, and an accurate spectrum of the $^{15}NO(v = 1) \to {}^{15}NO(v = 2)$ absorption is obtained. Vibrational-energy-transfer rates can also be derived.

F. Chemically Reactive Gases

Chemically reactive gases may produce transient intermediate chemical species, and the high sensitivity available with OA detection may be useful to identify some of the intermediate products, and hence provide important understanding of the reaction channel. In continuous photolysis, concurrent spectroscopic identification of photolysis products is generally very difficult since the steady-state concentration of intermediates is usually too small for spectroscopic measurements without using a matrix isolation procedure. However, Colles et al. (1976) have shown that in a gas-phase continuous-wave photolysis experiment, OA detection of intermediates is possible using a tunable cw dye laser. Their experiment was performed on CH_3NO_2, continuously photolyzed by a 250-W mercury lamp. A tunable dye laser beam, chopped at 1.4 kHz, passes through the stainless-steel cell in the opposite direction. The OA signal is detected by a sensitive microphone in a sidearm. The experiment of Colles et al. (1976) indicates that the continuous photolysis produces NO_2, which strongly absorbs the chopped dye laser beam, giving a characteristic OA spectrum. In a sense, this experiment is quite similar to the experiments on atmospheric pollutant detection discussed in Section II.D, except now the NO_2 is produced in situ by some photochemical reactions.

Information on photofragmentation products depending on the wavelength of excitation can also be gained by OA measurements. This was demonstrated in the case of CH_3I by Hunter and Kristjansson (1978). It was already known that photofragmentation of CH_3I caused I atoms to be produced, but the spectral dependence of the ratio ϕ of $I(^2P_{1/2})$ excited state to $I(^2P_{3/2})$ ground state being produced was uncertain. Hunter and

Kristjansson (1978) performed UV OA spectroscopy using modulated light from a high-pressure Hg arc at wavelengths from 240 to 270 nm. The phase of the OA signal depending on the wavelength of excitation was used to derive the branching ratio ϕ, which turns out to be highly dependent on the excitation wavelength.

G. Raman-Gain Spectroscopy

Up to now, we have considered the use of the OA method to detect linear absorptions; however, because of its high sensitivity, the OA method is also ideally suited to measure nonlinear absorption or nonlinear optical scattering effects, since some degree of heat deposition (often a small amount) in the gas usually occurs owing to these nonlinear optical interactions.

Photoacoustic Raman-gain spectroscopy (PARS) was first suggested by Nechaev and Ponomarev (1975), and was first observed by Barrett and Berry (1979). Raman-gain spectroscopic techniques are of strong current interest because of their potentially high sensitivity for measuring Raman frequencies and cross sections, and their applicability to cases of luminescent samples or hostile environments (flames, discharges, etc.) where spontaneous Raman scattering cannot be performed. To perform stimulated Raman scattering, a "pump beam" of intensity I_p and wavenumber v_p and a "signal beam" of intensity I_s and lower wavenumber v_s are incident on a Raman-active medium collinearly. One of the beams is tunable. When the difference wavenumber $v_p - v_s$ equals a Raman frequency v_R of the medium, Raman gain occurs, i.e., I_p is attenuated and I_s is amplified, photon for photon. Since a "pump photon" is of higher energy (being of larger wavenumber) than a "signal photon," energy is deposited in the medium owing to the stimulated Raman scattering, and the energy deposited usually appears as heat due to vibrational–translational relaxations in the medium.

For the case of weak Raman gain (i.e., the increase in signal intensity ΔI_s being much less than I_s), it can be shown that (see Barrett and Berry, 1979; Patel and Tam, 1979c) the energy W deposited in the medium due to a pulsed signal beam of energy E_s is

$$W = G_s l \bar{I}_p E_s v_R / v_s, \tag{32}$$

where G_s is the Raman gain coefficient that depends on the Raman scattering cross section, l is the gain pathlength, and $\bar{I}_p$ is the averaged pump intensity during the signal pulse.

Equation (32) is the basis of the PARS work of Barrett and co-workers. In their first experiments (Barrett and Berry, 1979), continuous laser

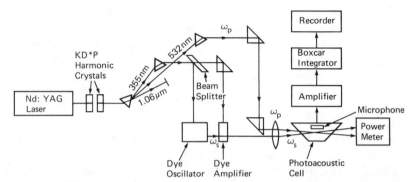

Fig. 8. Schematic of the apparatus for pulsed Photoacoustic Raman Spectroscopy (PARS). [After West and Barrett (1979).]

sources were used for both pump and signal beams. The pump beam was an Ar^+ laser mechanically chopped at 573 Hz, while the signal beam was a continuous dye laser beam tuned near the 6054-Å band of CH_4. With microphone detection, the symmetric stretch vibrational mode of methane near 2900 cm^{-1} was detected by the PARS method.

In the more recent experiments with PARS, West and Barrett (1979) reported that greatly enhanced sensitivity could be obtained using pulsed lasers for the pump and probe beams instead of cw lasers as previously. This is obvious from Eq. (32), which can be rewritten as

$$W \propto E_p E_s / (\tau_{\text{laser}} A_{\text{laser}}) \qquad (33)$$

where E_p is the total pump beam energy incident on the sample, τ_{laser} is the duration of the pump pulse, and A_{laser} is the cross-sectional area of the pump beam. Thus, for constant energies E_p and E_s, the PARS signal is inversely proportional to τ_{laser} and A_{laser}. The experimental arrangement for pulsed PARS of Barrett and co-workers is shown schematically in Fig. 8. With the sensitivity greatly enhanced by using pulsed lasers of 7-nsec duration, Siebert *et al.* (1980) reported that PARS trace analysis of many types of gases in N_2 is possible, with detection limits of a few parts per million possible for CH_4 or CO_2, for example. Also, PARS of rotational transitions (which are much weaker than the vibrational transition because ν_R is much smaller for rotational transition, and the PARS signal is proportional to ν_R) has been performed (West and Barrett, 1979; West *et al.* 1980).

H. Doppler-Free Spectroscopy

The OA spectroscopies performed so far are mostly of resolution at best equal to the Doppler width. In principle, the high sensitivity of detection implies that the technique of saturation spectroscopy (already shown for fluorescence monitoring, absorption monitoring, or optogalvanic monitoring) can be used with OA monitoring to allow a Doppler-free linewidth, i.e., with the spectral linewidth being limited by the molecular lifetimes. The idea of saturation spectroscopy is quite simple. If a highly monochromatic light beam (of bandwidth much less than the Doppler width) at frequency ν_L cm^{-1} is incident on a gas medium, then only a small velocity subgroup with velocity component u parallel to the light beam will be excited, as given by

$$\nu_L = \nu_0 + u\nu_0/c_L, \tag{34}$$

where ν_0 is the frequency of the absorption line center and c_L is the velocity of light. The lifetime-limited linewidth is assumed to be small. By similar argument, an oppositely directed light beam of the same wavelength will excite a velocity subgroup with velocity component $-u$ with respect to the original light beam. Thus, the two oppositely directed light beams do not interact with the same velocity subgroup unless $u = 0$, in which case Eq. (34) implies that ν_L must equal ν_0. The satisfaction of this condition can be detected by modulating the two opposite beams at two different frequencies f_1 and f_2 as shown in Fig. 9 for the Doppler-free OA spectroscopy by Marinero and Stuke (1979a). When the two beams are tuned to ν_0 so that they excite the same zero-velocity subgroup, the excited-state density N' will vary as

$$N' \sim \exp(i2\pi f_1 t)\exp(i2\pi f_2 t) \tag{35}$$
$$= \exp[i2\pi(f_1 + f_2)t].$$

Hence, the lock-in amplifier will detect an OA signal at the sum frequency $f_1 + f_2$. On the other hand, if the laser frequency is not at ν_0 so that different velocity subgroups are excited, then the excited-state density will vary as $\exp(i2\pi f_1 t) + \exp(i2\pi f_2 t)$, and no sum frequency will be generated. In the experiment of Marinero and Stuke (1979a), the P(193) line of the 11–0 band of the B $\leftarrow$ X transition of I_2 is measured by using a single-longitudinal-mode cw dye laser beam which is split into two opposite beams chopped at $f_1 = 757$ and $f_2 = 454$ Hz. Other work on Doppler-free OA spectroscopy has been reported by Lieto et al. (1979) and by Inguscio et al. (1979).

I. Multiphoton Absorption

Cox (1978) has used a high-power CO_2 TEA laser beam with intensity adjustable from 16 W/cm^2 to 5×10^6 W/cm^2 to excite a SF$_6$-Ar mixture. At low intensity, the pulsed OA signal S varies linearly with the beam inten-

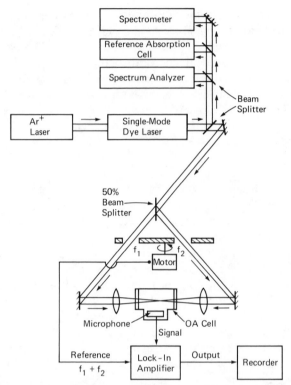

Fig. 9. Experimental arrangement for Doppler-free OA spectroscopy. [After Marinero and Stuke (1979a).]

sity I, as expected for linear absorption. At intermediate I, S is observed to change as $I^{1/2}$, indicating optical saturation of an inhomogeneously broadened absorption. At large I, rapid increase of S with I is observed, indicating the occurrence of multiphoton absorption. Furthermore, the addition of more buffer gas to increase the collision frequency is observed to quench the multiphoton excitation of SF_6.

Fukumi (1979) has also used CO_2 laser lines to observe multiphoton OA signals in ethylene. By measuring the dependence of the OA signal magnitude on the laser energy, Fukumi (1979) has shown that the absorption at 949 cm^{-1} is due to a single-photon transition in ethylene, while the absorptions at 953 and 939 cm^{-1} are due to two-photon transitions. At sufficiently high laser pulse intensity, higher-order multiple-photon transitions can also be detected.

Several other workers have performed similar IR multiphoton absorption spectroscopy with OA detection. For example, Brenner *et al.* (1980)

have studied multiphoton absorption in propynal in a collisionless environment, and performed both OA detection and laser-induced-fluorescence detection. Webb *et al.* (1980) have measured the two-photon OA spectroscopy of sym-triazine, and derived the vibronic assignments in the $^1E''$ state. Weulersse and Genier (1981) have obtained multiphoton absorption cross sections for CO_2 laser lines in CF_3I at a vapor pressure of 0.2 Torr.

III. PA SPECTROSCOPY IN CONDENSED MATTER

A. The Gas-Coupling Method: Theory

The use of a gas-phase microphone for detecting photoacoustic signals in condensed matter and obtaining photoacoustic spectra of solids and liquids began with the work of Hey and Gollnick (1968), Harshbarger and Robins (1973), and Rosencwaig (1973). The PA signal was generated by a sinusoidally modulated cw light beam incident on the condensed sample, and the periodic heating of the gas at the irradiated surface of the sample generated the acoustic wave, which was detected by a gas-phase microphone. The nature of the PA signal (magnitude and phase) in this gas-coupling method was first quantitatively investigated theoretically by Rosencwaig and Gersho (1976) with refined theories given later on by various authors such as Bennett and Forman (1977), Aamodt *et al.* (1977), Aamodt and Murphy (1978), McDonald and Wetsel (1978), McDonald (1979, 1980, 1981), Mandelis *et al.* (1979), Helander *et al.* (1980), Quimby and Yen (1980).

The basic idea of the theory of PA generation is represented in the block diagram of Fig. 10. The periodic heating of the sample occurs in the "absorption length" μ_α of the sample, but only the heat within a diffusion length μ_s from the interface can communicate with the gas and heat up a layer of gas of length μ_g (diffusion length in the gas) which expands periodically, producing acoustic waves. Here μ_α is the reciprocal of the optical absorption coefficient α in the sample, and

$$\mu_s = [D_s/(\pi f)]^{1/2}, \tag{36}$$

$$\mu_g = [D_g/(\pi f)]^{1/2}, \tag{37}$$

with D_s and D_g being the thermal diffusivities in the sample and in the gas, respectively, and f being the modulation frequency of the light beam.

The above theory of PA generation by the gas layer of thickness μ_g, first given by Rosencwaig and Gersho (1976), is usually referred to as the "gas-piston" model. Later, McDonald and Wetsel (1978) suggested that

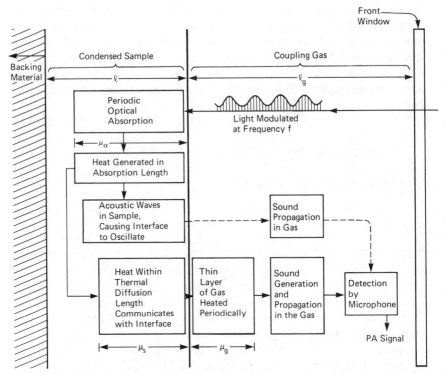

Fig. 10. Block diagram to explain the method of PA detection in condensed matter using gas coupling.

in some circumstances, additional effects may be generated owing to the acoustic wave produced in the sample causing the interface to vibrate as indicated in Fig. 10. This "surface vibration" effect is frequently small compared to the effect of the gas piston, except for the case when the sample is sufficiently transparent (e.g., $\alpha < 1$ cm^{-1}) and the modulation frequency is sufficiently large (e.g., $f > 10^3$ Hz). In this case, McDonald and Wetsel (1978) and McDonald (1979) have shown that a "composite piston" model must be used because the gas-piston effect and the sample vibration effect may both contribute significantly to the observed PA signal.

It is obvious that in the gas-piston effect, the PA signal generated will depend critically on the relative magnitudes of three lengths: l, μ_α, and μ_g (l is the sample thickness). Let us use the ordered bracket (a, b, c) to indicate the case of $a > b > c$, with a, b, and c each being one of the three lengths. There are 3! permutations for a, b, and c, and hence there are six cases for gas-piston PA generation. These six cases are listed in Table II, where the approximate expressions for the PA signal, according to Rosencwaig and Gersho (1976) are also given. In Table II, optically thin (thick) means that $\mu_\alpha > (<)l$, and thermally thin (thick) means that $\mu_s > (<)l$. The definitions of the symbols used are as follows: γ is the ratio of specific heats of the coupling gas, P_0 the ambient pressure of gas, I_0 the amplitude of the incident light intensity, l_g the thickness of the coupling gas, T_0 the ambient temperature of the gas, μ_b the thermal diffusion length of the backing material, and k the thermal conductivity, with subscripts b, s, and g standing for backing, sample and gas, respectively. In Table II, F is a function that is the same for all cases and depends only on properties of the coupling gas and of the light intensity,

$$F = \gamma P_0 I_0 \mu_g / 4\sqrt{2} l_g T_0. \tag{38}$$

Several points are worth noting for Rosencwaig and Gersho's results summarized in Table II: (i) There is always a phase lag of the sinusoidally modulated PA signal from the sinusoidally modulated light; if μ_s is shorter than μ_α and l, the phase lag is about 90°; otherwise, the phase lag is about 45°. (ii) The PA signal is proportional to the absorption coefficient α for all the optically thin cases 1–3, and for case 6 in Table II, i.e., the optically thick but thermally thicker case; it is only in these four cases that PA

TABLE II

Various Possible Cases for PA Detection by Gas-Coupling Method

Condition	Case	Approximate expression for the PA signal[a]
1. (μ_s, μ_α, l)	Thermally thin, optically thin	$(1 - i)\alpha l(\mu_b/k_b)F$
2. (μ_α, μ_s, l)	Optically thin, thermally thin	$(1 - i)\alpha l(\mu_b/k_b)F$
3. (μ_α, l, μ_s)	Optically thin, thermally thick	$-i\alpha\mu_s(\mu_s/k_s)F$
4. (μ_s, l, μ_α)	Thermally thin, optically thick	$(1 - i)(\mu_b/k_b)F$
5. (l, μ_s, μ_α)	Optically thick, thermally thick	$(1 - i)(\mu_s/k_s)F$
6. (l, μ_α, μ_s)	Thermally thick, optically thick	$-i\alpha\mu_s(\mu_s/k_s)F$

[a] Here $F = \gamma P_0 I_0 \mu_g / (4\sqrt{2} l_g T_0)$.

detection can be used for spectroscopy (in the other two cases, PA signal saturation occurs). (iii) Case 5 in Table II is the best case for measuring thermoelastic properties and depth profiling studies of the sample, since the PA signal is big and independent of the backing material; note that the signals in cases 3 and 6 are less than in case 5, and that the PA signals depend on the backing properties in all the other cases.

The results in Table II indicate that the PA signal is inversely proportional to the gas thickness l_g through the universal function F. This inverse proportionally can only hold for sufficiently large l_g. This is shown by Tam and Wong (1980) explicitly for case 5, which is the important case for depth profiling studies. Consider a light beam of radius r striking a flat opaque surface of radius R (see Fig. 11). The heat generated in the thin absorption layer of thickness μ_α is mainly conducted into the condensed sample (heat conduction into the gas is much smaller); the heat conduction equation is

$$k_s\theta_0/\mu_s \approx I_0, \tag{39}$$

where θ_0 is the amplitude of the surface temperature modulation and I_0 is the modulated light absorbed (i.e., reflection loss accounted for). θ_0 is coupled to an active gas volume V_{act} near the sample surface, given by

$$V_{act} \approx \pi r^2\mu_g \qquad \text{for } l_g > \mu_g$$

or   (40)

$$V_{act} \approx \pi r^2 l_g \qquad \text{for } l_g < \mu_g.$$

Using the ideal gas law, we obtain the amplitude δV of the volume change of V_{act}:

$$\delta V = V_{act}\theta_0/T_0 \tag{41}$$

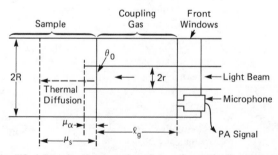

Fig. 11. Simplified PA signal generation theory for optically opaque sample due to modulated cw light beam and microphone detection. [After Tam and Wong (1980).]

with T_0 being the ambient temperature. The corresponding pressure change δP is obtained by considering an adiabatic expansion of an ideal gas

$$\delta P = \gamma P_0 \delta V/V, \tag{42}$$

where V is the total PA cell volume given by

$$V = \pi R^2 l_g + V_{res}. \tag{43}$$

Here V_{res} is the residual cell volume for $l_g = 0$, and can be due to the dead space in front of the microphone. Combining Eqs. (39)–(43), we have

$$\delta P \approx \gamma P_0 \begin{Bmatrix} \mu_g \\ l_g \end{Bmatrix} \mu_s I_0 r^2/[k_s(l_g + V'_{res})T_0 R^2], \tag{44}$$

where the upper (lower) quantity in the curly brackets should be used for $l_g > (<)\mu_g$, and $V'_{res} = V_{res}/\pi R^2$. Since V_{res} is typically negligible when $l_g > \mu_g$, Eq. (44) reproduces the result for case 5 in Table II, where only $l_g > \mu_g$ is considered. Thus, Eq. (44) agrees with Rosencwaig and Gersho's (1976) results that the PA signal varies as $l_g^{-1}f^{-1}$ at large l_g (i.e., $l_g \gg \mu_g$). However, for small l_g (i.e., $l_g \ll \mu_g$), Eq. (44) predicts that the PA signal varies as $l_g f^{-1/2}$. Thus, it is obvious that an optimum l_g exists, and the PA signal decreases for both larger and smaller l_g. Experimentally, the optimum l_g is found to be

$$l_{g,opt} \approx 1.8 \ \mu_g \tag{45}$$

as discussed in the next section.

B. The Gas-Coupling Method: Instrumentation

The instrumentation for PA spectroscopy with the gas-coupling method is quite similar to that for gas-phase OA spectroscopy indicated in Fig. 3. The only difference is that the OA cell now contains a condensed sample whose PA spectrum is to be determined, and the coupling gas in the PA cell does not absorb the light appreciably. The employment of Brewster windows for minimum light scattering, multipassing for increased sensitivity, recessed microphone for avoiding light scattered onto it, acoustic baffles for reduced effects of window absorption, etc., that are illustrated in Fig. 4 for the gas OA studies are also applicable in the present case. The idea of acoustic resonance and its consequent enhancement of the detected microphone signal, illustrated in Fig. 5 is also applicable here. The gas-coupling method for PA studies of condensed matter was first demonstrated by Hey and Gollnick (1968), and was subsequently extended by Harshbarger and Robin (1973) and by Rosencwaig (1973).

One of the simplest PA cells employing a flat microphone (e.g., General Radio Electret Microphone model 1961-9601) used by Rosencwaig (1977) is shown in Fig. 12. The microphone has high sensitivity, and may have flat frequency response up to about 100 kHz, thus making the nonresonant PA cell of Fig. 12 well suited for chopping-frequency-dependence study to obtain the depth-dependent PA spectrum.

Another type of PA cell with a narrow passage connecting the sample and the microphone has been used by Rosencwaig (1977), McClelland and Kniseley (1976a,b), Monahan and Nolle (1977), Aamodt and Murphy (1977), Bechthold et al. (1981), and others. An example is shown in Fig. 13. The passage serves several purposes: (i) Light scattering from the sample, sample holder, and window onto the microphone (causing spurious acoustic signals because of absorption at the microphone surface) may be reduced. (ii) By varying the volume of the sample or the microphone chamber, Helmholtz resonance can be obtained, thus enhancing the PA signal. (iii) By using a sufficiently long connection passage, the sample chamber can be kept at a very cold or very hot temperature to perform PA spectroscopy at these temperatures with the microphone being kept at room temperature (see, e.g., Aamodt and Murphy, 1977; Bechthold et al., 1981).

Similar to the idea of the dual-beam spectrophotometer, dual-channel PA cells have been designed, for example, by Cahen et al. (1978c), as shown in Fig. 14. Here PA signals from identical twin cells, one contain-

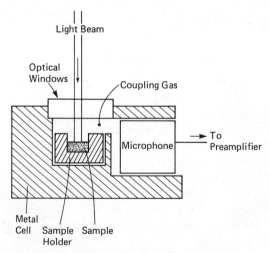

Fig. 12. Simple PA cell with flat microphone. A mirror at 45° to the vertical for directing a horizontal light beam onto the sample is sometimes attached to the upper part of the cell. [After Rosencwaig (1977).]

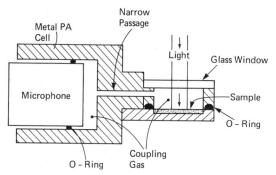

Fig. 13. Example of a PA cell with a sample compartment and a microphone compartment connected by a narrow passage. Such cells have Helmholtz resonances, and can be designed so that the sample is at low or high temperatures. [After Monahan and Nolle (1977).]

ing the sample and the other containing a reference material (e.g., carbon black), are recorded simultaneously, and ratioing of the signals would eliminate optical fluctuation and some other source of noise. Alternatively, differential signals can be obtained if one of the quartz tubes contains a solution and the second tube contains the pure solvent.

We have seen in Section III.A that the thickness of the coupling gas is an important parameter for PA signal optimization. For a maximum PA signal, the gas thickness should be about 1.8 μ_g and hence depends on the chopping frequency and on the type of gas used. The PA signal for helium

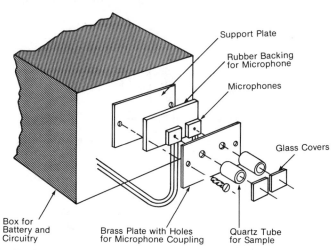

Fig. 14. Exploded view of a simple differential PA cell. [After Cahen *et al.* (1978c).]

coupling gas is a factor of 3.2 larger than that for N_2 coupling gas at the same chopping frequency (see Tam and Wong, 1980). Thus, to optimize OA signal, both the gas thickness and the type of coupling gas should be adjustable, as shown in Fig. 15. Here the coupling-gas thickness (between the sample and the front window) can be controlled accurately by the micrometer, and a "magnet coupling" is used so that the micrometer does not cause rotation of the sample. Also, various coupling gases can be introduced into the box enclosing the PA cell. A subminiaturized microphone (Knowles BT-1751) with built-in FET preamplifier is used to minimize dead space. With the apparatus of Fig. 15, Tam and Wong successfully verified that an optimum l_g exists as given in Eq. (45).

It should be noted that an enclosed sample chamber (used in Figs. 12–15) is not essential. An "open-membrane spectrophone" is also possible as first demonstrated by Kanstad and Nordal (1978a). The design is shown in Fig. 16. By placing the sample of powders, liquids, or thin films on a thin membrane (e.g., aluminum of 40 μm thickness), the PA signal can be recorded by the microphone situated on the opposite side of the membrane.

A "black" reference material is frequently needed in PA spectroscopy because the spectral intensity of the incident light may vary. Typically, lamp black is used as the black reference material, but "aging" and irreproducibility are very common for lamp black. A new type of PA reference material has been developed by Coufal (1982). This is a self-supporting amorphous carbon film. The advantages of such a film for PA spectroscopy reference compared to lamp black are (i) high reproducibility, stability, and dynamic range, and (ii) no backing materials are needed.

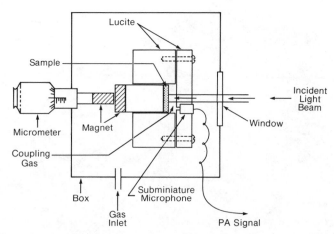

Fig. 15. PA cell with adjustable thickness of the coupling gas and possibility of using different coupling gases. [After Tam and Wong (1980).]

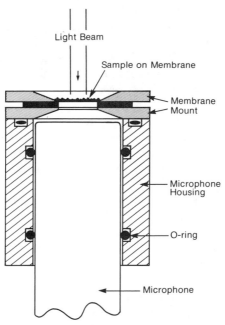

Fig. 16. Design of the open-membrane spectrophone. [After Kanstad and Nordal (1978a).]

C. The Direct-Coupling Method: Theory

As pointed out by McDonald and Wetsel (1978), the gas-microphone PA signal becomes more complex (i.e., the gas-piston model becomes inadequate) when the sample absorption coefficient is sufficiently small and the modulation frequency is sufficiently large. In this case, the sample vibration itself may contribute a significant portion of the microphone signal. As an example, McDonald and Wetsel's (1978) results indicated that for $\alpha = 0.1$ cm^{-1} and $f = 200$ Hz, the microphone signal due to the acoustic vibration in the sample is about twice as big as that due to the gas-piston contribution. On the other hand, gas-microphone detection of acoustic waves inside a condensed sample is not an efficient detection method because of the serious acoustic impedance mismatch at the sample–gas interface, causing little acoustic transmission (for example, only a factor of 10^{-5}) across the interface.

It is clear from McDonald and Wetsel's (1978) calculations that the gas-phase-microphone method for PA detection is extremely inefficient when dealing with a low-absorption sample, especially in many circumstance when the chopping frequency cannot be less than ~100 Hz because of line-frequency interference, mechanical resonances in the surroundings, etc. Thus, for low absorption, a direct-coupling method must be used. Direct coupling involves the insertion or attachment of a transducer (usually piezoelectric) into or onto the sample without the intervention of a gas medium. Thus, the serious acoustic impedance mismatch from condensed matter to gas can be avoided. Two general types of photoacoustic excitation are possible, as in the gas-phase OA case treated in Section II.A. These are (i) the use of a chopped or modulated cw excitation beam when the detected PA signal depends on the boundary conditions; (ii) the use of a pulsed excitation beam when the boundary conditions frequently have no effect on the detected optoacoustic signal, especially if short-duration pulses (for example, $\lesssim 1$ μsec) at low repetition rate (e.g., ~ 10 Hz) are used.

An example of case (i) is treated by Kohanzadeh et al. (1975). In that work, a liquid in a hollow cylinder of a solid excited by a modulated light beam propagating along the cylinder axis is considered; the algebra used resembles the theoretical considerations in Section II.A, and the PA signal is expressed in terms of Bessel functions to satisfy the cylindrical geometry. However, the boundary conditions are much more complex than the gas-phase OA case, because the cylindrical container surface cannot now be considered as a rigid surface as in the gas OA case; the acoustic impedance of the liquid is comparable to that of the container material so that acoustic waves can be transmitted into the solid. The boundary conditions now are the continuity of the normal displacement, traction, temperature, and heat flux at the liquid–container interface. If the container is a hollow cylinder of a solid in air, then additional boundary conditions exist at the solid–air interface. The results of Kohanzadeh et al. (1975) are not reproduced here. However, one simple special case of Kohanzadeh et al. is noted: the case in which the solid–liquid interface is at infinity. In this case, the PA pressure p produced by the light beam of power $\mathcal{P}_0$ modulated at angular frequency ω is given by

$$p(r) = (\alpha \mathcal{P}_0 \beta \omega / 4 C_p) \mathcal{H}_0(\omega r/c) \tag{46}$$

with the corresponding radial velocity V_a given by

$$V_a(r) = (-i\omega\alpha\mathcal{P}_0\beta/4\rho C_p c)\mathcal{H}_1(\omega r/c). \tag{47}$$

Here, α is the optical absorption coefficient (assumed small), β is the volume expansion coefficient, C_p is specific heat at constant pressure, r is the radial distance, c is sound velocity, ρ is the liquid density, and $\mathcal{H}_n$ is a Hankel function of the second kind, which represents asymptotically a wave propagating outward.

Equation (46) is a basis for performing liquid-phase PA spectroscopy using, for example, a small piezoelectric transducer immersed in a large volume of liquid and exciting the liquid by a modulated cw laser beam. It is valid for small absorption coefficient, and is intuitively appealing because the PA pressure signal is proportional to the light energy in one period $\mathcal{P}_0\omega$, the absorption coefficient α, the expansion coefficient β; and inversely proportional to the specific heat C_p. The time-averaged acoustic power is given by Kohanzadeh *et al.* (1975) as

$$P_{\text{acoust}} = \omega\beta^2\alpha\mathcal{P}_0^2/16\rho C_p^2, \tag{48}$$

which clearly shows that the integrated acoustic power is *quadratic* in the incident optical power. This does not violate the energy conservation principle because the absorbed optical energy is mainly converted into heat; for the case of 1-W blue light incident on water considered by Kohanzadeh *et al.* (1975), the conversion efficiency of light into sound is $\sim 10^{-15}$. Much higher conversion efficiency occurs for the pulsed laser case, as considered below.

The theory of pulsed OA spectroscopy has been considered by several workers, and we follow here the work of Patel and Tam (1981) because their work is most directly related to the use of pulsed OA spectroscopy for measuring weak absorptions in condensed matter. The basic semi-quantitative idea of pulsed OA generation is quite simple. If a light pulse of energy E_0 passes through an optically thin medium of length l and absorption coefficient α, the energy absorbed is

$$E_{\text{abs}} = E_0\alpha l. \tag{49}$$

If we assume that nonradiative relaxation predominates in the medium, then E_{abs} is also equal to the heat generated in the medium, given by

$$E_{\text{abs}} = \rho C_p V\Delta T \tag{50}$$

where V is the illuminated volume and ΔT is the temperature rise. If R is the radius of the light pulse, then

$$V = \pi R^2 l. \tag{51}$$

If we assume adiabatic, isobaric expansion, then the expansion ΔR of the illuminated volume is given by

$$\pi(R + \Delta R)^2 l - \pi R^2 l = \beta V\Delta T,$$

or

$$\Delta R \approx \tfrac{1}{2}R\beta\Delta T \tag{52}$$

if $\Delta R \ll R$, which is true in all the cases presently considered. Combining Eqs. (49), (50), and (52), we get

$$\Delta R = E_0\alpha\beta/2\pi RC_p\rho. \tag{53}$$

This expansion produces a pressure wave which travels radially from the illuminated cylinder. The excess pressure p is related to the frequency of the sound wave f and the displacement d by

$$p = 2\pi f c \rho d. \tag{54}$$

Since d at a certain radial distance r from the illuminated cylinder is proportional to ΔR [i.e., $d \approx \Delta R (R/r)^{1/2}$], we conclude from Eqs. (53) and (54) that

$$p = (\text{const})(\beta c/C_p)E_0\alpha \tag{55}$$

where the prefactor "const" is a constant for a fixed geometry and laser pulse shape.

Since the OA signal V_{OA} observable from a piezoelectric transducer is proportional to p, we have

$$V_{OA} = K(\beta c/C_p)E_0\alpha \tag{56}$$

where K is a constant for a fixed geometry, laser pulse shape, and transducer. Now we define the normalized OA signal S as V_{OA}/E. Thus

$$S = K(\beta c/C_p)\alpha. \tag{57}$$

In the above derivation of Eq. (57) for pulsed OA detection, we have assumed that the acoustic transit time $\tau_a = 2R/c$ is smaller than the laser pulse width τ_p. The opposite case of τ_a larger than τ_p has been considered by Nelson and Patel (1981), and the resultant equation is almost the same as Eq. (57) except c is replaced by c^2. Thus, without loss of generality, we can assume in the following discussion that $\tau_a < \tau_p$. Equation (57) is the basis of pulsed OA spectroscopy. We see that for a fixed material, so that $\beta c/C_p$ is a constant, S is proportional to the absorption coefficient α. By scanning the pulsed laser or other pulsed light source, an OA spectrum proportional to the absorption coefficient is produced. The proportionality constant can be obtained by independently determining α at one wavelength and measuring S at the same wavelength. The dye doping method for liquids (see Tam *et al.*, 1979) is an example of such an absolute calibration procedure for liquids. Once the proportionality constant K is obtained for one material, then Eq. (57) implies that absolute calibration is also achieved for another material if the geometry, laser pulse shape, and transducer remain unchanged from the old to the new material, and if $\beta c/C_p$ is known for the old and the new material. In other words, the absolute calibration can be scaled from sample to sample through the material-dependent factor $\beta c/C_p$.

The derivation of Eq. (57) is only semiquantitative because the time development of the acoustic pulse is not considered, only the final expansion is calculated. A more rigorous treatment of OA pulse generation is possible, following the fluid mechanics of Landau and Lifshitz (1959), as done by Patel and Tam (1981). A potential function $\phi(r, t)$ is used to define the acoustic field, with the acoustic velocity V_a and pressure p given by

$$V_a = \nabla \phi \tag{58}$$

and

$$p = -\rho \frac{\partial \phi}{\partial t}. \tag{59}$$

For the case of a long thin cylinder generating an acoustic wave due to the time-dependent expansion of the cylinder, Landau and Lifshitz (1959) have shown that the potential ϕ is given by

$$\phi(r, t) = -\frac{c}{2\pi} \int_{-\infty}^{t-r/c} \frac{\dot{S}_c(t')dt'}{[c^2(t - t')^2 - r^2]^{1/2}}, \tag{60}$$

where $\dot{S}_c(t')$ is the time derivative of the cross-sectional area $S_c(t')$ of the cylindrical source at the retarded time t', i.e.,

$$S_c(t') = \pi[R + \Delta R(t')]^2$$

$$= \pi R^2 + 2\pi R \Delta R(t'). \tag{61}$$

Patel and Tam (1981) have shown that Eq. (60) can be simplified to give

$$p_{peak}(r) = (\beta c/C_p)E_0\alpha(4/\pi^3\tau_p^3 cr)^{1/2} \tag{62}$$

where $p_{peak}(r)$ is the peak of the acoustic signal observable at a radial distance r from the line source; more specifically, a positive peak occurs at time $r/c - 0.35\tau_p$, and a negative peak of comparable magnitude occurs at the time $r/c + 0.35\tau_p$. Note that τ_p is the $1/e$ full width of the light pulse, assumed to have a Gaussian time envelope with peak intensity at time $t = 0$.

Equation (62) confirms the semiquantitative result of Eq. (55) except for one small difference, that the "const" in Eq. (55), equal to $(4/\pi^3\tau_p^3 cr)^{1/2}$, depends on c as well as laser pulse shape and separation r. This, of course, only affects the scaling of the proportionality constant for different materials as discussed after Eq. (57). In other words, the result of Eq. (62) indicates that the scaling factor should be $\beta c^{1/2}/C_p$ instead of $\beta c/C_p$ previously discussed. However, this difference is usually small if scaling from liquid to liquid samples is done.

Many theoretical papers on direct-coupling OA detection principles, and on the optical generation of acoustic waves and the subsequent

acoustic propagation (see Section V.A), have been published in the literature (e.g., Atalar, 1980, Liu, 1982, and Lai and Young, 1982). Some
papers are listed in Table IV for readers interested in the theories of
various cases of optoacoustic generation.

D. The Direct-Coupling Method: Instrumentation

The possibility of generating and detecting acoustic waves inside liquids or solids by a chopped or pulsed radiation beam has been known for
some time. For example, White (1963), Gournay (1966), Westervelt and
Larson (1973), Lyamshev and Naugol'nykh (1980), and others have calculated the effect, while Carome *et al.* (1964), Rentzepis and Pao (1966), Bell
and Landt (1967), Bushanam and Barnes (1975), and others have observed the effect. However, it was not until the work of Hordvik and
Schlossberg (1977) that the direct-coupling OA technique was first demonstrated as useful for detecting weak absorptions in condensed matter.

The experiment of Hordvik and Schlossberg (1977) is shown in Fig. 17.
In their experiment, a cw laser (CO_2, CO, Ar^+, etc.) is used; the chopped
beam is incident on a highly transparent solid sample (e.g., LiF, MgF_2,
etc.). A piezoelectric transducer is attached on one surface of the sample
with epoxy, while another similar transducer is positioned close to the
opposite surface of the sample without contact. The purpose of the latter
transducer is to measure the effect of laser light scattered onto the transducers. Using laser powers of several hundred milliwatts and chopping
frequencies of 0.15–3 kHz, Hordvik and Schlossberg (1977) achieved a
detection sensitivity in the range of $10^{-4}–10^{-5}$ cm^{-1}, depending on the
amount of light scattering in the solid. In a later experiment with a similar

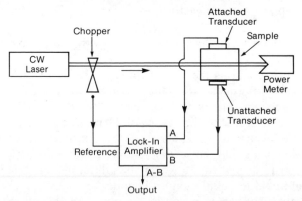

Fig. 17. Schematic of the experimental arrangement for the first OA detection of weak
absorption by direct coupling. [After Hordvik and Schlossberg (1977).]

apparatus, Hordvik and Skolnik (1977) reported that surface and bulk absorption in solids can be separately identified, because surface absorption produces a hemispherical wave, while bulk absorption produces a cylindrical wave. These two types of OA waves can be distinguished by their spatial and temporal differences.

The above study was further developed by Farrow *et al.* (1978a). Their sample is glued onto a front-surface mirror with an Eastman 910 cyanoacrylic adhesive. A barium titanate transducer (Endo-Western EC-64) is attached to the back of the mirror. A continuous light source, either a Xenon arc and monochromator assembly or a dye laser pumped by an Ar^+ laser, is used and chopped at 135 Hz. Simultaneous light intensity and PA signal measurements permit a normalized PA spectrum to be obtained using a DEC LSI-11 microcomputer. For weak absorption, the normalized PA signal is proportional to the absorption coefficient α, as seen in Eq. (46) for the chopped cw excitation beam case.

Besides the above direct-coupling OA spectroscopy of solids, similar experiments for liquids have been performed. For example, Lahmann *et al.* (1977) have used the apparatus of Fig. 18 to detect 0.012 ppb of β-carotene in chloroform. Here, an Ar ion laser producing simultaneously the 4880- and the 5145-Å lines is used. Modulation is performed with a dual chopper so that light of periodically alternating wavelengths is incident onto the liquid. This wavelength alternation permits a considerable reduction of the background signal due to light scattering and window absorptions, because absorptions by the solute at the two colors are quite differ-

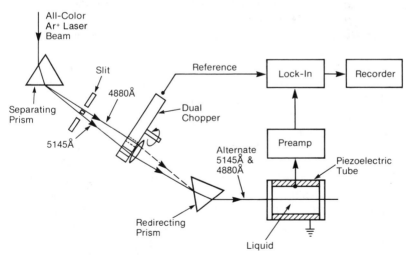

Fig. 18.   Direct-coupling OA technique for liquids, using wavelength switching in addition. [After Lahmann *et al.* (1977).]

ent, while the background signal is almost independent of the two colors. Lahmann *et al.* (1977) achieved a detection limit of 10^{-5} cm^{-1} absorption by the liquid. A similar OA experiment, without the wavelength alternation, has also been done by Oda *et al.* (1978) for liquids; their detection limit for a Cd-containing chloroform solution is 0.054 ppb, corresponding to an absorption coefficient of 2×10^{-5} cm^{-1} at 5145 Å.

All the above direct-coupling OA techniques for solids or liquids rely on the use of a chopped light beam, typically at chopping frequencies below a few kilohertz. Window absorptions and effects of light scattering onto the transducer are usually the origins of detection limitations, although other factors like heating of the liquid by the strong cw laser beam of power ~1 W (causing convection currents, self-defocusing, etc.), mechanical noise at low frequencies, electrical pickup noise at multiples of line frequencies, etc., also pose limitations. These limitations are avoided in the pulsed OA technique (Patel and Tam, 1979a) using pulsed laser beams of low duty cycle (e.g., 1-μsec, 1-mJ pulses at 10 Hz). Time-gating the desired OA signal (which travels ballistically from the illuminated region of the sample to the transducer) can be used to discriminate against window absorption (which usually arrives later than the desired OA signal) and against light scattering (which occurs almost instantaneously, i.e., earlier than the desired OA signal). Heating up of the liquid and associated noise effects are minimized by using low average laser power of ~10 mW. Mechanical noise and electrical pickup are noises peaking at frequencies ≲10 kHz, and can be eliminated by using high-frequency bandpass filters (e.g., passing between 0.1 and 0.5 MHz); this is possible because the pulsed OA signal produced by short laser pulses (≲1 μsec) is of high acoustic frequency (≳0.1 MHz), in contrast to the low-frequency OA signals produced by chopped cw excitation beams considered above.

The experimental arrangement used by Tam *et al.* (1979) to demonstrate the capability of pulsed OA spectroscopy technique is shown in Fig. 19. The transducer used is shown in Fig. 20. The lead zirconate titanate (PZT) cylinder used is a commercially available poled ceramic cylinder of 4 mm diameter and 4 mm height. The transducer is modeled after the commercial hydrophones with the important difference that the present transducer is rigidly spring loaded against a flat and polished stainless steel front diaphragm. Here a polished front steel diaphragm is used instead of a plastic or epoxy front seal as in most commercial hydrophones because it is desirable to achieve minimum contamination of the liquid due to corrosions at the diaphragm, and good reflectivity of the front diaphragm so that scattered light onto it tends to be reflected instead of absorbed. Also, electrical pickup is minimized by enclosing the PZT cylinder in a metal case. A low-noise "microdot" miniaturized connector

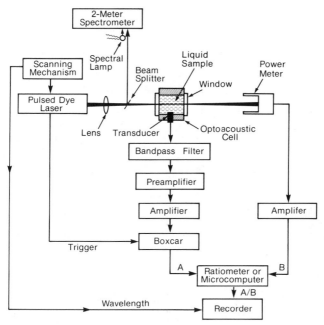

Fig. 19. Experimental arrangement to perform pulsed OA spectroscopy for measuring weak absorptions in liquids. [After Tam *et al.* (1979).]

is used as the output jack for the transducer shown in Fig. 20. An alternative way to construct a transducer with the same PZT cylinder is shown in Fig. 21, where a common BNC connector is used as the output jack.

The liquid OA cell shown in Fig. 19 is a stainless steel cell with a polished inside surface and with fused-silica windows attached to the cell

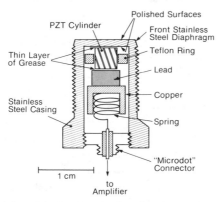

Fig. 20. Cross section of the homemade piezoelectric transducer utilizing a lead zirconate titanate cylinder (LTZ-2 or PZT-5A). [After Tam and Patel (1980).]

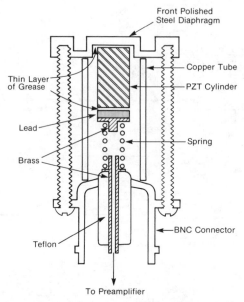

Fig. 21. Another method of mounting the PZT cylinder to make a transducer.

using Teflon O rings. The construction design aims at minimizing contamination of the liquid and light-scattering effects (for details, see Patel and Tam, 1981). Further optimization of the sensitivity for OA detection should be possible, for example, by multipassing the laser beam or by acoustic focusing. In fact, Patel and Tam (1980) have suggested elliptical OA focusing, with the pulsed dye laser excitation at a focus of an elliptical OA cell or substrate, and with the transducer located at the other focus. Such an elliptical OA focusing cell was first constructed and used by Heritier *et al.* (1982).

An observed pulsed OA signal is shown in Fig. 22. Such an undulating OA signal due to a laser pulse has been studied and explained by Burt (1979). The signal consists of a positive pulse followed by a negative pulse with multiple ringing thereafter. The ringing signal is reproducible if no geometry is changed and is due to the multiple reflections in the OA liquid cell as well as the internal ringing of the transducer; also a small amount of absorption (probably $\sim 10^{-3}$ or 10^{-4}) at the optical windows contributes to some of the OA signals at later times. The boxcar is usually adjusted to capture the first large pressure peak only, as shown in Fig. 22. This is done because all window absorption effects and light-scattering effects are thus minimized. The boxcar output in Fig. 19 is normalized by the laser pulse energy as the laser is scanned to obtain a normalized OA

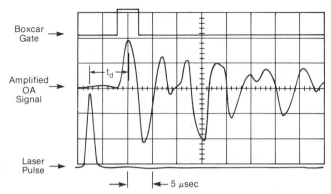

Fig. 22. This oscilloscope picture shows the laser pulse (lower trace), amplified OA signal (middle trace), and boxcar gate for the OA detection (upper trace) for a typical liquid. [After Patel and Tam (1981).]

spectrum, which is proportional to the absorption spectrum [see Eq. (62)]. Absolute calibration of the spectrum is performed as described in Section III.C.

The fact that the OA signal in Fig. 22 represents ballistic propagation of sound from the illuminated cylinder of liquid to the transducer has been demonstrated by Patel and Tam (1979a). In that work, the separation r between the laser beam and the transducer is varied by using a micrometer translation stage, and the corresponding delay time t_d (Fig. 22) is measured for each r. A plot of Δt_d versus Δr for the case of liquid benzene yields a straight line passing through the origin, with the gradient corresponding to a sound velocity of $(1.32 \pm 0.04) \times 10^5$ cm/sec, in good agreement with the known ultrasonic velocity in benzene at 20°C of 1.324×10^5 cm/sec (see Patel and Tam, 1979a). Thus, the pulsed OA technique is not only good for spectroscopy, but also good for many other nonspectroscopic applications like sound velocity measurements (see, e.g., Tam *et al.*, 1982).

The OA cell used in most of the studies of Patel and Tam (1981) is designed to minimize contamination or corrosion; thus only stainless steel, fused silica, and Teflon are exposed to the liquid. Such a cell works well for liquids like water (Tam and Patel, 1979c), benzene (Tam *et al.*, 1979a), and other organic liquids (Patel and Tam, 1979c). However, for highly reactive liquids like strong acids, the exposure of any metal part to the liquid becomes unacceptable because of chemical reactions. Thus, a highly corrosion-resistant OA cell suitable for pulsed OA spectroscopy is desired, as is shown in Fig. 23 (Tam and Patel, 1980). Here a silica or sapphire cuvette is used to contain the liquid, and the PZT transducer is

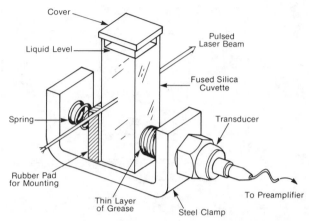

Fig. 23. Highly corrosion-resistant OA cell using a quartz cuvette with externally coupled PZT transducer. The cuvette can also be simply resting on the flat transducer surface. [After Tam and Patel (1980).]

externally coupled to the cuvette. Such an external coupling reduces the OA detection sensitivity compared to the direct-coupling case (Fig. 19) by about a factor of 2 (see Tam and Patel, 1980). However, the OA cell of Fig. 23 offers the highest achievable corrosion resistance as well as ease of sample change. This is in total contrast to the OA cell used by other workers (see Fig. 18) where the silvered surfaces of the piezoelectric tube is directly exposed to the liquid, contamination is likely, and sample change is laborious.

The external coupling technique of Tam and Patel (1980) has been improved and extended by other workers. For example, Voigtman *et al.* (1981) have recently demonstrated that a quartz cuvette resting on a PZT disk with glycerol coupling is ideally suited for routine testing of trace amounts of dyes, drugs, and biochemicals. More recently, Voigtman and Winefordner (1982) have developed a windowless flow cell that is suitable for simultaneous detections of fluorescence, PA, and photoconductivity signals with pulsed laser excitation of the flowing liquid column. Such flow-cell detections appear to have broad applications for trace detections and identifications.

E. Liquid-Coupling Method

The idea of the direct-coupling method using piezoelectric transducers can be extended to a liquid-coupling technique, which is illustrated schematically in Fig. 24. Two types of acoustic signals are possible. (i) For

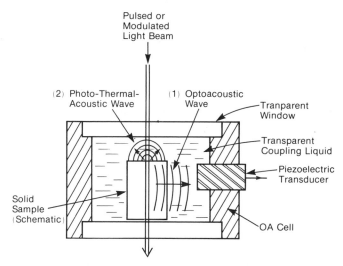

Fig. 24. Schematic representation of a liquid-coupling OA cell for solid sample, which can be a single intact piece as shown or can be in other forms like powder, long thin fiber, coating on a substrate, etc.

optically thin samples with negligible surface absorptions, only the optoacoustic wave (see Fig. 24) due to bulk absorption as discussed in Section III.C is important. A significant fraction of the OA wave is transmitted into the liquid, depending on the acoustic impedance mismatch. This method of detection is basically the same as in the direct-coupling case with the disadvantage that two solid–liquid interfaces are encountered so that the transmission factor due to the two interfaces may be as low as 1%. However, there are certain advantages with this liquid-coupling method: (a) change of sample is simpler and faster because the transducer is not attached to the sample; (b) for certain types of sample (e.g., powders, thin fibers, materials that react with transducer, etc.), the transducer cannot be directly attached to the sample, and thus the liquid-coupling method is the other feasible alternative. (ii) For optically thick samples, or for samples with significant surface absorption, a photo-thermal-acoustic wave (see Fig. 24) is generated because of the surface heating and thermal diffusion from a diffusion thickness layer of the solid into the liquid. Such a photo-thermal-acoustic wave has the same origin as that due to the PA signal for the gas-coupling case discussed in Sections III.A and III.B, and similar theories hold for the two cases with one exception: the sample vibration effect is frequently not significant compared to the gas-piston effect in the PA gas-coupling case, but the sample vibration effect should be significant in most cases for the liquid-coupling case. The advantage of liquid-coupling for detecting photo-thermal-acoustic waves is that most of the applications using the gas-microphone PA technique (like depth profiling using chopping-frequency dependence) can be done here using piezoelectric transducers, which have much higher frequency response than microphones.

Not many experiments using the liquid-coupling technique have been reported in the literature. Sam and Shand (1979) have reported the use of liquid coupling to detect the optoacoustic wave. In that work, a pulsed dye laser is used for excitation, and a lead zirconate titanate disk is used for OA detection. OA spectra of weakly absorbing Nd-doped crystals are obtained, and they pointed out that the liquid-coupling method should be useful for routine testing of laser-active materials like Nd-doped glasses, noting that the OA signal should be smaller for more fluorescent materials.

F. Measuring Weak Absorptions

Similarly to the gas-phase OA case, the high sensitivity available with the use of piezoelectric transducers in direct contact with a condensed sample has been utilized to measure weak absorptions. The gas-microphone method is seldom used for measuring weak absorptions because of the comparatively lower sensitivity, although McDavid et al. (1978) have used it in conjunction with high-intensity CO_2 laser sources to measure weak absorptions of alkali halide samples at 10.6 μm.

For OA monitoring of weak absorptions in solids, Hordvik and Schlossberg (1977) were the first to use the attached-transducer technique, and they have measured weak absorption by highly transparent solids like CaF_2, SrF_2, etc., with a detection sensitivity of 10^{-5} cm^{-1}. Weak absorptions in optical fibers have been measured by Burt et al. (1980), who obtained a large OA signal (2 mV) when a piezoelectric transducer was attached to a commercial step-index optical fiber of 200 μm diameter transmitting a ruby laser pulse of 1 J energy. However, no OA signal was produced when the transducer was not in contact with the fiber. This OA sensing of the absorption of light by optical fibers may be useful for determining fiber losses, for investigating the properties of fiber defects, and for determining the exact location of fiber damage inside a sheathed cable.

Weak absorptions due to some "forbidden" transitions in liquids have also been measured by OA techniques with immersed transducers. For example, Sawada et al. (1979) have measured some "forbidden" transitions of rare-earth ions in aqueous solutions. They pointed out that, together with absorption and emission measurements of the trivalent rare-earth solutions, the OA spectrum offers additional information on the radiationless decay processes. This is quite relevant for the rare-earth ions because they frequently have high fluorescence efficiencies. "Forbidden" absorption due to excitation of high vibrational overtones have been reported by Tam et al. (1979) and Patel et al. (1979). In that series of work, high overtones of the C–H stretching mode in benzene and other liquid hydrocarbons were measured by the pulsed OA technique (Fig. 19), and new results on the absorption position and line shape were measured up to the 8th harmonic of the C–H stretch. Also, high overtones in halobenzenes were measured by Tam and Patel (1980) using the external-coupling OA cell (Fig. 23).

An important application of OA methods in liquids is the measurement of the visible absorption spectrum of water by Patel and Tam (1979d) and Tam and Patel (1979b). The results of the OA spectra for light and heavy waters are shown in Fig. 25. Water is, of course, the most important liquid in many respects (environmental, biological, technological, geological, etc.), and hence many workers have previously measured its absorption spectrum, mostly by using long-pathlength absorption techniques. Some of these previous works (Hulburt, 1945; Sullivan, 1963; Irvine and Pollack, 1968; Hale and Querry, 1973; Kopelevich, 1976; Hass and Davisson, 1977; Querry et al., 1978) are shown in Fig. 25 for comparison with our data. Accuracies of the previous measurements were severely limited by scattering losses at windows and in the liquid, by refractive index effects causing small changes in the collimation of the light beam through the liquid onto the detector, by contamination of the liquid due to the containing vessel, etc. Although some of these sources of errors are corrected for in the previous work, the large scatter of the previous data indicates that accurate corrections are at best very difficult. Indeed, the previous data near the "green minimum" of light water near 5000 Å are scattered by about an order of magnitude. Long-pathlength extinction measurements using hollow optical fibers for accurate transmission monitoring of weak absorptions have recently been developed (Stone, 1978), but this has not been applied to water. Our pulsed OA results, shown in Fig. 25, have estimated accuracies of about ±10%, and are believed to be the most reliable data for waters up to date. Such data should be valuable as reference when weak absorptions of solute in aqueous solutions are considered. Also, our data should be valuable for modeling sunlight penetration in clear ocean waters, which have nearly the same absorption as for pure water.

The pulsed OA detection for liquids is not only limited to the room-

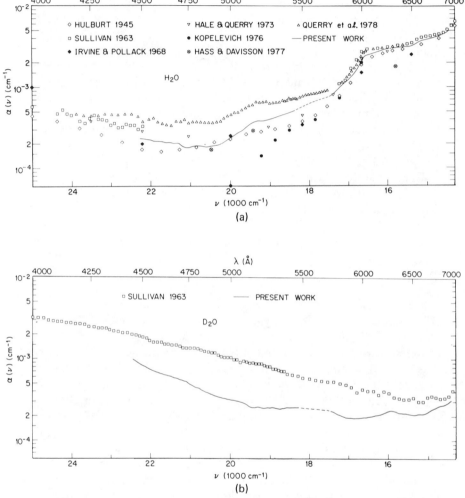

Fig. 25. Absorption spectra of triply distilled water [H_2O in (a) and D_2O in (b)] by optoacoustic determinations (shown by solid line). The dashed part is interpolation not covered by the tuning range of the laser dyes used. Also shown are selected experimental data from other authors for comparison. [From Tam and Patel (1979b).]

temperature liquids considered above. Piezoelectric materials such as lead zirconate titanate are usable for a broad temperature range (e.g., 1–600 K). For example, Patel *et al.* (1980) have used the pulsed OA method to measure the absorption spectrum of liquid methane at 94 K.

G. Trace Detection

Lahmann *et al.* (1977) have used the direct-coupling OA technique (in conjunction with wavelength switching) for detecting 0.012 ppb of carotene in a solution. Their work has been described in Section III.D (Fig. 18). Also Oda *et al.* (1978) have detected trace amount of cadmium extracted from a fungus, and they have achieved a detection limit of 0.054 ppb of cadmium in chloroform. The detectability of low concentrations relies on the prudent choice of a laser wavelength that is very strongly absorbed by the trace molecules to be detected. For example, consider an aqueous solution containing 10^{-12} g/cm^3 of a substance (i.e., 0.001 ppb by weight) of molecular weight say 100. If the substance has a molar absorption coefficient of 10^5 cm^{-1} L mol^{-1} at a certain wavelength, then the absorption coefficient of the solutions due to the substance is 10^{-6} cm^{-1}. This is within the detection capability of the pulsed OA technique as described in Section III.D and Fig. 19. Thus, the pulsed OA method should be capable of detecting a substance at the 0.001-ppb level if a laser wavelength that is very strongly absorbed by the substance is available.

For continuous measurement of a weak absorption (due, for example, to a trace constituent), a flow-through cell for laser OA detection is desirable. Such a cell has been designed by Sawada *et al.* (1981) (see Fig. 26),

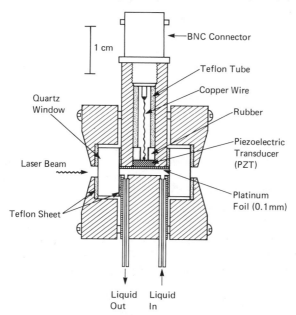

Fig. 26. PA flow cell for liquids suitable for laser excitation. [After Sawada *et al.* (1981).]

who have used it to detect carcinogenic dyes in solutions. Such flow-through OA cells, in conjunction with pulsed or suitably modulated cw lasers, should be quite valuable for continuous real-time sampling of liquids. Another type of flow cell, suitable for PA detection as well as other modes of detections simultaneously, has been described by Voigtman and Winefordner (1982).

H. Absorptions of Thin Films

The capability of laser OA techniques to detect weak absorptions implies that OA techniques should be useful for monitoring absorptions due to a thin film, which may be on a substrate or may be free standing. The reason, of course, is that weak absorption may be caused by a low absorption coefficient for the optical wavelength used (which are the cases considered so far), or may be caused by a short pathlength. Indeed, since the pulsed OA technique of Patel and Tam (1979a) can detect fractional absorptions of 10^{-6} (i.e., an absorption coefficient of 10^{-6} cm^{-1} for a 1-cm pathlength), it may be useful for detecting absorptions of thin films of 10^{-8} cm thickness if the absorption coefficient is 10^2 cm^{-1} or larger. In practice, it may be difficult to actually achieve such detectabilities for thin films because of difficulties associated with acoustic coupling to thin films, light scattering by interfaces causing spurious transducer signals, and absorption by the substrate.

Surface absorptions can, of course, be detected by the gas-coupling technique, by the direct-coupling technique, or by the liquid-coupling technique. Parker (1973) reported the observation of optical absorption at glass surfaces using a gas OA cell with a condenser microphone. He reported that a chopped light beam entering the high-pressure OA cell produces sound waves in the gas even though the gas does not absorb the light beam. The results can be quantitatively interpreted by assuming the existence of an optically absorbing surface layer on the glass window, and this assumption appears to be consistent with the effect of laser damage on glass windows at high intensities. Independently, Kerr (1973) reported a gas-microphone OA cell (the "alphaphone") for measuring thin-film absorption of laser beams. With a capacitance microphone, adequate sensitivity to detect a fractional absorption of 10^{-4} by a surface layer is achieved with 10 W incident CO_2 laser radiation and a chopping frequency of 0.3 Hz. Similar techniques have been used by Adams et al. (1978) and Fernelius et al. (1981) to examine absorption by laser mirror coatings. Rosencwaig and Hall (1975), Castleden et al. (1979), and Schneider et al. (1982a) have reported the use of gas-microphone PA techniques for analyzing the local distributions of dyes adsorbed on thin-layer chromatography plates. Such techniques should be very valuable for identifying separated compounds on chromatography plates because it is sometimes difficult to identify the compounds by conventional spectrophometric means because of optical opaqueness, light scattering, etc.

It may be noted that the gas-coupling method or liquid-coupling method has higher sensitivity for detecting absorption due to a thin film than for detecting an equal amount of absorption in a thick film since heat diffusion to the coupling medium is more efficient for a thin-film sample than for a thick-film sample. Kanstad and Nordal (1980a,b) have used a chopped cw tunable CO_2 laser beam and piezoelectric detection to measure a thin Al_2O_3 oxide film of various thickness that has been formed on an Al substrate by anodization at various voltages V_{anod}. They found that a characteristic OA absorption peak of Al_2O_3 at 10.55 μm, of magnitude S_{film}, can be observed (Fig. 27a), and that S_{film} is linear in V_{anod} (Fig. 27b), indicating that the OA absorption peak is proportional to the Al_2O_3 layer thickness. Their results indicate that they can easily detect a monolayer of Al_2O_3 on Al.

Pulsed OA detection has been used to measure absorptions by a thin film of powdered sample (Tam and Patel, 1979c), or by a liquid film of

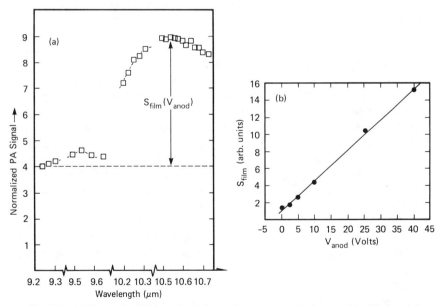

Fig. 27. (a) Normalized PA surface absorption spectrum (using tunable CO_2 laser) for an aluminum plate that has been anodized at 10 V. (b) Plot of height of the Al_2O_3 peak [marked S_{film} in (a)] as a function of the anodization potential V_{anod}. [After Kanstad and Nordal (1980a).]

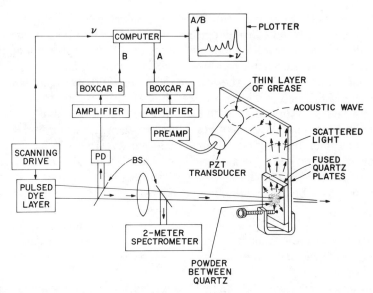

Fig. 28. Apparatus suitable for pulsed OA spectroscopy of a sample of a thin liquid film or a thin layer of powder in a transparent liquid film. The sample is clamped between two quartz substrates, and the OA signal generated in the sample is coupled to the substrate. The PZT transducer is acoustically coupled to a quartz substrate plate. PD denotes the pyroelectric detector and BS the 5% beam splitter. [From Tam and Patel (1979c).]

several microns thickness trapped between quartz (Patel and Tam, 1980). Their apparatus is shown in Fig. 28. The thin-film sample is clamped between quartz plates. One of the plates should have at least one bend to decrease the effect of light scattering. This is because a thin-film sample (especially of powders) frequently causes strong light scattering which induces spurious transducer signals if the scattered light is incident on the transducer. Sharp bends in the quartz substrate reduce light scattering greatly without attenuating the acoustic wave significantly because the acoustic wavelength (~1 cm for our case) is much longer than the optical wavelength, and acoustic diffraction is readily possible in the quartz substrate. Also, the great difference in acoustic and light velocities permits the use of time gating very effectively to discriminate against light-scattering effects. Using dye laser pulses of ~1 mJ energy at 10 Hz repetition rate and an integration time of 1 sec, Patel and Tam (1980) have demonstrated a detection sensitivity for a fractional absorption of 10^{-5} by liquid films between quartz plates.

Thin-film absorption is one of the most powerful applications of photoacoustic methods, and we can expect that many new cases of application will be reported in the future. In the above, we have noted several examples of absorptions by glass surfaces, laser mirror coating, and thin-film chromatography plates. Further noteworthy examples of surface absorption can be found in the series of papers by Nordal and Kanstad (1977, 1978). In their work, photoacoustic spectroscopy of various substances and complexes on metal surfaces is obtained. This may have important applications for surface chemistry and surface catalysis on metal surfaces. Such surface reactivity studies are best studied by infrared (IR) photoacoustic spectroscopy (to probe vibrational transitions) as shown by Low and Parodi (1980). To improve signal-to-noise ratios, surface studies can be done with Fourier transform IR spectroscopy as performed by several authors [e.g., Rockley and Devlin (1980) for cleaved coal surfaces]. More details will be given in Section III.J.

I. Absorptions of Aerosols and Particles

The optical absorptions by aerosols, colloids, powders, and all other forms of particles provide another example of the applicability of photoacoustic methods where conventional methods work poorly at best. It is very important to measure absorptions due to fine particles, both in basic science and in applied technologies like the motor industry, smog control, pigment manufacturing, coal conversions, etc. PA absorption measurements of powders using gas-coupling methods have been reported by several authors; examples of studies of inorganic powders (e.g., HO_2O_3, metal powders, etc.) can be found in the review article of Rosencwaig (1978) and in the book of Rosencwaig (1980a), and in Adams *et al.* (1977a,b), Eaton and Stuart (1978), Blank and Wakefield (1979), and Shaw (1979). Organic powders have also been measured; for example, Nordal and Kanstad (1977) have performed PA spectroscopy of glucose powders using a CO_2 laser.

An example of PA spectroscopy of colloidal suspensions in liquids is given by Oda *et al.* (1980). In that work, a chopped laser beam with piezoelectric detection is used. Fine particulates of $BaSO_4$ are produced chemically in a suspension, and the effect of particle size distributions on the PA signal and on the turbidimetric signal is investigated. Oda *et al.* (1980) reported that in comparison with a turbidimetric analysis, the PA signal was less affected by the particle size distribution, and the detection limit is about two orders of magnitude better for the PA detection case. Oda *et al.* (1980) suggested that laser-induced PA spectroscopy is well suited for quantitative determination of particulates in turbid solutions.

Absorbing species, including particulates, in the atmosphere is, of course, an area of great concern, and PA methods have shown great promise in this field. We have discussed in Section II.D the applications for pollution monitoring of molecular species. For the case of atmospheric aerosols, Yasa *et al.* (1979) have reported the use of an acoustically nonresonant cell using a Knowles microphone (BT-1759) for detecting the PA signals due to the aerosol absorptions. They show that the absorbing species in urban aerosols are mainly particulate carbon in the graphite form. McClenny and Rohl (1981) have shown that atmospheric aerosols can also be studied by first collecting them on a Teflon filter and subsequently studying the collected samples in a PA cell. They reported that the elemental carbon concentration in air can be estimated by a PA analysis technique. Similarly Rohl and Palmer (1981) have determined the concentration of airborne ammonium-salt particulates.

High-resolution optoacoustic spectroscopy of rare-earth powders was reported by Tam and Patel (1979c). Since most rare-earth oxides cannot be produced as macroscopic crystals, their spectral features cannot be easily studied by conventional transmission or reflection spectroscopy. The apparatus of Tam and Patel (1979c) is shown in Fig. 28, and some results are shown in Fig. 29. From these spectra, taken at room temperature, energy levels of the Stark sublevels both in the ground-state manifold and in an excited-state manifold are derived.

J. Fourier Transform Spectroscopy

Photoacoustic spectra using incoherent light sources (e.g., Xe lamps or glow bars) in conjunction with dispersive optical instruments (e.g., grating spectrometer) suffer from two disadvantages: (i) Data collection for the spectrum is point by point rather than simultaneous. (ii) Throughput of a dispersive spectrometer is small. These difficulties do not exist if a laser is used as the excitation source, but the limited tuning range or limited availability of lasers results in the fact that incoherent light sources are still being used in many PA measurements, especially in the infrared (IR). The technique of Fourier transform (FT) PA spectroscopy, analogous to the well-known Fourier transform absorption spectroscopy, was introduced by Farrow *et al.* (1978b) to overcome the difficulties mentioned above when using a lamp source. In the FT PA spectrometer, data at all spectral wavelengths emitted by the light source are simultaneously measured (Fellgett's advantage), and the throughput of the Michelson interferometer used in the FT PA technique is much larger than the throughput of a dispersive spectrometer (Jacquinot's advantage).

The schematic of a FT PA spectrometer is shown in Fig. 30. The top part of Fig. 30 is a Michelson interferometer with a fixed mirror at distance l_0 and a movable mirror at a distance $l_0 + X$ from the beam splitter. Constructive or destructive interference of the two beams from the mirrors recombining at the beam splitter occurs when $X = \frac{1}{2}\lambda, \lambda, \frac{3}{2}\lambda, \cdots$, or $X = \frac{1}{4}\lambda, \frac{3}{4}\lambda, \frac{5}{4}\lambda, \cdots$, respectively, for a given wavelength λ of light being

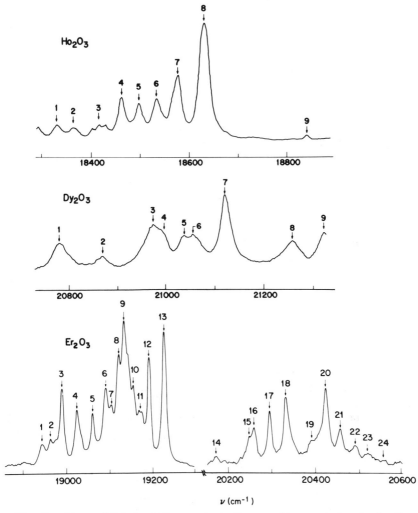

Fig. 29. Observed OA absorption spectra in the green–blue spectral region for three rare-earth-oxide powdered crystals at 20°C. Some prominent absorption lines are marked and numbered. [From Tam and Patel (1979c).]

considered. Hence, for the wavelength λ, the modulated part of the intensity $I(\lambda,X)$ incident on the sample is

$$I(\lambda,X) = A(\lambda) \cos (2\pi X/\tfrac{1}{2}\lambda),\qquad (63)$$

where $A(\lambda)$ is the spectral intensity emitted by the light source at λ. The corresponding PA signal, assumed to be proportional to the optical

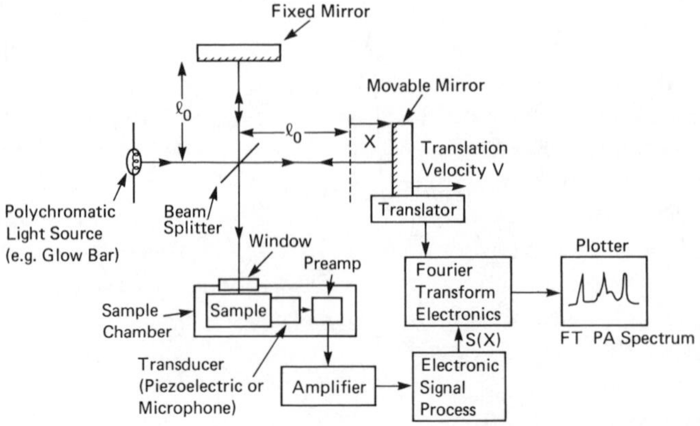

Fig. 30. Schematic diagram of a Fourier transform PA spectrometer.

absorption at λ (i.e., no signal saturation is present and full thermal relaxation of the excitation occurs), is

$$S(v,X) = A(v)\alpha(v) \cos (4\pi X v), \qquad (64)$$

where $v = 1/\lambda$. The PA signal $S(X)$ is due to all emitted wavelengths at position X.

$$S(X) = \int S(v,X)\,dv, \qquad (65)$$

where the integration is over the spectral range of the incident light. Hence $S(X)$ is a Fourier transform of the optical absorption $A(v)\alpha(v)$. By measuring $S(X)$ for an extended range of X, an inverse Fourier transform of $S(X)$ can be performed, obtaining back $A(v)\alpha(v)$ as a function of v. Since $A(v)$ is either known or can be measured by using a black absorber in place of the sample, the desired PA spectrum $\alpha(v)$ can be derived.

The FT PA spectroscopy was first used by Eyring and co-workers (Farrow *et al.*, 1978b; Lloyd *et al.*, 1980a,b). In their experiment, a white light (e.g., a 100-W tungsten iodide lamp) is used as the light source, and the transducer is either a piezoelectric transducer or a microphone. They demonstrated that FT PA spectra of various samples like Nd:glass materials, La_2O_3, whole blood, etc., can be obtained, and showed that the combined multiplexing and throughput advantages of the interferometry method significantly reduce the data collection time and/or improve the signal-to-noise ratio.

In the IR regime, Rockley and co-workers first demonstrated the applicability of FT PA spectroscopy (see Rockley, 1979, 1980a,b; Rockley and Devlin, 1980; Rockley *et al.*, 1980, 1981). In their work, a commercial FT IR spectrometer (Digilab FTS-20 vacuum spectrometer) is modified so that the sample chamber contains a small nonresonant PA cell of volume about 1 cm^3. A General Radio GR-1962 $\frac{1}{2}$-in. foil electret microphone is used as the sound transducer. The microphone signal is fed to an Ithaco 143L preamplifier, whose output is connected to the amplifier of the commercial instrument. Rockley has used this FT IR spectrometer to examine PA spectra of polystyrene films, of aged and freshly cleaved coal surfaces, of ammonium sulfate powders, and so on. The tremendous potential of the FT PA spectroscopy in the IR (where characteristic vibrational modes of various chemical bonds can be identified) is fully demonstrated.

A further advantage of FT PA spectroscopy in the IR is pointed out by the study of Royce and collaborators (Laufer *et al.*, 1980; Royce *et al.*, 1980; Teng and Royce, 1982). They pointed out that the conventional spectroscopy method of transmission or reflection monitoring frequently suffers from spectral distortions when powdered solids dispersed in a transparent matrix are examined. The reason is that light scattering decreases when the refractive index of the matrix equals that of the powdered solid, causing an increased transmission of the pellet (the Christiansen effect). In their work, a commercial FT IR spectrometer (Nicolet 7199) is used with the normal detector being replaced by a PA cell of volume about 1 cm^3, containing a $\frac{1}{2}$-in. B&K 4165 microphone of sensitivity about 50 mV/Pa. Spectra between 400 and 4000 cm^{-1} are obtained at a resolution of 4 cm^{-1}. The elimination of the Christiansen effect for powders such as AgCN is demonstrated.

K. Nonlinear Spectroscopy

Photoabsorption processes involving more than one photon (i.e., nonlinear spectroscopy) can be detected by OA methods: such processes may be multiphoton (i.e., two or more photons are absorbed and/or emitted simultaneously) or may be stepwise (i.e., two or more photons are absorbed and/or emitted at different times, e.g., absorptions of triplet excited states produced by optical excitation of the ground state). In nonlinear spectroscopy, the absorption is usually detectable only when high-intensity light sources (like pulsed lasers) are used. Hence, the pulsed OA technique developed by Patel and Tam (1979a) is ideally suitable for measuring weak nonlinear optical absorption/emission effects, as first pointed out by Tam *et al.* (1979).

The first demonstration of OA Raman-gain spectroscopy (OARS) for liquids was reported by Patel and Tam (1979c). Two synchronized flashlamp-pumped dye lasers are used to provide the pump and the signal pulses, and gated OA measurement with a boxcar integrator is used to detect the energy deposited in the liquid due to the stimulated Raman scattering (see Section II.G for the theory of Raman-gain spectroscopy and the gas-phase work of Barrett and co-workers). In the work of Patel and Tam (1979c), the pump wavelength λ_p is tuned in the green (480–560 nm) while the signal wavelength λ_s is fixed in the red (600 nm). An OARS signal is observed whenever $\lambda_p^{-1} - \lambda_s^{-1}$ equals a Raman frequency of the liquid. OARS spectra were obtained for several liquids such as benzene, acetone, trichloroethane, toluene, and hexane.

The sensitive detection of weak two-photon absorptions by OA method was first demonstrated by Tam and Patel (1979a). Previously, sound generation in liquids due to two-photon absorption was known, but the effect had never been applied for quantitative measurements of two-photon absorption cross sections. For example, Rentzepis and Pao (1966) used giant ruby-laser pulses in organic solutions of benzanthracene, and observed what they believed to be acoustic waves generated by two-photon absorption. More recently, Bonch-Bruevich et al.(1977) did succeed in detecting one-photon and two-photon absorptions in organic dye solutions using a pulsed OA technique, but the sensitivity achieved was rather poor; for example, the one-photon absorption sensitivity was not better than that attainable with a conventional spectrophotometer. The two-photon OA absorption spectra of Tam and Patel (1979a) were obtained for a "forbidden" $^1B_{2u} \leftarrow {}^1A_{1g}$ transition in liquid benzene. A lens of 5 cm focal length is used to focus the pulsed laser beam into the cell (in the linear absorption studies, much weaker focusing or no focusing was used). However, we have observed that a lens of even shorter focal length (i.e., tighter focusing) does not result in a stronger two-photon OA signal because tighter focusing results in a correspondingly shorter confocal length of the laser beam (i.e., the two-photon acoustic source becomes shorter for tighter focusing). The two-photon absorption spectrum of Tam and Patel (1979a) is shown in Fig. 31. From the signal-to-noise ratio of the result, we estimated that the pulsed OA technique, using laser pulses of ~1 mJ energy and ~1 μsec duration, can be used to detect two-photon absorption cross sections as small as 1×10^{-53} cm^4 molecule^{-1} photon^{-1} sec. The data of Fig. 31 are new quantitative two-photon absorption data for liquids, and represent substantial improvements on the previous two-photon absorption data for the same transition in liquid benzene, obtained by Twarowski and Kliger (1977) using a two-photon thermal blooming method. Other techniques to detect two-photon absorption include luminescence detection and photoconductivity detection. Voigtman and Winefordner (1982) have described a flow-cell apparatus suitable for simultaneous PA, luminescence, and photoconductivity detections, and they show that trace amount of drugs can be detected and identified.

Multiphoton absorption in solids is frequently a cause of laser damage

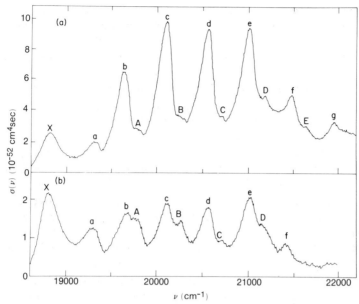

Fig. 31. Two-photon OA absorption spectra of liquid benzene for the $^1B_{2u} \leftarrow {}^1A_{1g}$ band. $\sigma(\nu)$ is the two-photon absorption cross section measured at the wavenumber ν for a single photon. The spectra obtained here for (a) linear laser polarization and (b) circular laser polarization have been normalized to the square of the excitation laser pulse energy. One of the absorption peaks shown here (at 18810 cm^{-1}) is *not* a two-photon absorption peak, but is a linear absorption peak of the seventh harmonic of the C–H stretch. [After Tam and Patel (1979a).]

of materials (e.g., laser-active crystals, windows, optical components, etc.). Hence, OA detection of two-photon (or higher-order) absorption processes should be very important for the development and application of intense lasers. Two-photon absorptions by OA detection in transparent materials such as glasses was reported by Munir *et al.* (1981). An optical glass disk (e.g., HO_2O_3-doped glass) was fitted tightly inside a hollow piezoelectric tube (Endo-Western EG64) with good acoustic coupling being achieved by using phenylbenzoate to glue the sample disk and tube together. A flash-lamp-pumped dye laser beam was focused with a lens of 10 cm focal length onto the center of the glass disk with a maximum focal intensity of 10 MW/cm^2. An OA signal that was quadratic to the laser pulse energy for certain wavelengths was observed, indicating the detection of two-photon absorption.

Two-photon absorption in semiconductor materials is important because semiconductor materials are frequently exposed to strong radiation (laser-active materials, laser annealing, etc.). Also, two-photon absorption spectra provide basic understanding of excited states in the materials. Van Stryland and Woodall (1980) have reported the observations of two-photon absorption measurements in CdTe and CdSe using pulsed dye lasers and piezoelectric transducers coupled to the sample with a suitable liquid.

OA detection of stepwise excitation instead of the two-photon absorption discussed above has been reported by Rockley and Devlin (1977). In their work, rose bengal dye was deposited as a thin film onto a substrate, and two pulsed dye lasers at 4732 and 4745 Å of power about 5 kW and duration 6 nsec were used to excite the sample. The PA signal was detected by a GR-1962 $\frac{1}{2}$-in. microphone in conjunction with an Ithaco 143L preamplifier and a PAR 162/164 boxcar integrator. They observed that the PA signal is strongly enhanced when the two pulsed lasers overlap spatially compared to the nonoverlapping case. This was interpreted as due to stepwise excitations, i.e., one laser pulse produces a lower excited state, which is excited by the other laser to a higher excited state. An implication of Rockley and Devlin's (1977) study is that excited-state lifetimes can be studied by providing a variable time delay between the two laser pulses and observing the corresponding variation in the PA signal.

IV. PA MONITORING OF DEEXCITATIONS

Photoacoustic methods have other applications besides the spectroscopies considered above. In the following, we will consider examples of nonspectroscopic applications. Figure 32 provides an idea of the origin of three general classes of applications of PA methods. (i) PA spectroscopy: in this class of application, the optical wavelength is varied to provide a PA spectrum; other factors like branching ratios, efficiency in the heat generation, efficiency in the acoustic wave generation, etc., are usually kept fixed while the PA spectrum is obtained. (ii) PA monitoring of deexcitation processes: in this class of application, branches complementary to the thermal decay branch are monitored. After optical excitation, four decay branches are possible: luminescence, photochemistry, photoelectricity, and heat (which may be generated directly or through energy transfer processes such as collisions). If there are only two competing branches, for example, luminescence and heat, then by PA monitoring of the heat branch, the quantum efficiency of luminescence under various conditions can be deduced. (iii) PA probing of thermoelastic and other physical properties of materials: in this class of application, heat generation, acoustic wave production, and acoustic propagation are the impor-

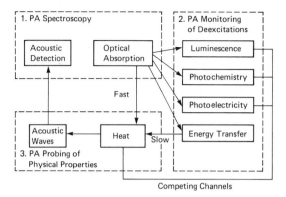

Competing Channels

Fig. 32. Block diagram showing the three different classes of applications of PA detection. Each dotted loop indicates the processes most relevant to the class of applications indicated.

tant processes considered. For example, sound velocity measurements, detecting changes in specific heat or thermal diffusivity, probing thickness of thin films, etc., belong to this class of application. Section IV deals with applications in class (ii), while Section V deals with applications in class (iii).

A. Fluorescence Quantum Yields

In many luminescent materials, the optically excited state can decay only by two channels: radiatively, or nonradiatively with heat generation. Under this circumstance, the measurement of the absolute optical energy absorbed W_{abs} and the absolute heat energy generated W_{heat} would give the fluorescence quantum efficiency ϕ for a simple two-level system as

$$\phi = (W_{abs} - W_{heat})/ W_{abs}. \tag{66}$$

However, a simple two-level system may be a poor representation of a luminescing system if intersystem crossing to a triplet state is efficient; in this case, two types of heat generated may be observable, a fast heat component due to nonradiative relaxation of the singlet excited state, and a slow heat component due to the nonradiative relaxation of the triplet state. Callis *et al.* (1969) have developed a flash calorimetry technique to examine quantum yields in such a system (for example, anthracene in ethanol, where triplet formation quantum yield is 66%); the slow and fast heat produced (only $\sim 10^{-5}$ cal) is detected by a capacitance microphone which measures directly the volume change of the liquid caused by heating. This work may be considered as the first experiment of PA monitoring of a triplet quantum yield.

A key issue for measuring fluorescence quantum yield as suggested in Eq. (66) is that absolute heat energy is involved. This is difficult to measure by PA means since the PA signal is only proportional to the modulated heat generated, with the proportionality constant being poorly known at best. The best way to solve this problem is to perform the PA measurement twice, once with the desired luminescence quantum yield ϕ_f, and once with the heat produced being altered in a known way. This provides two equations with two unknowns, and so ϕ_f can be solved. As concrete examples, consider a substance with the energy-level diagram of Fig. 33. For optical excitation to an S_1 excited state at energy E_1, the heat energy W_{heat} produced per unit volume by an incident photon flux Φ cm^{-2} sec^{-1}, pulse duration τ, and absorption coefficient α is given by

$$W_{heat} = \Phi\alpha\tau(E_1 - \phi_f E_1'), \qquad (67)$$

where E_1' is the energy of the lowest S_1 excited state, and relaxation from E_1 to E_1' is assumed to be fast and nonradiative. Now, since the PA signal V_{PA} is proportional to W_{heat}, we have

$$V_{PA} = KI_{in}\alpha\tau(E_1 - \phi_f E_1'), \qquad (68)$$

which contains two unknowns: the proportionality constant K and the luminescence quantum efficiency ϕ_f. Two methods to evaluate ϕ_f are indicated in Fig. 33. The first method is the quenching method (a). In essence, the method consists of the addition or doping of a low concentration of efficient "quencher" molecules into the sample so that ϕ_f in the doped sample becomes zero, without changing anything else. Thus

$$V_{PA}' = KI_{in}\alpha\tau E_1, \qquad (69)$$

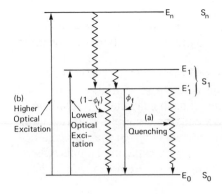

Fig. 33. Schematic of some methods for measuring the luminescence quantum yields: (a) by a quenching technique and (b) by a higher optical excitation technique. Straight lines indicate radiative transitions and sawtoothed lines indicate nonradiative transitions.

where V'_{PA} is the PA signal for the doped sample at the same wavelength. Thus, Eqs. (68) and (69) can be solved for ϕ_f. Such quenching methods have been used by Hall *et al.* (1976) for absolute fluorescence quantum yield studies of benzene vapor in the near ultraviolet, by Starobogatov (1977), Lahmann and Ludewig (1977) and Adams *et al.* (1977a,b) for measuring ϕ_f of dye solutions, by Quimby and Yen (1978) for measuring ϕ_f of Nd:glass materials, and by Murphy and Aamodt (1977a,b) for explaining the nonradiative processes affecting the concentration-dependent quenching rates in ruby and the subsequent effect on the PA signal at liquid-N_2 temperatures. The second method (b) in Fig. 33 relies on the possibility of optically pumping higher excited states S_n that nonradiatively decay to the lowest state in S_1. This higher excitation method was first proposed and demonstrated by Rockley and Waugh (1978). The mathematical basis of the method is that the PA signal V''_{PA} due to the higher excitation is given by

$$V''_{PA} = K\Phi''\alpha''\tau''(E_n - \phi_f E'_1), \qquad (70)$$

where Φ'' is the incident optical intensity to excite S_n with a pulse width of τ'' and absorption coefficient α''. Thus, Eqs. (68) and (70) provide two equations with two unknowns (K and ϕ_f), with all other quantities assumed to be measurable, and hence ϕ_f can be solved. Using this higher excitation method as "internal standard", Rockley and Waugh are able to obtain ϕ_f for a crystal violet solution.

Experiments analogous to PA monitoring of fluorescence quantum yield have been performed, using electrical excitation instead of optical excitation. As an example, semiconductor diode laser materials have been tested acoustically by Suemune *et al.* (1980); in that work, current-injection-induced acoustic signals are observed for GaAs–GaAlAs double-heterostructure lasers. The dependence of the acoustic intensity on the injection current shows an anomaly close to the threshold of lasing. Such work may have important applications in the research and manufacturing of semiconductor lasers.

B. Photochemical Effects

Photochemical changes can affect PA signals in several ways, some of which are listed in Table III. A simple reason why photochemistry affects the PA signal is that the quantum efficiencies for photochemistry and heat production are complementary (assuming the luminescence and photoelectricity branches to be zero) so that the increase of one branch must mean the decrease of the other. This complementary effect is present in all the photochemical monitoring studies discussed below, but may be

TABLE III

Examples of PA Monitoring of Photochemical Changes

Photochemical effect affecting the PA signal	Example	References
Complementary branch, i.e., photochemical energy + heat energy = light energy absorbed	PA spectroscopy of active and poisoned chloroplast membranes	Cahen et al. (1978a,b); Malkin and Cahen (1979)
Proton-transfer-induced volume changes	Volume change measurements of Bacteriorhodopsin-containing membrane fragments	Ort and Parson (1978, 1979)
Gas evolution or consumption	UV photodissociation of a diatomic gas;	Diebold (1980)
	Photocatalytic oxidation of acetic acid, or photo-oxidation of rubrene in oxygen	Gray and Bard (1978)
Photochemical chain reaction	Photopolymerization of diacetylenes	Chance and Shand (1980)
	Halogen–hydrogen reaction	Diebold and Hayden (1980)

negligible compared to other effects like gas evolution. The complementary effect is well demonstrated in the important photosynthesis work of Cahen et al. (1978a,b) and Malkin and Cahen (1979). In that work, lettuce chloroplast membrane in a liquid medium is contained in the sample chamber of a differential PA cell (Fig. 14). They observe that for samples with active photosynthesis, the PA spectrum and the optical absorption spectrum actually differ by an amount corresponding to the conversion into chemical energy (they called this effect the "photochemical loss"). Furthermore, "poisoned" chloroplasts give PA signals that are larger than those obtained with active chloroplasts because in the poisoned case there is no photochemical loss.

Energetics in the purple membrane of *Halobacterium halobium* represents another important photochemical effect that has been studied by PA techniques (Cahen *et al.*, 1978a,b; Ort and Parson, 1978, 1979). This membrane contains a protein, bacteriorhodopsin (bacteriopsin binds to retinal to form bacteriorhodopsin). The absorption of light causes a cyclic photochemical process, driving proton transfer from one side of the membrane to the other. Ort and Parson (1978, 1979) have used a capacitance microphone PA cell to detect the volume changes due to the proton transfer following an optical flash excitation. This is performed at 3.4°C in an aqueous buffered solution, where thermal expansion effects are negligible (water has zero expansion coefficient at 4°C). The rapid volume changes due to the flash-induced proton transfer can be detected.

Photochemical gas evolution or consumption is another important cause of the PA effect that may be very large compared to that caused by the thermal expansion effects only. Diebold (1980) has considered the case of a periodic ultraviolet photodissociation of a diatomic gas. Since the equilibrium point of the photodissociation depends on the slow reverse process, namely a three-body recombination, a phase lag is introduced in the PA signal. The amplitude and phase lag of the PA signal provide important information on the photodissociation process. Gray and Bard (1978) have studied both photochemical gas generation and consumption by PA detection. Gas evolution was demonstrated for the heterogeneous photocatalytic oxidation of acetic acid to methane and CO_2 at a platinized TiO_2 catalyst. Gas consumption was shown for the case of O_2 depletion due to photo-oxidation of rubrene.

In photochemical chain reactions, large "gain factors" for the PA generation are possible. Chance and Shand (1980) have demonstrated the PA detection of light-induced chain reaction by observing the solid-state photopolymerization of diacetylenes. The thermal energy evolved is found to be four times greater than the light energy absorbed; this corresponds to about eight polymerized diacetylene units being produced per photon absorbed. Independently, Diebold and Hayden (1980) reported the PA detection of the photoinduced chain reaction in a halogen (X_2) and hydrogen mixture

$$X_2 \xrightarrow{\text{light}} X \xrightarrow{H_2} H \xrightarrow{X_2} X \xrightarrow{H_2} H \xrightarrow{X_2} \cdots. \tag{71}$$

Diebold and Hayden (1980) reported that the PA signal is directly proportional to the photochemical chain length in reaction (71), i.e., chemical amplification of the PA effect occurs. They estimated that a "chemical amplification factor" of several orders of magnitude is possible, and showed that the PA measurements provide a determination of the chain length without prior knowledge of the photon flux or detailed rate constants.

C. Photoelectrical Effects

When part of the light energy is converted into electrical energy (e.g., in a photovoltaic or photoconductive device), the thermal energy produced in the optical excitation will be correspondingly reduced from the full light energy. Under circumstances such that the released free carriers do not cause any volume change in the sample (e.g., by electrostriction or by Joule heating), the observed OA signal from the sample should be smaller when the sample is photoelectrically active than when it is inactive. Under the opposite case when electrostriction or Joule heating by the carriers produced is important, the OA signal is still affected by the photoelectrical effect, but the interpretation of the results becomes more complex.

Cahen (1978) was the first to demonstrate the optoacoustic monitoring of photoelectrical effects. The OA signal V_{OA} from a solar cell device depends on the photoelectrical generation efficiency $\phi_e(R)$, which is dependent on the external load R:

$$V_{OA}(R) = KI_0[1 - \phi_e(R)] \tag{72}$$

where K is a constant for a fixed OA cell geometry and a fixed optical wavelength, and I_0 is the modulation amplitude of the incident light intensity. Under open-circuit conditions, the OA signal V'_{OA} can again be measured, and is now given by

$$V'_{OA} = KI_0 \tag{73}$$

since $\phi_e = 0$ at open circuit. Combining Eqs. (72) and (73), we have

$$\phi_e(R) = 1 - V_{OA}(R)/V'_{OA}. \tag{74}$$

Cahen's (1978) experiment was performed on a Si solar cell; he measured simultaneously $V_{OA}(R)$ and the electrical power output $P_{out}(R)$ as R was varied from a small value ($\sim 0.1\ \Omega$) to a large value ($\sim 100\ \Omega$). His results showed that $V_{OA}(R)$ is at a minimum while $P_{out}(R)$ is at a maximum when $R = 6\ \Omega$; which is presumably the internal resistance of the solar cell. By using Eq. (74), Cahen obtained a value of 18% for ϕ_e (6 Ω) at the wavelength used. Cahen's experiment may have important applications in the testing and quality control of solar cell conversion efficiency averaged over the whole cell surface or scanned over different parts of the surface. Furthermore, efficiencies of individual elements of a solar cell array can be measured without any disconnection.

Tam (1980a) has extended Cahen's (1978) idea of OA monitoring of photoelectricity to other systems. Tam (1980a) used an OA method for the first time to study the photoconductive quantum efficiency ϕ_e of a thin organic dye film. Here, ϕ_e can be defined as the averaged number of mobile electrons or holes generated per photon absorbed. The experimental arrangement of Tam (1980a) is shown in Fig. 34. The photoconductive sample used is a multilayer film suitable for electrophotographic applications. The special, highly photoconductive dye (see Tam, 1980a) is coated onto aluminized Mylar, and the dye film thickness ranges from 0.1 to 1 μm. It is overcoated with a 20-μm-thick "transport layer," which is a doped polycarbonate film that transports holes. A piezoelectric trans-

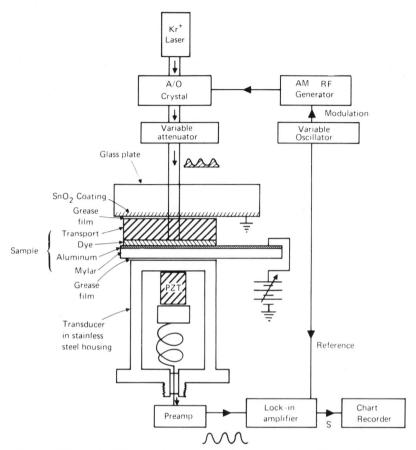

Fig. 34. Schematic of the experimental arrangement to measure photoconductive properties of dye films at various applied voltages. The thickness of the sample film is not drawn to scale; the thicknesses of the transport, dye, aluminum, and Mylar layers are 20, 1, 0.1, and 75 μm, respectively. A spring clamp to compress the transducer, sample, and glass plate together is not shown. [From Tam (1980a).]

ducer (Fig. 20) is spring loaded against the Mylar with a thin layer of grease used between the piezoelectric transducer and the Mylar for acoustic coupling. The sample is excited by a modulated cw Kr$^+$ laser beam sufficiently attenuated so that the light intensity at the sample never exceeds 0.1 mW/cm^2 to avoid material damage. The OA signal V_{OA} is found to be very much dependent on the modulation frequency f; strong enhancement (by a factor of $\sim$100) of V_{OA} occurs when f equals a fundamental mechanical resonance frequency of the sample–transducer assembly (about 40 kHz in this experiment). To measure ϕ_e at a laser wavelength (e.g., 676 nm), the modulation frequency is fixed at a resonance frequency, and $V_{OA}(\epsilon)$ is measured for various electric fields ϵ across the sample. It is well known (see Tam, 1980a) that ϕ_e in organic dyes is very dependent on ϵ, and ϕ_e is zero at $\epsilon = 0$. Hence, we may write the following equations, which are analogous to Cahen's (1978) equations:

$$V_{OA}(\epsilon) = KI_0[1 - \phi_e(\epsilon)], \tag{72'}$$

$$V_{OA}(0) = KI_0, \tag{73'}$$

which can be combined to give

$$\phi_e(\epsilon) = 1 - V_{OA}(\epsilon)/V_{OA}(0). \tag{74'}$$

Using Eq. (74'), Tam (1980a) obtained ϕ_e as a function of voltage applied across the sample (electric field ϵ = voltage/thickness, with thickness being about 20 μm). The results are shown in Fig. 35, where ϕ_e obtained by Eq. (74') is compared to the ϕ_e measured directly, i.e., by measuring the number of carriers flowing through the sample for a known amount of light energy absorbed. At low voltage V (i.e., small ϵ), the direct measurement becomes very unreliable because of the low mobility and high trapping of carriers at low fields. The present OA method avoids these difficulties and offers a better method for measuring ϕ_e at low fields. At high fields, the results from OA measurements approach the results from direct measurements.

We may note that the use of Eq. (74) or (74') to obtain photoelectrical quantum efficiencies is not unconditional. Luminescence and photochemistry are assumed to be absent. The photon absorption should deposit a negligible amount of kinetic energy on the generated carrier. Joule heating and electrostriction effects should be negligible. These conditions should be examined before Eq. (74) or (74') is used.

There are several other recent papers on PA monitoring of photoelectrical carrier generation or related effects in semiconductors and organic dyes. Thielemann and Neumann (1980) have applied a PA technique similar to Cahen's (1978) described above to determine the photocarrier generation quantum efficiency of a Schottky diode. Wasa et al. (1980b) have investigated nonradiative states in GaAs and InP by PA spectroscopy. Tokumoto et al. (1981) have used PA spectroscopy to study Si, Ge, InSb, GaAs, and GaP in the region above the fundamental absorption edge. Iwasaki et al. (1979) have used laser-induced PA spectroscopy to examine voltage-dependent electron recombination processes at a semiconductor electrode–dye solution interface. Iwasaki et al. (1981) have used PA methods to study spectral sensitizations of dyes on ZnO powders.

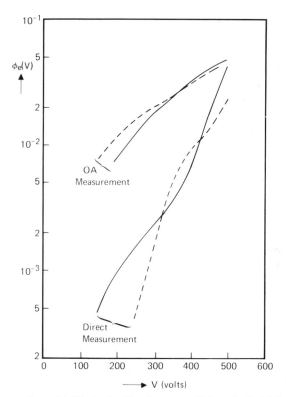

Fig. 35. Comparison of $\phi_e(V)$ obtained by the present OA method with that obtained by a direct method, i.e., (number of carrier collected)/(number of photon absorbed). The full lines are for dye I (chlorodiane blue) at a wavelength of 676 nm, and the broken lines are for dye II (methyl squarylium) at a wavelength of 799 nm. [From Tam (1980a).]

D. Energy Transfer

The simple mathematical basis of the OA monitoring of energy transfer processes has been given in Eqs. (5), (8), and (9), which can be combined to give

$$H = [\Phi_0 \alpha A_n \tau E'/(1 + \omega^2 \tau^2)^{1/2}] \exp[i(\omega t - \psi)], \tag{75}$$

where Φ_0 is the photon flux, α the absorption coefficient, A_n the nonradiative decay rate, τ the lifetime, E' the average heat energy released due to nonradiative deexcitation of the excited state, $\omega = 2\pi \times$ chopping frequency, and $\psi = \tan^{-1}(\omega\tau)$. Usually E' equals the total excited-state energy. Equation (75) is valid only for a simple two-level system, but it provides a qualitative understanding of the techniques used in OA monitoring of energy transfer.

There are at least two ways to achieve energy transfer: collisions or optically stimulated transitions. The effects of energy transfer on Eq. (75) are twofold: (i) The branching ratio $A_n \tau$ for nonradiative decay is altered; e.g., it is increased by more quenching collision so that the magnitude of the OA signal increases. (ii) The net lifetime τ is changed by energy transfer processes; e.g., with increased quenching collisions, τ decreases, which causes a decrease in the phase lag ψ and an increase in the magnitude of the OA signal through the factor $(1 + \omega^2\tau^2)^{1/2}$. Conversely, by measuring the dependence of the magnitude and phase of the OA signal as functions of gas pressures and composition, chopping frequencies, etc., quantities of interest like A_n and τ can be derived. This is for the case of a modulated cw beam being used for excitation. In the case of pulsed excitation, the idea is the same, namely, a time-resolved detection of the OA signal provides information on the excited-state radiationless processes as discussed by Wrobel and Vala (1978).

Parker and Ritke (1973) performed some of the pioneering work on the OA measurement of collisional deactivation time of the first vibrationally excited level of the $^1\Delta_g$ electronic state of O_2, and obtained a pressure $\times$ lifetime product of 0.05 sec atm for pure oxygen.

Hunter and co-workers (Hunter *et al.*, 1974; Hunter and Stock, 1974; Hall *et al.*, 1976) have performed a series of experiments on energy transfer processes in vapors of organic materials (e.g., biacetyl) using OA monitoring. For example, Hunter and Stock (1974) derived a triplet yield of 97% in a biacetyl vapor excited at a wavelength of 436 nm. This large triplet yield is due to the fast intersystem crossing in biacetyl, where S_1 and T_1 are only 3000 cm^{-1} apart. Hunter *et al.* (1974) also obtained a much more refined theory [compared to Eq. (75)] taking into account both fast heat release and slow heat release (e.g., due to nonradiative decays of S_1 and T_1, respectively) after optical excitation.

Robin and collaborators (Kaya *et al.*, 1974, 1975; Robin and Kuebler, 1975; Robin, 1976; Robin *et al.*, 1980) performed a series of experiments on organic vapors which beautifully demonstrate the use of OA spectroscopy to measure energy transfer processes when various triplet or singlet excited states are involved. An example of their results in the study of biacetyl is shown in Fig. 36. In this work, Kaya *et al.* (1975) demonstrated that radiationless decay from S_1 of biacetyl when excited at wavelengths longer than 4430 Å involves a very slow $T_1 \rightarrow S_0$ step with a decay rate of $\sim 10^3$ sec^{-1}. However, at shorter wavelengths, a second fast channel opens, namely $T_2 \rightarrow S_0$. Addition of O_2 makes the apparent $T_1 \rightarrow S_0$ relaxation very fast.

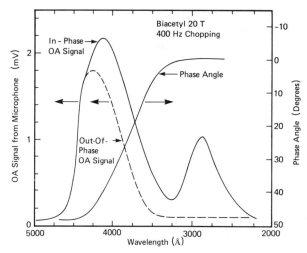

Fig. 36. In-phase and out-of-phase OA spectra of biacetyl, taken with a phase angle chosen to maximize the S_2 signal for the in-phase case. The curve labeled "phase angle" gives the angle at which the OA signal is nulled at each wavelength. [After Kaya *et al.* (1975).]

Vibrational relaxations of CO_2 molecules in collisions with CO_2 or with other molecules like N_2, He, CO, etc., represent an important field of research, both scientifically and technologically. Huetz-Aubert and Tripodi (1971) first used a spectrophone (i.e., an OA cell) to investigate such vibrational relaxation rates in CO_2. Subsequently, Lepoutre and co-workers (Huetz-Aubert and Lepoutre, 1974; Lepoutre *et al.*, 1979; Taine and Lepoutre, 1980) have investigated in detail the collisional deactivation rates of vibrationally excited CO_2 with a host of other molecules at temperatures ranging from 170 to 400 K. Gelfand and co-workers (Gelfand *et al.*, 1979; Rohlfing *et al.*, 1980, 1981) have used tunable-dye-laser OA spectroscopy to examine the pressure broadening and relaxation rates of high vibrational overtones in simple molecules like CH_4 and HD.

Relaxation rates in solids have also been measured by monitoring the phase and magnitude of the OA signal, similarly to the gas-phase measurements described above. Theories of such OA measurements for condensed matter have been given by Mandelis *et al.* (1979) and Mandelis and Royce (1980a,b). Powell *et al.* (1980a,b) have applied OA methods to study lifetimes in the laser materials, e.g., NdP_5O_{14}, and Peterson and Powell (1978) have studied the radiationless relaxations in $Cr^{3+} : MgO$, which is shown to be quite different from the case of $Cr^{3+} : Al_2O_3$, i.e., ruby.

Up to now, energy transfer due to collisions or intermolecular interactions have been discussed. Another class of energy transfer (or more accurately, state transfer) is due to transitions or emissions stimulated by light. Allen *et al.* (1977) and Anderson *et al.* (1981) have ingeniously applied such a technique to obtain a quenching rate in a flame. The OA technique is, of course, ideally suited for studies in a flame because of its noncontact nature [except if a microphone is used for detection, it must be located outside the flame, as in Allen *et al.* (1977)]. Probe-beam deflection methods can be used for detection instead of a microphone, making the OA probing method totally remote-sensing and suitable for studies in hostile environments (see Tam *et al.*, 1982). In the experiment of Allen *et al.* (1977), the flame is doped with Na atoms, and a pulsed-dye-laser beam tuned to the resonant transition (Na–Na*) is used to excite the Na in the flame. The laser intensity is varied so that the excitation rate (Na $\rightarrow$ Na*) and the simulated emission rate (Na* $\rightarrow$ Na) are correspondingly changed. The stimulated emission Na* $\rightarrow$ Na competes with the other two decay process of Na*, namely radiative decay and quenching collision. Since the stimulated emission rate is known, as given by an Einstein coefficient, and the radiative decay rate is known, the quenching rate Q can be derived. Allen *et al.* (1977) obtain $Q = 2.6 \times 10^{10}$ sec^{-1} under their flame conditions. The use of stimulated emission to cause state transfer has also been applied to condensed matter, as performed by Razumova and Starobogatov (1977) for organic dye solutions.

V. PA PROBING OF PHYSICAL PROPERTIES

We have indicated in Fig. 32 that the processes of heating up of a sample due to optical absorption and the generation and propagation of acoustic waves in the sample depend critically on the thermoelastic and physical properties of the sample. By monitoring the PA signal we may be able to probe or measure these properties, e.g., acoustic velocities, elasticity, density, thickness, specific heat, electrostrictive coefficients, material discontinuities, crystallinity, and phase transitions. Large-amplitude waves can also be generated for studying nonlinear acoustics or high-pressure effects. By focusing the light beam, some of these physical properties may be measured locally, and hence by raster-scanning the beam over the sample, new types of PA imaging of the property concerned can be obtained. Also, the use of light beams is not necessary, since any other types of beams (electron, ion, or other types of radiations) which can generate heat in a sample can be used to do the probing. Applications where beams other than light are used are discussed in Section VI.

A. Acoustic Generation and Propagation

Theories of PA generation (e.g., caused by pulsed heating by a short-duration laser beam) have been given in the literature. White (1963) gave the first theory of generation of elastic waves due to transient surface heating by electrons or electromagnetic waves. Gournay (1966) extended White's (1963) work, and studied in detail the stress evolving from a thermally shocked liquid. The radiation pattern of the acoustic waves generated by one or more light beams have been studied, for example, by Westervelt and Larson (1973), Gorodetskii *et al.* (1978), and Dunina *et al.* (1979). The shapes and magnitudes of the acoustic pulses produced under different absorption conditions (surface absorption or bulk absorption), different dimensions of the excited region, and different constraint conditions (free surface or rigid boundary) have been calculated by Bunkin and Komissarov (1973). Lyamshev and Naugol'nykh (1976, 1980), Kasoev and Lyamshev (1977), and Naugol'nykh (1977).

Many cases of strong PA generation have been reported since giant pulsed lasers were invented in the early 1960s. For example, Askaryan *et al.* (1963) made the first qualitative experimental investigation of strong acoustic wave generation via boiling in a liquid by pulsed ruby lasers. Carome *et al.* (1964) have used a Q-switched ruby laser of 0.1 J energy and 50 nsec duration to excite a ferrocyanide aqueous solution contacting a glass backing plate and reported the observation of peak acoustic pressures of a few atmospheres. Silberg (1965) has reported that a 50-J ruby laser pulse incident on a mercury surface causes rapid boiling, sending a strong "reaction force" into the liquid metal. Observations of high-pressure shock waves in liquid, usually water, have been described by Bell and Landt (1967), Bushanam and Barnes (1975), Emmony *et al.* (1976), Sigrist and Kneubuhl (1978), and others. Such shock waves are important not only for fundamental understanding, but also for high-pressure applications (transient pressures in the Mbar range can be generated) and underwater communications [for example, Kohanzadeh *et al.* (1975) have estimated that a ruby laser pulse can produce an acoustic pulse in water observable many kilometers away]. Attempts have been made to understand the time development of the shock wave generated by pulsed lasers. For example, Felix and Ellis (1971) used Schlieren photographic techniques to provide a step-by-step account of the expanding cavitation due to laser-induced breakdown in liquids. Gordienko *et al.* (1978) have used a stroboscope to examine OA generation by a Nd–YAG laser of 50 mJ energy and 10 nsec duration in a $CuCl_2$ aqueous solution.

Teslenko (1977) has investigated the efficiency of conversion of radiation into mechanical energy as a result of breakdown in liquids. He defined a "PA coefficient" as the ratio of acoustic shock wave energy to the incident optical pulse energy. Using a ruby laser of 1 J energy and 20 nsec duration, Teslenko (1977) reported a PA coefficient as large as 30% due to laser-induced breakdown in water. This large photoacoustic coefficient may be compared with typical values of less than 10^{-10} in most PA spectroscopy experiments discussed in Sections II and III. If a pulsed laser is incident on a liquid containing a dissolved gas, strong acoustic generation may also occur due to effervescence or gas evolution, as studied by Askar'yan et al. (1963) and by Egerev and Naugol'nykh (1977).

PA generation can also produce surface elastic waves besides the bulk waves described above. For example, Lee and White (1968) have used a spatially periodic illumination of an aluminum film on a fused silica substrate to generate surface acoustic waves of ~1 MHz frequency.

Intense OA generation is possible in gases as well as in condensed matter. An example is the recent work of Tam et al. (1982), who demonstrated that strong blast waves are produced in a metal vapor due to transient ionization and plasma formation by a pulsed dye laser. Their experimental arrangement is shown in Fig. 37. A flash-lamp-pumped dye laser of energy ~1 mJ and duration 1 μsec tuned to an atomic transition in the Cs atom is used to cause transient plasma formation in a Cs metal vapor (Tam and Happer, 1977; Tam, 1980b). The plasma formation results in a rapid expansion of the illuminated volume, causing a blast wave to be generated. This cylindrical acoustic wave propagates outward radially at a supersonic speed initially. The acoustic propagation can be probed by a weak cw Kr^+ laser at 676 nm, which is parallel to but displaced from the pulsed laser by a radial distance R. As the acoustic pulse passes through the weak probe beam, the refractive index change due to the atomic density variation in the pulse causes a corresponding deflection of the probe beam, which can be detected by using a knife edge and a photodiode as shown in Fig. 37, where a typical photodiode signal is also indicated. The acoustic pulse arrival time t after the laser pulse is found to be dependent on R, and a plot of R versus t for different energies of the pulsed dye laser is shown in Fig. 38. We see that the R versus t plot is characterized by an initial nonlinear part followed by a linear part. The initial nonlinear part indicates supersonic propagation of the blast wave, which is stronger if a more energic dye laser pulse is used. The linear part indicates that the blast wave has become sufficiently weak and propagates as a sound wave. Tam et al. (1982) showed that the data of Fig. 38 not only provide new understandings of blast wave propagation (Vlases and Jones, 1966), but also provide new data on acoustic velocities in a hot metal vapor, which is a nonideal gas. Recently, Zapka and Tam (1982b)

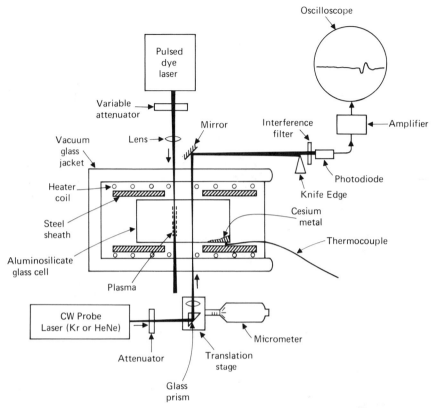

Fig. 37. Experimental arrangement for acoustic studies of hot metal vapors by photoacoustic pulse generation via breakdown and monitoring the acoustic pulse propagation by the transient deflection of the probe beam. [From Tam *et al.* (1982).]

have shown that the gas-phase measurement method indicated in Fig. 37 can be extended to condensed matter; this opens up many new possibilities of noncontact ultrasonic measurements in matter via PA pulse generation and optical detections (e.g., probe-beam deflection and laser interferometry).

The PA technique of Tam *et al.* (1982) is a totally noncontact method, and is hence nonintrusive and suitable for use in hostile environments. It differs from the photorefractive techniques discussed in Section VI.C, where the diffusive thermal refractive-index gradient is probed, not the propagating acoustic refractive-index gradient. Extension of the technique of Tam *et al.* (1982) for nonintrusive flow-velocity and temperature measurements in a flowing fluid has been demonstrated by Tam and Zapka (1982) and Zapka and Tam (1982c). Furthermore, applicability of

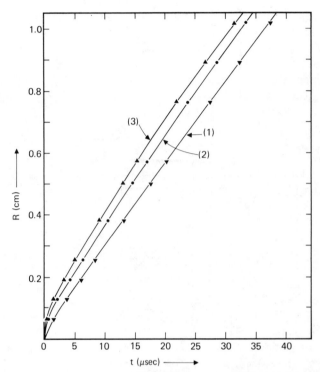

Fig. 38. Plot of R vs t for three different incident laser pulse energies of 106, 876, and 2420 µJ for (1), (2), and (3), respectively. The corresponding energies absorbed by the vapor are 21, 464, and 2001 µJ, respectively. The saturated vapor is at 402°C. [From Tam *et al.* (1982).]

the technique in highly hostile environments is clearly exemplified by the work of Zapka *et al.* (1982), where temperature profile and flow-velocities in a propane flame are mapped out by using a Nd–YAG laser for pulsed PA generation and two parallel HeNe probe beams for acoustic detection. Earlier, PA studies in flames have been reported by Allen *et al.* (1977), Anderson *et al.* (1981), and Tennal *et al.* (1981); however, no temperature profile or flow data were provided in these earlier measurements, where the PA pulse is detected outside the flame by a microphone, and the transmission of the PA pulse through the turbulent flame–air boundary as well as the slow microphone risetime may limit the accuracy of the determination of the acoustic arrival time.

Besides the ultrasonic pulse propagation measurements and intense acoustic pulse production indicated above, PA generation may have other intriguing and unique applications in industry (acoustic devices, material characterizations, quality controls, etc.). For example, Tam and Gill

(1982) have shown that PA pulse generation in a small ink reservoir can be used to controllably eject small ink drops from a nozzle. This "drop-on-demand" Photoacoustic Ejection from a Nozzle (PEN) requires only small optical pulse energies (obtainable from GaAs diode lasers or from LED's); hence, PEN should be useful for printing applications, and for other cases where precisely controlled emission of droplets is important (e.g., in laser ignition of liquid fuel drops, described by Dabora, 1981).

B. Electrostriction

There is another mechanism of OA generation which is distinct from the generation due to heating, gas evolution, chemical reactions, or dielectric breakdown discussed in Section V.A. This is electrostriction, which has received considerable attention recently because Brueck *et al.* (1980) have suggested that electrostriction may limit the OA detection sensitivity to no better than $\sim 10^{-5}$ cm^{-1} in condensed matter. Brueck *et al.* (1980) further suggested that electrostriction poses a much less severe limit on the photorefractive (PR) detection technique (see Section VI.C). These suggestions are not totally valid as discussed below.

Electrostriction (Forsbergh, 1956) is basically due to the electric polarizability of molecules in the sample so that they tend to move into (or out of) regions of higher light intensity if they have positive (or negative) polarizability. The atomic motions produce a density gradient and hence cause OA and PR effects. This has been noted by Bebchuk *et al.* (1978). Hence, electrostriction may produce spurious OA signals. Bebchuk *et al.* (1978) have also pointed out that the fractional volume change $(\delta V/V)_{ES}$ due to electrostriction is given by

$$(\delta V/V)_{ES} \propto I/B, \tag{76}$$

where I is the intensity of the excitation beams and B is the compressibility of the liquid. Thus, electrostriction-produced acoustic amplitude depends linearly on the incident optical intensity, as for the thermal-expansion-produced acoustic amplitude (Eqs. 53 or 62). However, the magnitude of the former depends on the "electrostrictive coefficient" (Brueck *et al.*, 1980), while that of the latter depends on the absorption coefficient for a transparent sample. Furthermore, the temporal profile of the electrostrictive acoustic signal is quite different from that of the thermal acoustic signal; Lai and Young (1982) have shown theoretically that by using suitable gated detection (Patel and Tam, 1981) of the acoustic signal, strong suppression of the electrostrictive component is possible. Thus, electrostrictive limitations of OA sensitivities need not be as serious as Brueck *et al.* (1980) indicated.

Experimental OA signals apparently due to electrostriction effects only (but not optical absorption) in liquid N_2 have been reported by Brueck *et al.* (1980). Such electrostrictive OA signals may be unusually large in their experiment, which is performed in liquid N_2 close to the critical point. The electrostrictive constant under such circumstances has unusually large values and is very dependent on the pressure applied to the liquid. Indeed, Brueck *et al.* (1980) observed that the electrostrictive OA signal decreases by a factor of 8 for a doubling of pressure above the liquid.

The present conclusion is that although electrostrictive signals may be observable under certain circumstances (e.g., close to a boiling point in a liquid), they seldom pose any limit to OA detection capabilities. Indeed, in the experiment of Tam and Patel (1979a), who used a focused dye laser beam in liquid benzene at room temperature to perform OA two-photon-absorption spectroscopy, no electrostrictive signals can be observed. This is because electrostrictive effects are in general fairly insensitive to wavelength; hence a large electrostrictive signal should correspond to a large nonzero "baseline," which is not the case in Tam and Patel's (1979a) two-photon OA spectra.

C. Radiation Damage, Crystallinity Change, and Phase Transitions

The PA signal obtained from a sample may change for several reasons. For the direct-coupling case, the generation of acoustic waves by thermal expansion is determined by the expansion $(\Delta V)_{th}$ (see Eq. 53):

$$(\Delta V)_{th} = \beta H / C\rho \tag{77}$$

where β is the volume expansion coefficient, H is the heat deposited in the volume V, C is a specific heat per unit mass, and ρ is the density. Thermal losses from V are neglected in Eq. (77). Thus, the acoustic generation efficiency is characterized by a PA "figure of merit" M_{PA}

$$M_{PA} = \beta / C\rho. \tag{78}$$

M_{PA} has appeared in several equations describing PA generation [e.g., Eqs. (47), (53)]. If M_{PA} of a sample varies (with position or with time), or if H changes because absorption or reflectivity changes, the acoustic signal produced in the condensed sample would vary correspondingly. The technique of high-spatial-resolution imaging with PA detection is usually referred to as "PA microscopy," and will be treated in detail in Section V.E. In this section, we are more concerned with time-dependent changes of the PA signal due to some treatment of the sample (e.g., ion bombardment, laser annealing, structural change, etc.).

An interesting application of monitoring time-dependent PA signals has been suggested by McClelland and Kniseley (1979a,b) and by McFarlane and Hess (1980). They suggested the application of PA methods to monitor laser annealing of semiconductors where it is important to control the semiconductor surface heating during irradiation by the annealing laser, to monitor dopant concentrations and their diffusions, to know the degree of recrystallization, and to detect any possible damage associated with the intense irradiation. In an actual application, the PA signals can be obtained either from the annealing laser or with the use of an auxiliary probe laser. Real-time monitoring of the laser annealing is possible by observing the change in the PA signal when crystallinity changes, which causes changes in the value of M_{PA}, the amount of thermal coupling with a gas, and/or the magnitude of the absorption.

In the work of McClelland and Kniseley (1979a), optical surface damage (e.g., due to thermal etching and oxidation after surface melting) caused by cw laser irradiation of Ge in a gas-microphone PA cell is studied. That work shows that PA detection is potentially useful in the control of processing conditions during annealing of semiconductors, where laser-induced damage should be avoided. In their later work, McClelland and Kniseley (1979b) demonstrated the additional potential of PA detection of crystallinity changes in Si due to annealing. They mounted a Si sample with regions of different crystallinity (by sputtering amorphous Si onto part of the surface of a single crystal) in a gas-microphone PA cell, and irradiated the sample surface with a cw Ar ion laser chopped at 400 Hz. When the laser beam was scanned from disordered to single-crystal material, a decrease in the PA signal amplitude was observed. They suggested that their observation provides a means of rapidly evaluating the degree of crystallinity achieved after laser annealing irradiation by monitoring the PA signal using the same annealing laser beam but set at a much lower power level for PA excitation.

Other authors have also observed that crystallinity or structural changes (without chemical changes) of materials cause the PA signals to vary. Florian *et al.* (1978) have demonstrated PA detection of phase transitions in Ga, H_2O, and K_2SnCl_6 samples. Luukkala and Askerov (1980) have observed that plastic deformation of a polycrystalline Al disk results in a larger PA signal obtained with a gas-microphone cell and a cw laser beam chopped at a few hundred hertz; they suggested that the thermal conductivity k of plastic-deformed Al is smaller than that of polycrystalline Al, causing a larger microphone signal. Schneider *et al.* (1982b) have reported that photoisomerization results in an intensity-dependent change in the PA spectrum, as observed for the case of the reversible photoisomerization of DODCI (3,3'-diethyldicarbocyanide iodide).

D. Depth-Profiling Studies

Depth profiling is the investigation of a layered sample to find out the properties of the various layers. Depth profiling by PA means may be destructive or nondestructive.

An example of destructive depth profiling is provided by the work of Yeack *et al.* (1982), who monitored the PA signal due to laser ablation on a composite layered sample. By continuously monitoring the PA signal obtained from successive laser pulses incident on the same region of a layered sample, they show that they can control the degree of local thermal ablation of the composite material. The PA pulse is detected by a piezoelectric ring positioned close to the ablation spot, and no special acoustic cell is used. Their technique is actually used as a PA monitoring technique of optical ablation of composite materials rather than as a destructive depth-profiling tool. Their method of PA monitoring of stepwise ablation by laser pulses may be extended to many other novel technological or medical applications where optical ablation, evaporation, coagulation, polymerization, or other chemical or physical changes can be performed in steps with light pulses. In these applications, the PA pulse signal can be continuously monitored and the completion of the desired operation can be indicated by a characteristic change in the pulsed PA signal.

Nondestructive depth-profiling studies are of much more general applicability. Usually, depth profiling is obtained by a "chopping-frequency-dependence" measurement of the PA signal with a gas-coupling detection (see Section III.B). The qualitative idea is simple: the thermal coupling at the gas–sample interface is between a sample diffusion length μ_s and a gas diffusion length μ_g. Since the diffusion lengths depend on the chopping frequency f as $f^{-1/2}$, we see that higher chopping frequency corresponds to probing the sample closer to the surface. If the sample is also optically thick with an optical absorption length μ_α much smaller than the sample thickness l, then another interesting effect occurs as the chopping frequency is varied. When f is small so that $\mu_\alpha < \mu_s$, then all the heat generated in μ_α due to optical absorption can communicate with the gas, and so PA signal saturation occurs, i.e., the signal is independent of the sample absorption coefficient α. On the other hand, when f is large enough so that $\mu_\alpha > \mu_s$, only the heat generated within μ_s can communicate with the gas, and so the PA signal magnitude is proportional to α; i.e., PA spectroscopy can be performed for the totally opaque sample. This effect has been pointed out by Rosencwaig and Gersho (1976) and has also been treated in Section III.A (see Table II).

A good example of depth profiling by chopping-frequency dependence

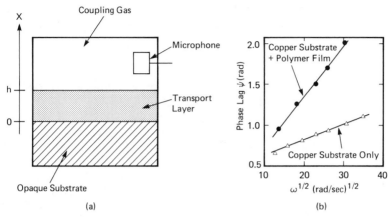

Fig. 39. PA depth-profiling measurement for a simple layered structure. (a) Schematic of the apparatus; the excitation beam can be incident on the opaque substrate from the top or from the bottom. (b) Variation of phase lag ψ with the angular chopping frequency ω for a bare copper substrate and a coated copper substrate. [After Adams and Kirkbright (1977b).]

is the visible PA spectra of an apple peel or of a spinach leaf as performed by Rosencwaig (1978), by Adams and Kirkbright (1976), and by others. In these studies, gas-microphone detection is used. At a comparatively high chopping frequency (e.g., $f \approx 300$ Hz, $\mu_s \approx 10$ μm), the PA spectrum corresponds to the optical absorption due to the top waxy layer of the plant matter, while at a lower chopping frequency (e.g., $f \approx 30$ Hz, $\mu_s \approx 33$ μm), the PA spectrum corresponds to absorption by the plant pigment below the waxy top layer as well.

The PA depth profiling for a simple layered sample (see Fig. 39a) composed of a transparent thin layer of thickness h on an opaque substrate can be semiquantitatively understood as follows. Since no heat is generated in the transparent layer, the temperature T within this layer satisfies the diffusion equation

$$D\ \partial^2 T/\partial x^2 = \partial T/\partial t, \tag{79}$$

where D is the thermal diffusivity and only one-dimensional conduction is considered. Assume that the optical excitation beam has the sinusoidal time dependence $\exp(i\omega t)$. The solution to Eq. (79) is

$$T(x) = \theta_0 \exp(-ax + i\omega t), \tag{80}$$

where

$$a = (i\omega/D)^{1/2}. \tag{81}$$

Here θ_0 is the temperature at the interface and depends on the *absorbed* light amplitude I_0 (assumed strongly absorbed by the opaque substrate) and on the chopping period $2\pi/\omega$ (Helander *et al.*, 1981):

$$\theta_0 = (\text{real const})(I_0/\omega^{1/2})\exp(-\tfrac{1}{4}i\pi). \tag{82}$$

Equations (80)–(82) obviously satisfy Eq. (79). Combining Eqs. (80)–(82) (only the real parts have physical meaning), we may write the temperature at the top of the transparent layer, i.e., $T(h)$, as

$$T(h) = (\text{real const})(I_0/\omega^{1/2})\exp(-h/\mu)\cos(\omega t - h/\mu - \tfrac{1}{4}\pi), \tag{83}$$

where

$$\mu = (2D/\omega)^{1/2} \tag{84}$$

is the thermal diffusion length.

Equation (83) is the basis of PA depth profiling for the simple layered sample. Similar equations have been given by Adams and Kirkbright (1977a,b) and by Helander *et al.* (1981). Equation (83) implies that (for gas-microphone detection) the magnitude and phase of the PA signal denoted by A_{PA} and ψ_{PA}, respectively, are given by

$$A_{PA} \propto \omega^{-1} \exp[-h\omega^{1/2}/(2D)^{1/2}], \tag{85}$$

$$\psi_{PA} = -h\omega^{1/2}/(2D)^{1/2} - \tfrac{1}{4}\pi. \tag{86}$$

because the PA signal is determined by $T(h) \times$ (gas diffusion length), assuming that the gas diffusion length is smaller than the gas thickness [see Eqs. (40) and (41)]. Thus, $h/D^{1/2}$ of the top layer can be obtained either by plotting a semilog plot of $A_{PA}\omega$ versus $\omega^{1/2}$, or by plotting a linear plot of ψ_{PA} versus $\omega^{1/2}$. Adams and Kirkbright (1977b) have verified the latter by obtaining phase-lag plots for various thicknesses of polymer coatings on copper substrates (see Fig. 39b). They show that h obtained from such plots (with D being known) agrees with the coating thickness measured directly.

Refinement of the above depth-profiling theory and extension of the theory to multiple layers have been attempted by Helander *et al.* (1981). They show that the linear dependence of ψ_{PA} on the layer thickness h [Eq. (86)] is only a first approximation. In general, ψ_{PA} and A_{PA} are given by linear combinations of $\cosh(ah)$ and $\sinh(ah)$ functions, with the quantity a being given by Eq. (81). However, the experimental results of Helander *et al.* (1981) performed on a photographic emulsion do indicate that Eq. (86) is approximately valid for h/μ not exceeding 0.3.

Depth-profiling techniques are uniquely suitable for certain *in vivo* studies in medicine and biology, where optical reflection or extinction measurements usually cannot be performed without much sample preparation. For example, Campbell *et al.* (1979) have reported some remarkable applications of PA techniques to research in dermatology. They show that both drug diffusion, water content, and thermal properties of skin can be measured. They have developed a multilayer model of the PA effect to account for the nonuniform thermal properties of the intact skin due to the gradient of water content.

Various PA cells using gas microphones for depth-profiling studies of solid samples have been described in the literature. For example, Adams and Kirkbright (1976, 1977a,b) have described the apparatus in detail. Tam and Wong (1980) have described a PA cell (Fig. 15) with the type and the thickness of the coupling gas being variable so that the PA signal for the depth-profiling study can be optimized. Depth profiling with the use of direct coupling (Section III.D) is also possible, and the PA signal with piezoelectric detection is usually larger than the signal with microphone detection. Some of these studies will be discussed in conjunction with PA microscopy (Section V.E). Also, depth profiling with photoradiometric detection (see Section VI.B) or with photorefractive detection (Section VI.C) have also been demonstrated.

E. PA Microscopy and Imaging

PA microscopy is one of the most important nonspectroscopic application of PA detection because it is a unique method for providing subsurface imaging of any irregularities, flaws, doping concentrations, etc. Von Gutfeld and Melcher (1977) were the first to demonstrate that subsurface holes in an Al cylinder can affect the pulsed PA signal detected by a piezoelectric transducer. Wong *et al.* (1978) first reported actual PA images of subsurface structures in solids. They used an Ar ion laser chopped at a frequency between 50 and 2000 Hz, and they raster-scanned the focused beam across a silicon carbide ceramic sample which is used for the manufacturing of turbine blades. In addition, Wong *et al.* (1978) observed that the PA signal can detect visible surface microstructures as well as subsurface inhomogeneities that are not visible with an optical microscope. In a subsequent paper, Wong *et al.* (1979) showed that for the optically thick case (which is true for their SiC or Si_3N_4 samples excited by 4880- or 5145-Å beams), the PA microscopy detection depth is approximately given by the thermal diffusion length, which is ~ 100 μm in their cases with chopping frequencies ~ 100 Hz.

The principle of the PA microscopy technique is shown in Fig. 40. The best optical excitation source for PA microscopy is a highly collimated laser beam focused to a small spot at a wavelength that is strongly ab-

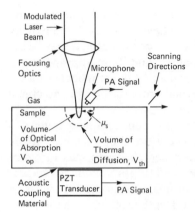

Fig. 40. Principle of the PA microscopy technique. A laser beam that is strongly absorbed by the sample is shown here. Any irregularities within the volume V_{op} affecting the optical absorption, or within the volume V_{th} affecting the thermal properties, will result in a change in the PA signal as the focused laser spot is scanned across the sample surface. The PA imaging has a resolution defined by V_{th}. The PA signal can be detected by a gas microphone, or by a piezoelectric transducer, as indicated, or even by radiometry or photorefractive effects (see Section VI).

sorbed by the sample (i.e., the absorption length is less than the diffusion length). If the beam is not strongly absorbed, then both PA signal magnitude and spatial resolution would suffer. Note that electron beam or other types of radiation can also be used for excitation (see Section VI). As shown in Fig. 40, the surface defining the volume of thermal diffusion V_{th} is outside the surface defining the volume of optical absorption V_{op} at a separation μ_s, which is the thermal diffusion length of the sample, of thermal diffusivity D. If a sinusoidally modulated laser beam of frequency f is used, then

$$\mu_s = (D/\pi f)^{1/2}, \tag{87}$$

while if a pulsed laser beam of duration τ is used, then

$$\mu_s \approx (4D\tau)^{1/2}. \tag{88}$$

The volume V_{th} (approximately a hemisphere of radius μ_s) defines the PA microscopy resolution. For highly thermally conductive materials such as Al, Au, or Si, $\mu_s \approx 50\ \mu m$ at $f = 10$ kHz. For moderately conductive materials like Al_2O_3, $\mu_3 \approx 15\ \mu m$ at the same chopping frequency. Three kinds of PA imaging are possible, as pointed out by Rosencwaig (1979, 1980b): (i) Any foreign materials or voids included in V_{th} causing a change in the thermal properties of V_{th} would change the amplitude and the phase of the PA signal; this is the origin of PA microscopy by *thermal wave*. (ii) Any foreign materials or voids included in V_{op} causing a change in the optical absorption in V_{op} would also change the amplitude of the PA signal; this is the origin of PA microscopy by *optical absorption*. (iii) If PZT transducers are used as shown in Fig. 40, the PA ultrasonic wave generated in the sample may be scattered by some foreign materials or voids before reaching the transducer; this is the origin of PA microscopy by *ultrasonic scattering*. Typically, ultrasonic scattering is negligible because the ultrasonic wavelength at 10 kHz exceeds 10 cm for almost all solids, and is much larger than the irregularities of interest. (Also, ultrasonic scattering is not observable with gas-microphone detection.) One notable exception is the PA microscopy experiment of Wickramasinghe *et al.* (1978). They have used a mode-locked Q-switched Nd : YAG laser to excite a metal film with optical pulse trains of durations 0.2 nsec at 210 MHz repetition rate. The PA ultrasonic wave is coupled through a water film and an Al_2O_3 acoustic lens to a SnO_2 thin-film piezoelectric transducer responding to sound waves at 840 MHz (wavelength $\lesssim 10$ μm). Deep subsurface imaging of gross features or discontinuities by PA ultrasonic reflections with mode conversions have been reported by Tam and Coufal (1983).

Numerous PA microscopy experiments to demonstrate new applications of the technique have been reported in the literature after the pioneering work of von Gutfeld and Melcher (1977), Wong *et al.* (1978, 1979). Rosencwaig (1979), and Wickramasinghe *et al.* (1978). For example, Wang *et al.* (1979) have used a CO_2 laser and a gas-microphone cell to map out localized absorption regions in KCl windows. Luukkala and Penttinen (1979) have used a HeNe laser chopped at 1 kHz and focused by a microscope for high-resolution imaging of a photolithographic mask. Freese and Teegarden (1979) have demonstrated the use of PA microscopy for nondestructive detection of inhomogeneities (that preferentially cause damage under intense laser-irradiation) in layered samples. Petts and Wickramasinghe (1980) have used a train of narrow laser pulses for PA microscopy to minimize sample heating by the focused laser beam. Thomas and collaborators (Thomas *et al.*, 1980; Favro *et al.*, 1980; Pouch *et al.*, 1980) and also Busse and co-workers (Busse and Rosencwaig, 1980; Busse and Ograbeck, 1980; Perkowitz and Busse, 1980; Rosencwaig and Busse, 1980) have performed PA imaging on devices such as integrated circuits, using either the *magnitude* or (often preferably) the *phase* of the PA signal. McClelland *et al.* (1980) have obtained PA images of compositional variations of a semiconducting alloy, $Hg_{1-x}Cd_xTe$, which is important for far-infrared detection. McFarlane *et al.* (1980) have done PA mapping of the damage due to ion implantation, and the subsequent recrystallization due to laser annealing.

PA imaging experiments have been performed with spot-by-spot laser excitation and data taking, so far. Coufal *et al.* (1982b) have shown that PA imaging can also be performed by a spatial multiplexing method, e.g., Hadamard transformation technique. This has two advantages for PA imaging: (i) The power density at the sample is reduced, thus lowering the risk of sample damage or spurious effects due to sample heating; (ii) Improvement of signal-to-noise ratio is achievable because of the multiplexing advantage.

VI. PHENOMENA CLOSELY RELATED TO THE PA EFFECT

In the photoacoustic effect, the excitation source is a light beam, and the detected quantity is an acoustic wave generated directly or indirectly by heat. There are several other effects closely related to it. For example, the "beam-acoustic" effect is similar to the PA effect except that the excitation source is an energetic beam (e.g., electron beam, ion beam, meson beam, x ray, rf, etc.). The "photothermal" effect is similar to the PA effect except that some other direct consequence of the heat deposited in the sample is monitored. The "photorefractive effect" is similar to the PA effect except that the detected quantity is a refractive index gradient generated directly or indirectly by heat (this effect is discussed in Chapter 3, by Fang and Swofford, in this book).

A. "Beam-Acoustic" Effect

The name "beam-acoustic" (BA) is used here to mean the method of "exciting a sample with a beam, and detecting by acoustic monitoring." Thus the BA effect includes the PA effect. It is probably impossible to trace who first observed a BA effect. Throwing a stone into water and hearing a sound may be called a BA effect. Also, it may be impossible to delineate all the different fields of science and technology in which a BA effect may be useful. For example, the BA effect is useful even in high-energy physics and cosmic-ray research. This is discussed by Rothenberg (1979), Learned (1979), Hunter *et al.* (1981), and by others; a cosmic-ray project called DUMAND is considered (Deep Underseas Muon and Neutrino Detector), utilizing hydrophone arrays under the sea to "hear" energetic particles in cosmic rays instead of detecting their radiation or ionization. Hence, the present discussions of application of BA effects must be regarded as sketchy at best.

An excellent example of the BA effect of relevance to PA microscopy described in Section V.E is the electron-acoustic (EA) microscopy work reported by Cargill (1980a,b) and independently by Brandis and Rosencwaig (1980). Cargill (1980a) modified a commercial scanning electron microscope to obtain an electron beam modulated at 6 MHz and focused to 1 μm spot size at a sample surface and detected the ultrasonic wave at the same frequency with a piezoelectric transducer attached to the opposite surface of the sample. The transducer output is used to form a scanned, magnified image of the sample. Image contrast comes primarily from spatial variations in the thermal and elastic properties within a resolution depth, which is typically a few microns for the sample studied (Au, Cu, Al, and Si). Cargill (1980a,b) showed that images of integrated circuits can be obtained with ~4-μm resolution, and suggested that 0.1-μm resolution is possible in thin-film specimens.

Brandis and Rosencwaig (1980) have also reported EA imaging of ceramic substrates. Their work is independent of that of Cargill (1980a), but the technique is similar. Brandis and Rosencwaig (1980) succeeded in identifying subsurface cracks in alumina substrates with soldering pads by thermal-wave imaging with phase optimization. More recently, Rosencwaig and White (1981) have demonstrated that EA thermal-wave microscopy can be used at 640 kHz modulation frequency to detect and to image phosphorus doping in a Si wafer.

Acoustic detection of the absorption of electromagnetic radiation far from the visible region has also been demonstrated. Nunes *et al.* (1979) first demonstrated this by their acoustic detection of ferromagnetic resonance in a thin iron foil in the microwave region. Melcher (1980) has reported that electron paramagnetic resonance can be detected by acoustic means; i.e., in magnetic resonance, the modulated radio-frequency field (acting like a modulated light beam in PA experiments) is absorbed, producing a detectable acoustic wave. Melcher (1980) found that magnetic field modulation can also be used instead of rf amplitude modulaton. Futher work on the acoustic detection of magnetic resonances has been reported by Coufal (1981), Hirabayashi *et al.* (1981), and DuVarney *et al.* (1981).

B. Photothermal Effects

The photothermal (PT) effect is the generation of heat by optical absorption in a sample (see Fig. 32). Subsequently, the heat can generate acoustic waves by thermal expansion of the sample, by thermal leak to an adjacent gas, or by some less common means like boiling, ablation, chemical decomposition, polymerization, phase transition, etc.; this is the PA effect, which is thus a consequence of the PT effect. The PT effect can produce other consequences as well.

The technique of direct measurement of PT temperature rise is also called optical calorimetry or laser calorimetry if a laser beam is used for the optical excitation. The technique is simply based on the conservation of energy, i. e., the optical energy absorbed equals the thermal energy produced, if no luminescence, photochemistry, or photoelectric effects occur. To prevent heat leaks, the PT calorimetry measurement must be carefully performed in a well-insulated vacuum Dewar, with the sample being in good thermal contact with a temperature sensor (of low thermal capacity) but in poor thermal contact with anything else. PT spectroscopy can be performed by scanning the wavelength of the incident optical radiation on the sample and measuring the corresponding temperature rise. However, because of the slow response, data taking can be extremely slow; it may take many minutes to take one data point.

Several examples of careful PT calorimetry have been reported. For example, Hass and Davisson (1977) performed a laser-calorimetry measurement of the absolute optical absorption coefficients of H_2O at 20°C at two Ar ion laser wavelengths, 4880 and 5145 Å. Their results, shown in Fig. 25a, compare very favorably with the pulsed OA measurement of Tam and Patel (1979b). Brilmyer et al. (1977) made a detailed study of PT calorimetry. They positioned a thermistor in close proximity to the solid or liquid sample, and monitored the temperature-dependent thermistor resistance due to the irradiation of the sample by intense monochromatic light. They have used their experimental arrangement to study absorbing samples, including crystalline solids (CdS, TiO_2) and solutions of dyes (rose bengal, methylene blue, and aniline yellow).

In the microscopic scale, the PT effect may cause vibrational heating of molecules. If the PT source is a pulsed, localized heat source (produced by a focused pulsed laser beam), heat diffusion causes a time-dependent spreading of the vibrational heating. Zapka and Tam (1982a) have observed this heat diffusion in a metal vapor by probing the molecular absorption of a continuous laser beam outside the pulsed PT source. Thus, Zapka and Tam (1982a) have shown that thermal conductivities can be measured in a noncontact manner.

"PT radiometry" is the measurement of the increase in IR radiation of a sample after PT heat deposition and represents a totally noncontact means to study a sample. Both this noncontact advantage and the high sensitivity attainable with PT radiometry (PTR) imply that PTR may be superior to PA monitoring in some applications, especially when remote sensing is necessary (high-temperature sample, corrosive sample, on-line product inspection, too large or too small a sample so that attachment of transducers or enclosure with a PA cell becomes difficult, etc.).

The normalized PTR signal S, defined as the amplitude of the modulation of the IR emission divided by the amplitude of the excitation radiation, depends on the sample absorption coefficient α. However, S may not be proportional to α because the PTR signal also depends on several other factors: the initial temperature profile produced in the sample, thermal diffusion into the sample, heat loss at sample surface, radiation loss, reabsorption and scattering of the far-IR radiation emitted from the interior of the sample, and so on. As an example, PTR signal saturation may occur when all the heat deposited within the optical absorption length α^{-1} can contribute uniformly to the radiometry signal, as in the photoacoustically thick case 5 in Table II.

Kanstad and Nordal (1978b, 1980a,b) and Nordal and Kanstad (1979) have pioneered the investigation of the PTR technique. A representative experimental arrangement used by them is shown in Fig. 41. They have used a tunable CO_2 laser to examine the PTR spectrum of various powdered samples excited in the wavelength region of 9–11 μm. The variation in the thermal radiation from the sample is detected with a far-IR sensor. They have studied samples like powered K_2SO_4 at temperatures of 295–942 K. They showed that due to Stefan–Boltzmann's law (radiant energy varies as the fourth power of absolute temperature) the signal-to-noise ratio of the PTR spectrum may be enhanced at higher temperature, although at very high temperature, the increased background radiation may reduce the signal-to-noise ratio. Kanstad and Nordal's work clearly demonstrates the greatest advantage of PTR detection, i. e., remote sensing capability so that high-temperature samples can be examined, and opaque, highly-scattering powders can be measured without the necessity of sample preparation or transducer attachment.

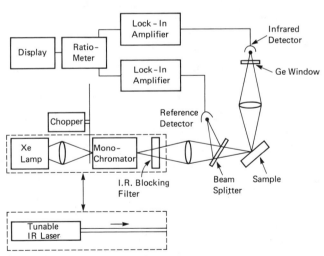

Fig. 41.   Experimental arrangement for PTR spectroscopy. Either the visible incoherent light source or the tunable IR laser (both shown in dotted rectangles) has been used for excitation. [After Nordal and Kanstad (1981).]

In a very recent work, Nordal and Kanstad (1981) demonstrated that PTR spectra of a variety of samples (Nd_2O_3 powder, whole blood, green leaf) can be measured with an inexpensive incoherent light source. This is simply a high-pressure Xe lamp in conjunction with a low-resolution spectrometer with an output beam power $\lesssim$ mW. They used a PbSnTe far-IR detector to measure the thermal radiation from the sample. They showed that their experimental results agree with their theory on the PTR signals.

Busse (1982) and Stadler *et al.* (1981) have recently extended the application of PTR detection to imaging, similar to the PA microscopy treated in Section V.E. Here again, the noncontact nature of the PTR method results in several distinct advantages of PTR imaging compared to PA imaging, namely, rapid imaging is possible, the sample can be at a distance, contamination of the sample is minimized, and large samples and minute samples can be easily studied. An important application possible for medical diagnostic is *in vivo* imaging (of parts of the human body) by scanning PTR.

C. Photorefractive Techniques

The photorefractive (PR) effect is the production of a refractive index gradient (RIG) Due to the heat gradient generated by the absorption of a light beam (called the excitation beam here) that may be modulated or not. The RIG generated by the excitation beam affects the propagation of the excitation beam itself, causing the well-known effect of "self-defocusing" or "thermal blooming" of the beam. Self-defocusing usually occurs instead of self-focusing because the derivative of the refractive index with respect to temperature is usually negative, so that the temperature gradient produces a negative lens. On the other hand, the RIG by the excitation beam also affects the propagation of another weak "probe" beam in the vicinity of the excitation beam. Thus, the RIG can be monitored either by self-defocusing (SD) or by probe-beam refraction (PBR). Leite *et al.* (1964) were the first to show that monitoring the SD of the excitation beam can be used as a sensitive spectroscopic tool. Solimini (1966) has provided a quantitative theory for this SD spectroscopy method. Swofford *et al.* (1976) have shown that the PBR method with an additional collinear probe beam provides higher sensitivity than the single-beam SD method; this is called "thermal lensing" (TL) of the probe beam. Further refinements and applications of PR methods are given by Murphy and Aamodt (1980, 1981), Boccara *et al.* (1980), Jackson *et al.* (1981a), and Fournier *et al.* (1982). Details on TL techniques are given in Chapter 3 in this book and also given in reviews by Whinnery (1974) and Kliger (1980).

VII. CONCLUSIONS

This chapter has presented a review of the theories, experiments, and applications of PA detections with emphasis on the most recent results (obtained within the past three years). With the theoretical and experimental principles of many different PA techniques presently understood, many more applications should be forthcoming. Examples of applications that are expected to expand rapidly are real-time PA pollutant, drug, and trace-impurity detection; PA particle or aerosol detection; miniaturized PA cell using small scannable solid-state lasers; PA monitoring of thin films, surfaces, and powders; PA Fourier transform spectroscopy for material identification; PA imaging; noncontact PA techniques; PA material characterizations and quality control techniques; and so on. These applications may well become standard procedures in industrial production or in medical diagnostics.

PA detection methods are currently competing with other PT methods like PT radiometry and PT refraction. These other methods usually have the distinct advantage of remote sensing, but are usually somewhat more complicated and vulnerable to errors than PA methods.

Table IV provides a list of references that are considered to be the most relevant to the subject matters covered in the present chapter. Some of the references have not been discussed in the text because of space limitations. It is hoped that a reasonably comprehensive collection of the most recent PA papers may be of use to the reader.

TABLE IV

Publications that Are Relevant to the Present Review in Order Presented

I. GENERAL REVIEW

Colles *et al.* (1979)	Hordvik (1977)	Patel (1978a)
Coufal *et al.* (1982)	Kanstad and Nordal	Patel and Tam (1981)
Dewey (1974)	(1980a,b)	Robin (1976)
Harshbarger and Robin	Kirkbright and Castle-	Rosencwaig (1978,
(1973)	den (1980)	1980a)
	Lyamshev and Sedov	Somoano (1978)
	(1981)	
	Pao (1977)	

II. PA SPECTROSCOPY IN GASES

A. Theory

Dewey (1977)	Farrow and Richton	Morse and Ingard (1968)
Dewey and Flint (1979)	(1977)	Wrobel and Vala (1978)
	Kreuzer (1971, 1977)	

continued

TABLE IV (*continued*)

B. Instrumentations for OA Studies of Gases

Barrett and Berry (1979)
Bray and Berry (1979)
Dewey (1977)
Farrow and Richton (1977)
Gelbwachs (1977)
Gerlach and Amer (1980)
Kavaya *et al.* (1979)
Kelly (1977)
Kerr and Atwood (1968)

Kimble and Roessler (1980)
Koch and Lahmann (1978)
Kreuzer (1971, 1977)
Kreuzer and Patel (1971)
Kreuzer *et al.* (1972)
Leo *et al.* (1980)
McClenny *et al.* (1981)
Marinero and Stuke (1979b)
Miles *et al.* (1980)

Parker (1973)
Patel (1978a)
Patel and Kerl (1977)
Rosengren (1975)
Smith and Gelfand (1980)
Tam *et al.* (1982)
Thomas *et al.* (1978)
Wake and Amer (1979)

C. Measuring Weak Absorption Lines

Bray and Berry (1979)
Cox and Gnauck (1980)
Dalby and Vigue (1979)
Fang and Swofford (1982)

Reddy (1980)
Reddy and Berry (1981)
Reddy *et al.* (1978)

Smith and Gelfand (1980)
Stella *et al.* (1976)
Wong and Moore (1981)

D. High-Sensitivity Trace Detection

Adamowicz and Koo (1979)
Angus *et al.* (1975)
Claspy *et al.* (1976, 1977)
Crane (1978)
Dewey (1976)
Dewey *et al.* (1973)
Gerlach and Amer (1978)

Ioli *et al.* (1979)
Kerr and Atwood (1968)
Koch and Lahmann (1978)
Konjevic and Jovicevic (1979)
Kozubovskii *et al.* (1980)
Kreuzer (1971)
Kritchman *et al.* (1978)

Loper *et al.* (1980)
McClenny (1981)
Nodov (1978)
Patel (1978a)
Shtrickman and Slatkine (1977)
Trautmann *et al.* (1981)
Vansteenkiste *et al.* (1981)

E. Absorption by Excited States

Patel (1978b)

Patel *et al.* (1977)

F. Chemically Reactive Gases

Colles *et al.* (1976)

Hunter and Kristjansson (1978)

G. Raman-Gain Spectroscopy

Barrett and Berry (1979)
Siebert *et al.* (1980)

West and Barrett (1979)

West *et al.* (1980)

H. Doppler-Free Spectroscopy

Inguscio *et al.* (1979)

Lieto *et al.* (1979)

Marinero and Stuke (1979a)

I. Multiphoton Absorption

Brenner *et al.* (1980)
Chin *et al.* (1982)

Cox (1978)
Fukumi (1979)

Webb *et al.* (1980)
Weulersse and Genier (1981)

TABLE IV (*continued*)

III. PA SPECTROSCOPY IN CONDENSED MATTER

A. The Gas-Coupling Method: Theory

Aamodt and Murphy (1978)
Aamodt et al. (1977)
Afromowitz et al. (1977)
Bennett and Forman (1977)
Chow (1980)

Helander et al. (1980)
McDonald (1979, 1980, 1981)
McDonald and Wetsel (1978)
Mandelis et al. (1979)

Poulet et al. (1980)
Quimby and Yen (1980)
Rosencwaig and Gersho (1976)
Tam and Wong (1980)

B. The Gas-Coupling Method: Instrumentation

Aamodt and Murphy (1977)
Bechthold et al. (1981)
Blank and Wakefield (1979)
Cahen and Garty (1979)
Cahen et al. (1978c)
Castleden et al. (1981)
Coufal (1982)
Crowley et al. (1980)
Ducharme et al. (1979)

Eaton and Stuart (1978)
Fournier et al. (1978)
Gray et al. (1977)
Hey and Gollnick (1968)
Kanstad and Nordal (1977, 1978a)
Lin and Dudek (1979)
McClelland and Knise-ley (1976a,b,c)
McDavid et al. (1978)
Malkin and Cahen (1981)

Monahan and Nolle (1977)
Murphy and Aamodt (1977a,b)
Nordhaus and Pelzl (1981)
Quimby et al. (1977)
Rosencwaig (1973, 1975, 1977)
Shaw (1979)
Tam and Wong (1980)
Wetsel and McDonald (1977)

C. The Direct-Coupling Method: Theory

Atalar (1980)
Bonch-Bruevich et al. (1977)
Bunkin and Kormis-sarov (1973)
Burt (1979, 1980)
Egerev and Naugol'nykh (1977)
Farrow et al. (1978a)

Gorodetskii et al. (1978)
Gournay (1966)
Jackson and Amer (1980)
Kasoev and Lyamshev (1977)
Kohanzadeh et al. (1975)
Liu (1982)
Lyamshev and Naugol'nykh (1976, 1980)

McQueen (1979)
Naugol'nykh (1977)
Nelson and Patel (1981)
Patel and Tam (1981)
Westervelt and Larson (1973)
White (1963)

D. The Direct-Coupling Method: Instrumentation

Bell and Landt (1967)
Bushnam and Barnes (1975)
Callis (1976)
Carome et al. (1964)
Farrow et al. (1978a)
Heritier et al. (1982)
Hordvik and Schloss-berg (1977)

Hordvik and Skolnik (1977)
Kohanzadeh et al. (1975)
Lahmann et al. (1977)
Oda et al. (1978)
Patel and Tam (1979a–d, 1980)
Patel et al. (1979, 1980)
Rentsch (1979)

Rentzepis and Pao (1966)
Sawada and Oda (1981)
Tam (1980a,c)
Tam and Patel (1979a,b,c, 1980)
Tam et al. (1979)
Voigtman et al. (1981)
Voigtman and Wine-fordner (1982)

continued

TABLE IV (*continued*)

E. Liquid-Coupling Method

    Sam and Shank (1979)

F. Measuring Weak Absorptions

Burt *et al.* (1980)	Kuo *et al.* (1982)	Starobogatov (1979)
Hordvik (1977)	McDavid *et al.* (1978)	Tam and Patel (1979b,
Hordvik and Schloss-	Patel *et al.* (1979, 1980)	1980)
berg (1977)	Sawada *et al.* (1979)	Tam *et al.* (1979)
Huard and Chardon		
(1981)		

G. Trace Detection

Gomenyuk and	Oda *et al.* (1978)	Sawada *et al.* (1981)
Shaidurov (1979)	Raine and Brown (1981)	Voigtman and Wine-
Lahmann *et al.* (1977)		fordner (1982)

H. Absorptions of Thin Films

Adams *et al.* (1978,	Kanstad and Nordal	Rockley and Devlin
1979)	(1980a,b)	(1980)
Castleden *et al.* (1979)	Kerr (1973)	Rosencwaig and Hall
Fernelius (1980a)	Low and Parodi (1980)	(1975)
Fernelius *et al.* (1981)	Nordal and Kanstad	Sander *et al.* (1981)
Fishman and Bard	(1977, 1978)	Schneider *et al.* (1981)
(1981)	Parker (1973)	Tam (1980c)
Hordvik and Skolnik	Patel and Tam (1980)	Tam and Patel (1979c)
(1977)		
Ikeda *et al.* (1981)		

I. Absorptions of Aerosols and Particles

Adams *et al.* (1977a,b)	Nordal and Kanstad	Shaw (1979)
Blank and Wakefield	(1977)	Szkarlat and Japar
(1979)	Oda *et al.* (1980)	(1981)
Eaton and Stuart (1978)	Roessler and Faxvog	Tam and Patel (1979c)
Helander and Lund-	(1980)	Vejux and Bae (1980)
ström (1980)	Rohl and Palmer (1981)	Yasa *et al.* (1979, 1982)
Japar and Killinger	Rosencwaig (1978,	
(1979)	1980a)	
McClenny and Rohl		
(1981)		

J. Fourier Transform Spectroscopy

Chalmers *et al.* (1981)	Rockley (1979, 1980a,b)	Royce *et al.* (1980)
Farrow *et al.* (1978b)	Rockley and Devlin	Teng and Royce (1982)
Laufer *et al.* (1980)	(1980)	Vidrine (1980)
Lloyd *et al.* (1980a,b)	Rockley *et al.* (1980,	
	1981)	

TABLE IV (*continued*)

K. Nonlinear Spectroscopy

Bae *et al.* (1982)	Patel and Tam (1979c)	Tam and Patel (1979a)
Bonch-Bruevich *et al.* (1977)	Rentzepis and Pao (1966)	Tam *et al.* (1979)
Munir *et al.* (1981)	Rockley and Devlin (1977)	Van Stryland and Woodall (1980)

IV. PA MONITORING OF DEEXCITATIONS

A. Fluorescence Quantum Yields

Aamodt and Murphy (1977)	Lahman and Ludewig (1977)	Rockley and Waugh (1978)
Adams *et al.* (1980, 1981)	Merkle and Powell (1977)	Rosencwaig and Hildrum (1981)
Auzel *et al.* (1979)	Murphy and Aamodt (1977b)	Starobogatov (1977)
Cahen *et al.* (1980)	Quimby and Yen (1978)	Suemune *et al.* (1980)
Callis *et al.* (1969)	Razumova and Starobogator (1977)	Wasa *et al.* (1980a)
Hall *et al.* (1976)		

B. Photochemical Effects

Bults *et al.* (1981)	Diebold (1980)	Malkin and Cahen (1979)
Cahen *et al.* (1978a,b)	Diebold and Hayden (1980)	Ort and Parson (1978, 1979)
Chance and Shand (1980)	Gray and Bard (1978)	

C. Photoelectrical Effects

Cahen (1978)	Tam (1980a)	Tokumoto *et al.* (1981)
Iwasaki *et al.* (1979, 1981)	Thielemann and Neumann (1980)	Wasa *et al.* (1980b)

D. Energy Transfer

Allen *et al.* (1977)	Hunter and Stock (1974)	Peterson and Powell (1978)
Anderson *et al.* (1981)	Hunter *et al.* (1974)	Powell *et al.* (1980a,b)
Farrow and Richton (1977)	Kaya *et al.* (1974, 1975)	Razumova and Starobogatov (1977)
Frank and Hess (1979)	Lepoutre *et al.* (1979)	Robin (1976)
Gelfand *et al.* (1979)	Mandelis and Royce (1980a,b)	Robin and Kuebler (1975, 1977)
Hall *et al.* (1976)	Mandelis *et al.* (1979)	Robin *et al.* (1980)
Huetz-Aubert and Lepoutre (1974)	Merkle *et al.* (1978)	Rohlfing *et al.* (1980, 1981)
Huetz-Aubert and Tripodi (1971)	Parker and Ritke (1973)	Taine and Lepoutre (1980)
Huetz-Aubert *et al.* (1980)		

continued

TABLE IV (*continued*)

V. PA PROBING OF PHYSICAL PROPERTIES

A. Acoustic Generation and Propagation

Aindow *et al.* (1981)
Akhmanov *et al.* (1979)
Allen *et al.* (1977)
Ash *et al.* (1980)
Askaryan *et al.* (1963)
Bell and Landt (1967)
Bonch-Bruevich *et al.*
  (1975)
Bunkin and Komissarov
  (1973)
Bushanam and Barnes
  (1975)
Carome *et al.* (1964)
Cottet and Romain
  (1982)
Dunina *et al.* (1979)
Egerev and
  Naugol'nykh (1977)
Emmony *et al.* (1976)

Fairand and Cluer
  (1979)
Felix and Ellis (1971)
Gordienko *et al.* (1978)
Gorodetskii *et al.* (1978)
Gourney (1966)
Karabutov (1979)
Kasoev and Lyamshev
  (1977)
Kolomenskii (1979)
Lee and White (1968)
Lyamshev and
  Naugol'nykh (1976,
  1980)
Lyamshev and Sedov
  (1979)
Naugol'nykh (1977)
Nelson and Fayer (1980)
Onokhov *et al.* (1980)
Scruby *et al.* (1980)

Sigrist and Kneubuhl
  (1978)
Silberg (1965)
Sladky *et al.* (1977)
Tam and Gill (1982)
Tam and Zapka (1982)
Tam *et al.* (1982)
Tennal *et al.* (1981)
Teslenko (1977)
Viertl (1980)
von Gutfeld and Budd
  (1979)
von Gutfeld and Mel-
  cher (1977)
Westervelt and Larson
  (1973)
White (1963)
Zapka and Tam (1982b)
Zapka *et al.* (1982)

B. Electrostriction

Bebchuk *et al.* (1978)

Brueck *et al.* (1980)

Forsbergh (1956)
Lai and Young (1982)

C. Radiation Damage, Crystallinity Change, and Phase Transitions

Cleves-Nunes *et al.*
  (1979)
Evora *et al.* (1980)
Florian *et al.* (1978)
Hoh (1980)

Jackson *et al.* (1981b)
Luukkala and Askerov
  (1980)
McClelland and Knise-
  ley (1979a,b)
McFarlane and Hess
  (1980)

Pichon *et al.* (1979)
Schneider *et al.* (1982)
Siqueira *et al.* (1980)
Tam (1982)

D. Depth-Profiling Studies

Adams and Kirkbright
  (1976, 1977a,b)
Bennett and Patty (1981)
Campbell *et al.* (1977,
  1979)
Fernelius (1979, 1980b)

Fujii *et al.* (1981)
Helander *et al.* (1981)
Morita (1981)
Parpal *et al.* (1981)

Stadler *et al.* (1981)
Yamagichi *et al.* (1980)
Yeack *et al.* (1982)

TABLE IV (*continued*)

E. PA Microscopy and Imaging

Busse and Ograbeck (1980)	McClelland *et al.* (1980)	Sawada *et al.* (1981)
Busse and Rosencwaig (1980)	McFarlane *et al.* (1980)	Tam and Coufal (1983)
Coufal *et al.* (1982)	Perkowitz and Busse (1980)	Thomas *et al.* (1980)
Favro *et al.* (1980)	Petts and Wickrama- singhe (1980)	von Gutfeld and Mel- cher (1977)
Freese and Teegarden (1979)	Pouch *et al.* (1980)	Wang *et al.* (1977)
Khandelwal *et al.* (1980)	Rosencwaig (1979, 1980b, 1981)	Wang *et al.* (1979)
Luukkala (1979)	Rosencwaig and Busse (1980)	Wickramasinghe *et al.* (1978)
Luukkala and Penttinen (1979)		Wong *et al.* (1978, 1979)

VI. PHENOMENA CLOSELY RELATED TO THE PA EFFECT

A. "Beam-Acoustic" Effect

Brandis and Rosencwaig (1980)	DuVarney *et al.* (1981)	Nunes *et al.* (1979)
Cargill (1980a,b)	Hirabayashi *et al.* (1981)	Rosencwaig and White (1981)
Coufal (1981)	Hunter *et al,* (1981)	Rothenberg (1979)
Coufal and Pacansky (1979)	Learned (1979)	
	Melcher (1980)	

B. Photothermal Effects

Brilmyer *et al.* (1977)	Hass and Davisson (1977)	Nordal and Kanstad (1979, 1980, 1981)
Busse (1980, 1982)	Kanstad and Nordal (1978b, 1980a,b)	Stadler *et al.* (1981)
Hartung *et al.* (1980)	Luukkala (1980)	Zapka and Tam (1982a)

C. Photorefractive Techniques

Boccara *et al.* (1980)	Jackson *et al.* (1981a,b)	Murphy and Aamodt (1980, 1981)
Fang and Swofford (1979)	Kliger (1980)	Solimini (1966)
Fournier *et al.* (1982)	Leite *et al.* (1964)	Swofford *et al.* (1976)

ACKNOWLEDGMENTS

Some of the pulsed OA work using piezoelectric transducers described in the present review has been performed by the author in collaboration with C. K. N. Patel of Bell Laboratories. The noncontact PA monitoring methods using probe-beam deflections have been studied in collaboration with W. Zapka. I greatly appreciate these collaborations. I thank the following scientists for letting me know of their work or letting me have preprints or reprints: M. A. Afromowitz, N. M. Amer, R. C. Bray, D. Cahen, P. C. Claspy, H. Coufal, D. F. Dewey, G. Diebold, E. M. Eyring, R. P. Freese, J. Gelfand, M. J. Kavaya,

G. F. Kirkbright, F. LePoutre, J. F. McClelland, R. A. McFarlane, R. L. Melcher, N. Mikoshiba, R. B. Miles, J. C. Murphy, P. E. Nordal, S. Oda, D. R. Ort, C. K. N. Patel, R. C. Powell, M. G. Rockley, A. Rosencwaig, B. S. H. Royce, G. J. Salamo, M. L. Shand, R. L. Thomas, M. Vala, R. J. von Gutfeld, J. Winefordner, and C. Yeack. The critical comments and helpful suggestions on this review offered by D. Cahen, H. Coufal, S. O. Kanstad, G. F. Kirkbright, E. E. Marinero, J. F. McClelland, and E. Voigtman are greatly appreciated.

REFERENCES

Aamodt, L. C., and Murphy, J. C. (1977). *J. Appl. Phys.* **48,** 3502.
Aamodt, L. C., and Murphy, J. C. (1978). *J. Appl. Phys.* **49,** 3036.
Aamodt, L. C., Murphy, J. C., and Parker, J. G. (1977). *J. Appl. Phys.* **48,** 927.
Adamowicz, R. F., and Koo, K. P. (1979). *Appl. Opt.* **18,** 2938.
Adams, M. J., and Kirkbright, G. F. (1976). *Spectros. Lett.* **9,** 255.
Adams, M. J., and Kirkbright, G. F. (1977a). *Analyst* **102,** 281.
Adams, M. J., and Kirkbright, G. F. (1977b). *Analyst* **102,** 678.
Adams, M. J., Highfield, J. G., and Kirkbright, G. F. (1977a). *Anal. Chem.* **49,** 1850.
Adams, M. J., Beadle, B. C., and Kirkbright, G. F. (1977b). *Analyst* **102,** 569.
Adams, M. J., Beadle, B. C., Kirkbright, G. F., and Menon, K. R. (1978). *Appl. Spectrosc.* **32,** 430.
Adams, M. J., Kirkbright, G. F., and Menon, K. R. (1979). *Anal. Chem.* **51,** 508.
Adams, M. J., Highfield, J. G., and Kirkbright, G. F. (1980). *Anal. Chem.* **52,** 1260.
Adams, M. J., Highfield, J. G., and Kirkbright, G. F. (1981). *Analyst* **106,** 850.
Afromowitz, M. A., Yeh, P. S., and Yee, S. (1977). *J. Appl. Phys.* **48,** 209.
Aindow, A. M., Dewhurst, R. J., Hutchins, D. A., and Palmer, S. B. (1981). *J. Acoust. Soc. Am.* **69,** 449.
Akhmanov, S. A., Rudenko, O. V., and Fedorchenko, A. T. (1979). *Sov. Tech. Phys. Lett.* (*Engl. Transl.*) **5,** 387.
Allen, J. F., Anderson, W. R., and Crosley, D. R. (1977). *Opt. Lett.* **1,** 118.
Anderson, W. R., Allen, J. E., and Crosley, D. R. (1981). Technical Report ARBRL-TR-02336. U.S. Army Aberdeen Proving Ground, Maryland (unpublished).
Angus, A. M., Marinero, E. E., and Colles, M. J. (1975). *Opt. Commun.* **14,** 223.
Ash, E. A., Dieulesaint, E., and Rakouth, H. (1980). *Electron. Lett.* **16,** 470.
Askar'yan, G. A., Prokhorov, A. M., Chanturiya, G. F., and Shipulo, G. P. (1963). *Sov. Phys.—JETP* (*Engl. Transl.*) **17,** 1463.
Atalar, A. (1980). *Appl. Opt.* **19,** 3204.
Auzel, F., Meichenin, D., and Michel, J. C. (1979). *J. Lumin.* **18/19,** 97.
Bae, Y., Song, J. J., and Kim, Y. B. (1982). *Appl. Opt.* **21,** 35.
Barrett, J. J., and Berry, M. J. (1979). *Appl. Phys. Lett.* **34,** 144.
Bebchuk, A. S., Mizin, V. M., and Salova, N. Ya. (1978). *Opt. Spectrosc.* (*Engl. Transl.*) **44,** 92.
Bechthold, P. A., Campagna, M., and Chatzipetros, J. (1981). *Opt. Commun.* **36,** 369.
Bell, A. G. (1880). *Am. J. Sci.* **20,** 305.
Bell, C. E., and Landt, J. A. (1967). *Appl. Phys. Lett.* **10,** 46.
Bennett, C. A., and Patty, R. R. (1981). *Appl. Opt.* **20,** 911.
Bennett, H. S., and Forman, R. A. (1977). *Appl. Opt.* **16,** 2834.
Blank, R. E., and Wakefield, T., II, (1979). *Anal. Chem.* **51,** 50.
Boccara, A. C., Fournier, D., Jackson, W., and Amer, N. M. (1980). *Opt. Lett.* **5,** 377.

Bonch-Bruevich, A. M., Razumova, T. K., and Starobogatov, I. O. (1975). *Sov. Tech. Phys. Lett.* (*Engl. Transl.*) **1**, 26.

Bonch-Bruevich, A. M., Razumova, T. K., and Starobogatov, I. O. (1977). *Opt. Spectrosc.* (*Engl. Transl.*) **42**, 45.

Brandis, E., and Rosencwaig, A. (1980). *Appl. Phys. Lett.* **37**, 98.

Bray, R. G., and Berry, M. J. (1979). *J. Chem. Phys.* **71**, 4909.

Brenner, D. M., Brezinsky, K., and Curtis, P. M. (1980). *Chem. Phys. Lett.* **72**, 202.

Brilmyer, G. H., Fujishima, A., Santhanam, K. S. V., and Bard, A. J. (1977). *Anal. Chem.* **49**, 2057.

Brueck, S. R. J., Kidal, H., and Bélanger, L. J. (1980). *Opt. Commun.* **34**, 199.

Bults, G., Horwitz, B. A., Malkin, S., and Cahen, D. (1981). *FEBS Lett.* **129**, 44.

Bunkin, F. V., and Komissarov, V. M. (1973). *Sov. Phys.—Acoust.* (*Engl. Transl.*) **19**, 203.

Burt, J. A. (1979). *J. Acoust. Soc. Am.* **65**, 1164.

Burt, J. A. (1980). *J. Phys. D* **13**, 1185.

Burt, J. A., Ebeling, K. J., and Efthimiades, D. (1980). *Opt. Commun.* **32**, 59.

Bushanam, G. S., and Barnes, F. S. (1975). *J. Appl. Phys.* **46**, 2074.

Busse, G. (1980). *Infrared Phys.* **20**, 419.

Busse, G. (1982). *Appl. Opt.* **21**, 107.

Busse, G., and Ograbeck, A. (1980). *J. Appl. Phys.* **51**, 3576.

Busse, G., and Rosencwaig, A. (1980). *Appl. Phys. Lett.* **36,**. 815.

Cahen, D. (1978). *Appl. Phys. Lett.* **33**, 810.

Cahen, D., and Garty, H. (1979). *Anal. Chem.* **51**, 1865.

Cahen, D., Garty, H., and Caplen, S. R. (1978a). *FEBS Lett.* **91**, 131.

Cahen, D., Malkin, S., and Lerner, E. J. (1978b). *FEBS Lett.* **91**, 339.

Cahen, D., Lerner, E. I., and Auerback, A. (1978c). *Rev. Sci. Instrum.* **49**, 1206.

Cahen, D., Garty, H., and Becker, R. S. (1980). *J. Phys. Chem.* **84**, 3384.

Callis, J. B. (1976). *J. Res. Natl. Bur. Strand., Sect. A* **80A**, 413.

Callis, J. B., Gouterman, M., and Danielson, J. D. S. (1969). *Rev. Sci. Instrum.* **40**, 1599.

Campbell, S. D., Yee, S. S., and Afromowitz, M. A. (1977). *J. Bioeng.* **1**, 185.

Campbell, S. D., Yee, S. S., and Afromowitz, M. A. (1979). *IEEE Trans. Biomed. Eng.* **BME-26**, 220.

Cargill, C. S. (1980a). *Nature* (*London*) **286**, 691.

Cargill, C. S. (1980b). *In* "Scanned Image Microscopy" (E. A. Ash, ed.), p. 319. Academic Press, New York.

Carome, E. F., Clark, N. A., and Moeller, C. E. (1964). *Appl. Phys. Lett.* **4**, 95.

Castleden, S. L., Elliott, C. M., Kirkbright, G. F., and Spillane, D. E. M. (1979). *Anal. Chem.* **51**, 2152.

Castleden, S. L., Kirkbright, G. F., and Spillane, D. E. M. (1981). *Anal. Chem.* **53**, 2228.

Chalmers, J. M., Stay, B. J., Kirkbright, G. F., Spillane, D. E. M., and Beadle, B. C. (1981). *Analyst* **106**, 1179.

Chance, R. R., and Shand, M. L. (1980). *J. Chem. Phys.* **72**, 948.

Chin, S. L., Evans, D. K., McAlpine, R. D., and Selander, W. N. (1982). *Appl. Opt.* **21**, 65.

Chow, H. C. (1980). *J. Appl. Phys.* **51**, 4053.

Claspy, P. C., Pao, Y. H., Kwong, S., and Nodov, E. (1976). *Appl. Opt.* **15**, 1506.

Claspy, P. C., Ha, C., and Pao, Y. H. (1977). *Appl. Opt.* **16**, 2972.

Cleves-Nunes, O. A., Monteiro, A. M. M., and Skeff-Neto, K. (1979). *Appl. Phys. Lett.* **35**, 656.

Colles, M. J., Angus, A. M., and Marinero, E. E. (1976). *Nature* (*London*) **262**, 681.

Colles, M. J., Geddes, N. R., and Mehdizadeh, E. (1979). *Contemp. Phys.* **20**, 11.

Cottet, F., and Romain, J. P. (1982). *Phys. Rev. A* **25**, 576.
Coufal, H. (1981). *Appl. Phys. Lett.* **39**, 215.
Coufal, H. (1982). *Appl. Opt.* **21**, 104.
Coufal, H., and Pacansky, J. (1979). *IBM Tech. Discl. Bull.* **22**, 2875.
Coufal, H., Korpiun, P., Lüscher, E., Schneider, S., and Tilgner, R. (1982a). *In* "Photo-acoustics—Principles and Applications" (H. Coufal, ed.), Vieweg Verlag, Braun-schweig.
Coufal, H., Möller, U., and Schneider, S. (1982b). *Appl. Opt.* **21**, 116.
Cox, D. M. (1978). *Opt. Commun.* **24**, 336.
Cox, D. M., and Gnauck, A. (1980). *J. Mol. Spectrosc.* **81**, 207.
Crane, R. A. (1978). *Appl. Opt.* **17**, 2097.
Crowley, T. P., Faxvog, F. R., and Roessler, D. M. (1980). *Appl. Phys. Lett.* **36**, 641.
Dabora, E. K. (1981). *Prog. Astronaut. Aeronaut.* **76**, 119.
Dalby, F. W., and Vigue, J. (1979). *Phys. Rev. Lett.* **43**, 1310.
Dewey, C. F. (1974). *Opt. Eng.* **13**, 483.
Dewey, C. F. (1976). U.S. Patent 3,938,365.
Dewey, C. F. (1977). *In* "Optoacoustic Spectroscopy and Detection" (Y. H. Pao, ed.), p. 47. Academic Press, New York.
Dewey, C. F., and Flint, J. H. (1979). Technical Digest, Photoacoustic Spectroscopy Con-ference, Ames, Iowa (unpublished).
Dewey, C. F., Kamm, R. D., and Hackett, C. E. (1973). *Appl. Phys. Lett.* **23**, 633.
Diebold, G. J. (1980). *J. Phys. Chem.* **84**, 2213.
Diebold, G. J., and Hayden, J. S. (1980). *Chem. Phys.* **49**, 429.
Ducharme, D., Tessier, A., and Leblanc, R. M. (1979). *Rev. Sci. Instrum.* **50**, 1461.
Dunina, T. A., Egerev, S. V., Lyamshev, L. M., and Naugol'nykh, K. A. (1979). *Sov. Phys.—Acoust. (Engl. Transl.)* **25**, 32.
DuVarney, R. C., Garrison, A. K., and Busse, G. (1981). *Appl. Phys. Lett.* **38**, 675.
Eaton, H. E., and Staurt, J. D. (1978). *Anal. Chem.* **50**, 587.
Egerev, S. V., and Naugol'nykh, K. A. (1977). *Sov. Phys.—Acoust. (Engl. Transl.)* **23**, 422.
Emmony, D. C., Siegrist, M., and Kneubuhl, F. K. (1976). *Appl. Phys. Lett.* **29**, 547.
Evora, C., Landers, R., and Vargas, H. (1980). *Appl. Phys. Lett.* **36**, 864.
Fairand, B. P., and Cluer, A. H. (1979). *J. Appl. Phys.* **50**, 1497.
Fang, H. L., and Swofford, R. L. (1979). *J. Appl. Phys.* **50**, 6609.
Fang, H. L., and Swofford, R. L. (1981). *Appl. Opt.* **21**, 55.
Farrow, L. A., and Richton, R. E. (1977). *J. Appl. Phys.* **48**, 4962.
Farrow, M. M., Burnham, R. K., Auzanneau, M., Olsen, S. L., Purdie, N., and Eyring, E. M. (1978a). *Appl. Opt.* **17**, 1093.
Farrow, M. M., Burnham, R. K., and Eyring, E. M. (1978b). *Appl. Phys. Lett.* **33**, 735.
Favro, L. D., Kuo, P. K., Pouch, J. J., and Thomas, R. L. (1980). *Appl. Phys. Lett.* **36**, 953.
Felix, M. P., and Ellis, A. T. (1971). *Appl. Phys. Lett.* **19**, 484.
Fernelius, N. C. (1979). *NBS Spec. Pub. (U.S.)* **568**, 301.
Fernelius, N. C. (1980a). *Appl. Surf. Sci.* **4**, 401.
Fernelius, N. C. (1980b). *J. Appl. Phys.* **51**, 650.
Fernelius, N. C., Dempsey, D. V., and Quinn, D. B. (1981). *Appl. Surf. Sci.* **7**, 32.
Fishman, V. A., and Bard, A. J. (1981). *Anal. Chem.* **53**, 102.
Florian, R., Pelzl, H., Rosenberg, M., Vargas, H., and Wernhardt, R. (1978). *Phys. Status Solidi A* **48**, K35.
Forsbergh, P. W. (1956). *In* "Handbuch der Physik" (S. Flügge, ed.), Vol. 17, p. 264. Springer-Verlag, Berlin and New York.
Fournier, D., Boccara, A. C., and Bodoz, J. (1978). *Appl. Phys. Lett.* **32**, 640.

Fournier, D., Boccara, A. C., and Badoz, J. (1981). *Appl. Opt.* **21**, 74.

Frank, K., and Hess, P. (1979). *Chem. Phys. Lett.* **68**, 540.

Freese, R. P., and Teegarden, K. J. (1979). *NBS Spec. Publ.* (*U.S.*) **568**, 313.

Fujii, Y., Moritani, A., and Nakai, J. (1981). *Jpn. J. Appl. Phys.* **20**, 361.

Fukumi, T. (1979). *Opt. Commun.* **30**, 351.

Gelbwachs, J. A. (1977). *In* "Optoacoustic Spectroscopy and Detection" (Y. H. Pao, ed.), p. 79. Academic Press, New York.

Gelfand, J., Hermina, W., and Smith, W. H. (1979). *Chem. Phys. Lett.* **65**, 201.

Gerlach, R., and Amer, N. M. (1978). *Appl. Phys. Lett.* **32**, 228.

Gerlach, R., and Amer, N. M. (1980). *Appl. Phys.* **23**, 319.

Gomenyuk, A. S., and Shaidurov, V. G. (1979). *Sov. J. Opt. Technol.* (*Engl. Transl.*) **46**, 315.

Gordienko, V. M., Reshilov, A. B., and Shmal'gauzen, V. I. (1978). *Sov. Phys.—Acoust.* (*Engl. Transl.*) **24**, 73.

Gorodetskii, V. S., Egerev, S. V., Esipov, I. B., and Naugol'nykh, K. A. (1978). *Sov. J. Quantum Electron.* (*Engl. Transl.*) **8**, 1345.

Gournay, L. S. (1966). *J. Acoust. Soc. Am.* **40**, 1322.

Gray, R. C., and Bard, A. J. (1978). *Anal. Chem.* **50**, 1262.

Gray, R. C., Fishman, V. A., and Bard, A. J. (1977). *Anal. Chem.* **49**, 697.

Hale, G. M., and Querry, M. R. (1973). *Appl. Opt.* **12**, 555.

Hall, L. M., Hunter, T. F., and Stock, M. G. (1976). *Chem. Phys. Lett.* **44**, 145.

Harshbarger, W. R., and Robin, M. B. (1973). *Acc. Chem. Res.* **6**, 329.

Hartung, C., Jurgeit, R., and Ritze, H. H. (1980). *Appl. Phys.* **23**, 407.

Hass, M., and Davisson, J. W. (1977). *J. Opt. Soc. Am.* **67**, 622.

Helander, P., and Lundström, I. (1980). *J. Appl. Phys.* **51**, 3841.

Helander, P., Lundström, I., and McQueen, D. (1980). *J. Appl. Phys.* **51**, 3841.

Helander, P., Lundström, I., and McQueen, D. (1981). *J. Appl. Phys.* **52**, 1146.

Henry, B. R. (1977). *Acc. Chem. Res.* **10**, 207.

Henry, B. R., and Siebrand, W. (1968). *J. Chem. Phys.* **49**, 5369.

Heritier, J. M., Fouquet, J. E., and Siegman, A. E. (1982). *Appl. Opt.* **21**, 90.

Hey, E., and Gollnick, K. (1968). *Ber. Bunsenges. Phys. Chem.* **72**, 263.

Hirabayashi, I., Morigaki, K., and Sans, Y. (1981). *Jpn. J. Appl. Phys.* **20**, L208.

Hoh, K. (1980). *Electron. Lett.* **16**, 931.

Hordvik, A. (1977). *Appl. Opt.* **16**, 2827.

Hordvik, A., and Schlossberg, H. (1977). *Appl. Opt.* **16**, 101.

Hordvik, A., and Skolnik, L. (1977). *Appl. Opt.* **16**, 2919.

Huard, S., and Chardon, D. (1981). *Opt. Commun.* **39**, 59.

Huetz-Aubert, M., and Lepoutre, F. (1974). *Physica (Amsterdam)* **78**, 435.

Huetz-Aubert, M., and Tripodi, R. (1971). *J. Chem. Phys.* **55**, 5724.

Huetz-Aubert, M., Perrin, M. Y., and Lepoutre, F. (1980). *Chem. Phys. Lett.* **76**, 498.

Hulbert, E. O. (1945). *J. Opt. Soc. Am.* **35**, 698.

Hunter, S. D., *et al.* (1981), *J. Acoust. Soc. Am.* **69**, 1557.

Hunter, T. F., and Kristjansson, K. S. (1978). *Chem. Phys. Lett.* **58**, 291.

Hunter, T. F., and Stock, M. G. (1974). *J. Chem. Soc., Faraday Trans. 2* **70**, 1022.

Hunter, T. F., Rumbles, D., and Stock, M. G. (1974). *J. Chem. Soc., Faraday Trans. 2* **70**, 1010.

Hurst, G. S., Payne, M. G., Kramer, S. D., and Young, J. P. (1979). *Rev. Mod. Phys.* **51**, 767.

Ikeda, S., Murakami, Y., and Akatsuka, K. (1981). *Chem. Lett.* **3**, 363.

Inguscio, M., Moretti, A., and Strumia, F. (1979). *Opt. Commun.* **30**, 355.

Ioli, N., Violino, P., and Meucci, M. (1979). *J. Phys. E* **12**, 168.

Irvine, W. M., and Pollack, J. B. (1968). *Icarus* **8**, 324.

Iwasaki, T., Sawada, T., Kamada, H., Fujishima, A., and Honda, K. (1979). *J. Phys. Chem.* **83**, 2142.

Iwasaki, T., Oda, S., Sawada, T., and Honda, K. (1981). *Photogr. Sci. Eng.* **25**, 6.

Jackson, W. B., and Amer, N. M. (1980). *J. Appl. Phys.* **51**, 3343.

Jackson, W. B., Amer, N. M., Boccara, A. C., and Fournier, D. (1981a). *Appl. Opt.* **20**, 1333.

Jackson, W. B., Amer, N. M., Fournier, D., and Boccara, A. C. (1981b). Technical Digest, Second International Conference on Photoacoustic Spectroscopy, Berkeley (unpublished).

Japar, S. M., and Killinger, D. K. (1979). *Chem. Phys. Lett.* **66**, 207.

Kanstad, S. O., and Nordal, P. E. (1977). *Int. J. Quantum. Chem.* **12**, Suppl. 2, 123.

Kanstad, S. O., and Nordal, P. E. (1978a). *Opt. Commun.* **26**, 367.

Kanstad, S. O., and Nordal, P. E. (1978b). *Powder Technol.* **22**, 133.

Kanstad, S. O., and Nordal, P. E. (1980a). *Appl. Surf. Sci.* **5**, 286.

Kanstad, S. O., and Nordal, P. E. (1980b). *Appl. Surf. Sci.* **6**, 372.

Karabutov, A. A. (1979). *Sov. Tech. Phys. Lett. (Engl. Transl.)* **5**, 174.

Kasoev, S. G., and Lyamshev, L. M. (1977). *Sov. Phys.—Acoust. (Engl. Transl.)* **23**, 510.

Kavaya, M. J., Margolis, J. S., and Shumate, M. S. (1979). *Appl. Opt.* **18**, 2602.

Kaya, K., Harshbarger, W. R., and Robin, M. B. (1974). *J. Chem. Phys.* **60**, 4231.

Kaya, K., Chatelain, C. L., Robin, M. B., and Kuebler, N. A. (1975). *J. Am. Chem. Soc.* **97**, 2153.

Kelly, P. L. (1977). In "Optoacoustic Spectroscopy and Detection" (Y. H. Pao, ed.), p. 113. Academic Press, New York.

Kerr, E. L. (1973). *Appl. Opt.* **12**, 2520.

Kerr, E. L., and Atwood, J. G. (1968). *Appl. Opt.* **7**, 915.

Khandelwal, P. K., Heitman, P. W., Silversmith, A. J., and Wakefield, T. D. (1980). *Appl. Phys. Lett.* **37**, 779.

Kimble, H. J., and Roessler, D. M. (1980). U.S. Patent 4,200,399.

Kirkbright, G. F., and Castleden, S. L. (1980). *Chem. Br.* **16**, 661.

Kliger, D. S. (1980). *Acc. Chem. Res.* **13**, 129.

Koch, K. P., and Lahmann, W. (1978). *Appl. Phys. Lett.* **32**, 289.

Kohanzadeh, Y., Whinnery, J. R., and Carroll, M. M. (1975). *J. Acoust. Soc. Am.* **57**, 67.

Kolomenskii, A. A. (1979). *Sov. Phys.—Acoust. (Engl. Transl.)* **25**, 312.

Konjevic, N., and Jovicevic, S. (1979). *Spectrosc. Lett.* **12**, 259.

Kopelevich, O. V. (1976). *Opt. Spectrosc. (Engl. Transl.)* **41**, 391.

Kozubovskii, V. R., Perchi, Z. I., and Romanko, G. D. (1980). *Kvantovaya Elektron. (Kiev)* **18**, 86.

Kreuzer, L. B. (1971). *J. Appl. Phys.* **42**, 2934.

Kreuzer, L. B. (1977). In "Optoacoustic Spectroscopy and Detection" (Y. H. Pao, ed.), p. 1. Academic Press, New York.

Kreuzer, L. B., and Patel, C. K. N. (1971). *Science* **173**, 45.

Kreuzer, L. B., Kenyon, N. D., and Patel, C. K. N. (1972). *Science* **177**, 347.

Kritchman, E., Shtrickman, S., and Slatkine, M. (1978). *J. Opt. Soc. Am.* **68**, 1257.

Kuo, C.-Y., Vieira, M. M. F., and Patel, C. K. N. (1982). *Phys. Rev. Lett.* **49**, 1284.

Lahmann, W., and Ludewig, H. J. (1977). *Chem. Phys. Lett.* **45**, 177.

Lahmann, W., Ludewig, H. J., and Welling, H. (1977). *Anal. Chem.* **49**, 549.

Lai, H. M., and Young, K. (1982). *J. Acoust. Soc. Am.* (to be published).

Landau, L. D., and Lifshitz, E. M. (1959). "Fluid Mechanics" (transl. by J. B. Sykes and W. H. Reid), Chapter VIII. Pergamon, Oxford.

Laufer, G., Huneke, J. T., Royce, B. S. H., and Teng, Y. C. (1980). *Appl. Phys. Lett.* **37,** 517.

Learned, J. G. (1979). *Phys. Rev. D.,* **19,** 3293.

Lee, R. E., and White, R. M. (1968). *Appl. Phys. Lett.* **12,** 12.

Leite, R. C. C., Moore, R. S., and Whinnery, J. R. (1964). *Appl. Phys. Lett.* **5,** 141.

Leo, R. M., Hawkins, H. L., John, P., and Harrison, R. G. (1980). *J. Phys. E* **13,** 658.

Lepoutre, F., Louis, G., and Taine, J. (1979). *J. Chem. Phys.* **70,** 2225.

Lieto, A. D., Minguzzi, P., and Tonelli, M. (1979). *Opt. Commun.* **31,** 25.

Lin, J. W., and Dudek, L. P. (1979). *Anal. Chem.* **51,** 1627.

Liu, G. (1982). *Appl. Opt.* **21,** 955.

Lloyd, L. B., Riseman, S. M., Burnham, R. K., and Eyring, E. M. (1980a). *Rev. Sci. Instrum.* **51,** 1488.

Lloyd, L. B., Burnham, R. K., Chandler, W. L., Eyring, E. M., and Farrow, M. M. (1980b). *Anal. Chem.* **52,** 1595.

Loper, G. L., Calloway, A. R., Stamps, M. A., and Gelbwachs, J. A. (1980). *Appl. Opt.* **19,** 2726.

Low, M. J. D., and Parodi, G. A. (1980). *Appl. Spectrosc.* **34,** 76.

Luukkala, M. (1979). *IEEE Ultrason. Symp. Proc.* p. 412.

Luukkala, M. (1980). *In* "Scanned Image Microscopy" (E. A. Ash, ed.), p. 273. Academic Press, New York.

Luukkala, M., and Askerov, S. G. (1980). *Electron. Lett.* **16,** 84.

Luukkala, M., and Penttinen, A. (1979). *Electron. Lett.* **15,** 326.

Lyamshev, L. M., and Naugol'nykh, K. A. (1976). *Sov. Phys.—Acoust.* (*Engl. Transl.*) **22,** 354.

Lyamshev, L. M., and Naugol'nykh, K. A. (1980). *Sov. Phys.—Acoust.* (*Engl. Transl.*) **26,** 351.

Lyamshev, L. M., and Sedov, L. N. (1979). *Sov. Tech. Phys. Lett.* (*Engl. Transl.*) **5,** 403.

Lyamshev, L. M., and Sedov, L. N. (1981). *Sov. Phys.—Acoust.* (*Engl. Transl.*) **27,** 4.

McClelland, J. F., and Kniseley, R. N. (1976a). *Appl. Opt.* **15,** 2658.

McClelland, J. F., and Kniseley, R. N. (1976b). *Appl. Opt.* **15,** 2967.

McClelland, J. F., and Kniseley, R. N. (1976c). *Appl. Phys. Lett.* **28,** 467.

McClelland, J. F., and Kniseley, R. N. (1979a). *Appl. Phys. Lett.* **35,** 121.

McClelland, J. F., and Kniseley, R. N. (1979b). *Appl. Phys. Lett.* **35,** 585.

McClelland, J. F., Kniseley, R. N., and Schmit, J. L. (1980). *In* "Scanned Image Microscopy" (E. A. Ash, ed.), p. 353.

McClenny, W. A. (1981). Environmental Protection Agency Project Summary EPA-600/S7-81-026 (unpublished).

McClenny, W. A., and Rohl, R. (1981). Technical Digest, Second International Conference on Photoacoustic Spectroscopy, Berkeley (unpublished).

McClenny, W. A., Bennett, C. A., Russwurm, G. M., and Richmond, R. (1981). *Appl. Opt.* **20,** 650.

McDavid, J. M., Lee, K. L., Yee, S. S., and Afromowitz, M. A. (1978). *J. Appl. Phys.* **49,** 6112.

McDonald, F. A. (1979). *Appl. Opt.* **18,** 1363.

McDonald, F. A. (1980). *Appl. Phys. Lett.* **36,** 123.

McDonald, F. A. (1981). *J. Appl. Phys.* **52,** 1462.

McDonald, F. A., and Wetsel, G. C., Jr. (1978). *J. Appl. Phys.* **49,** 2313.

McFarlane, R. A., and Hess, L. D. (1980). *Appl. Phys. Lett.* **36,** 137.

McFarlane, R. A., Hess, L. D., and Olson, G. L. (1980). *IEEE Ultrason. Symp. Proc.* p. 628.

McQueen, D. H. (1979). *J. Phys. D* **12**, 1673.

Malkin, S., and Cahen, D. (1979). *Photochem. Photobiol.* **29**, 803.

Malkin, S., and Cahen, D. (1981). *Anal. Chem.* **53**, 1426.

Mandelis, A., and Royce, B. S. H. (1980a). *J. Appl. Phys.* **51**, 610.

Mandelis, A., and Royce, B. S. H. (1980b). *J. Opt. Soc. Am.* **70**, 474.

Mandelis, A., Teng, Y. C., and Royce, B. S. H. (1979). *J. Appl. Phys.* **50**, 7138.

Marinero, E. E., and Stuke, M. (1979a). *Opt. Commun.* **30**, 349.

Marinero, E. E., and Stuke, M. (1979b). *Rev. Sci. Instrum.* **50**, 241.

Melcher, R. L. (1980). *Appl. Phys. Lett.* **37**, 895.

Merkle, L. D., and Powell, R. C. (1977). *Chem. Phys. Lett.* **46**, 303.

Merkle, L. D., Powell, R. C., and Wilson, T. M. (1978). *J. Phys. C* **11**, 3103.

Miles, R. B., Gelfand, J., and Wilczek, E. (1980). *J. Appl. Phys.* **51**, 4543.

Monahan, E. M., and Nolle, A. W. (1977). *J. Appl. Phys.* **48**, 3519.

Morita, M. (1981). *Jpn. J. Appl. Phys.* **20**, 835.

Morse, P. M., and Ingard, K. V. (1968). "Theoretical Acoustics." McGraw-Hill, New York.

Moses, E. I., and Tang, C. L. (1977). *Opt. Lett.* **1**, 115.

Munir, Q., Winter, E., and Schmidt, A. J. (1981). *Opt. Commun.* **36**, 467.

Murphy, J. C., and Aamodt, L. C. (1977a). *Appl. Phys. Lett.* **31**, 728.

Murphy, J. C., and Aamodt, L. C. (1977b). *J. Appl. Phys.* **48**, 3502.

Murphy, J. C., and Aamodt, L. C. (1980). *J. Appl. Phys.* **51**, 4580.

Murphy, J. C., and Aamodt, L. C. (1981). *Appl. Phys. Lett.* **38**, 196.

Naugol'nykh, K. A. (1977). *Sov. Phys.—Acoust.* (*Engl. Transl.*) **23**, 98.

Nechaev, S. Yu., and Ponomarev, N. Yu. (1975). *Sov. J. Quantum Electron* (*Engl. Transl.*) **5**, 752.

Nelson, E. T., and Patel, C. K. N. (1981). *Opt. Lett.* **6**, 354.

Nelson, K. A., and Fayer, M. D. (1980). *J. Chem. Phys.* **72**, 5202.

Nodov, E. (1978). *Appl. Opt.* **17**, 1110.

Nordal, P. E., and Kanstad, S. O. (1977). *Opt. Commun.* **22**, 185.

Nordal, P. E., and Kanstad, S. O. (1978). *Opt. Commun.* **24**, 95.

Nordal, P. E., and Kanstad, S. O. (1979). *Phys. Scr.* **20**, 659.

Nordal, P. E., and Kanstad, S. O. (1980). *In* "Scanned Image Microscopy" (E. A. Ash, ed.), p. 331. Academic Press, New York.

Nordal, P. E., and Kanstad, S. O. (1981). *Appl. Phys. Lett.* **38**, 486.

Nordhaus, O., and Pelzl, J. (1981). *Appl. Phys.* **25**, 221.

Nunes, O. A. C., Monteiro, A. M. M., and Neto, K. S. (1979). *Appl. Phys. Lett.* **35**, 656.

Oda, S., Sawada, T., and Kamada, H. (1978). *Anal. Chem.* **50**, 865.

Oda, S., Sawada, T., Moriguchi, T., and Kamada, H. (1980). *Anal. Chem.* **52**, 650.

Onokhov, A. P., Razumova, T. K., and Starobogatov, I. O. (1980). *Sov. J. Opt. Technol.* (*Engl. Transl.*) **47**, 36.

Ort, D. R., and Parson, W. W. (1978). *J. Biol. Chem.* **253**, 6158.

Ort, D. R., and Parson, W. W. (1979). *Biophys. J.* **25**, 355.

Pao, Y.-H., ed. (1977). "Opto-acoustic Spectroscopy and Detection." Academic Press, New York.

Parker, J. G. (1973). *Appl. Opt.* **12**, 2974.

Parker, J. G., and Ritke, D. N. (1973). *J. Chem. Phys.* **59**, 3713.

Parpal, J. L., Monchalin, J. P., Bertrand, L., and Gagne, J. M. (1981). Technical Digest,

Second International Conference on Photoacoustic Spectroscopy, Berkeley (unpublished).

Patel, C. K. N. (1978a). *Science* **202,** 157.

Patel, C. K. N. (1978b). *Phys. Rev. Lett.* **40,** 535.

Patel, C. K. N., and Kerl, R. J. (1977). *Appl. Phys. Lett.* **30,** 578.

Patel, C. K. N., and Tam, A. C. (1979a). *Appl. Phys. Lett.* **34,** 467.

Patel, C. K. N., and Tam, A. C. (1979b). *Chem. Phys. Lett.* **62,** 511.

Patel, C. K. N., and Tam, A. C. (1979c). *Appl. Phys. Lett.* **34,** 760.

Patel, C. K. N., and Tam, A. C. (1979d). *Nature (London)* **280,** 302.

Patel, C. K. N., and Tam, A. C. (1980). *Appl. Phys. Lett.* **36,** 7.

Patel, C. K. N., and Tam, A. C. (1981). *Rev. Mod. Phys.* **53,** 517.

Patel, C. K. N., Kerl, R. J., and Burkhardt, E. G. (1977). *Phys. Rev. Lett.* **38,** 1204.

Patel, C. K. N., Tam, A. C., and Kerl, R. J. (1979). *J. Chem. Phys.* **71,** 1470.

Patel, C. K. N., Nelson, E. T., and Kerl, R. J. (1980). *Nature (London)* **286,** 368.

Perkowitz, S., and Busse, G. (1980). *Opt. Lett.* **5,** 228.

Peterson, R. G., and Powell, R. C. (1978). *Chem. Phys. Lett.* **53,** 366.

Petts, C. R., and Wickramasinghe, H. K. (1980). *Ultrason. Symp. Proc.* p. 636.

Pichon, C., LeLiboux, M., Fournier, D., and Boccara, A. C. (1979). *Appl. Phys. Lett.* **35,** 435.

Pouch, J. J., Thomas, R. L., Wong, Y. H., Schuldies, J., and Srinivasan, J. (1980). *J. Opt. Soc. Am.* **70,** 562.

Poulet, P., Charbron, J., and Unterreiner, R. (1980). *J. Appl. Phys.* **51,** 1738.

Powell, R. C., Neikirk, D. P., Flaherty, J. M., and Gualtien, J. C. (1980a). *J. Phys. Chem. Solids* **41,** 345.

Powell, R. C., Neikirk, D. P., and Sardar, D. (1980b). *J. Opt. Soc. Am.* **70,** 486.

Querry, M. R., Cary, P. A., and Waring, R. C. (1978). *Appl. Opt.* **17,** 3587.

Quimby, R. S., and Yen, W. M. (1978). *Opt. Lett.* **3,** 181.

Quimby, R. S., and Yen, W. M. (1980). *J. Appl. Phys.* **51,** 1252.

Quimby, R. S., Selzer, P. M., and Yen, W. M. (1977). *Appl. Opt.* **16,** 2630.

Raine, B. C., and Brown, S. D. (1981). Technical Digest, Second International Conference on Photoacoustic Spectroscopy, Berkeley (unpublished).

Razumova, T. K., and Starobogatov, I. O. (1977). *Opt. Spectrosc. (Engl. Transl.)* **42,** 274.

Reddy, K. V. (1980). *J. Mol. Spectrosc.* **82,** 127.

Reddy, K. V., and Berry, M. J. (1981). Technical Digest, Second International Conference on Photoacoustic Spectroscopy, Berkeley (unpublished).

Reddy, K. V., Bray, R. G., and Berry, M. J. (1978). *In* "Advances in Laser Chemistry" (A. Zewail, ed.), p. 48. Springer-Verlag, Berlin and New York.

Rentsch, S. (1979). *Exp. Tech. Phys.* **27,** 571.

Rentzepis, P. M., and Pao, Y. H. (1966). *J. Chem. Phys.* **44,** 2931.

Robin, M. B. (1976). *J. Lumin.* **13,** 131.

Robin, M. B., and Kuebler, N. A. (1975). *J. Am. Chem. Soc.* **97,** 4822.

Robin, M. B., and Kuebler, N. A. (1977). *J. Chem. Phys.* **66,** 169.

Robin, M. B., Kuebler, N. A., Kaya, K., and Diebold, G. J. (1980). *Chem. Phys. Lett.* **70,** 93.

Rockley, M. G. (1979). *Chem. Phys. Lett.* **68,** 455.

Rockley, M. G. (1980a). *Appl. Spectrosc.* **34,** 405.

Rockley, M. G. (1980b). *Chem. Phys. Lett.* **75,** 370.

Rockley, M. G., and Devlin, J. P. (1977). *Appl. Phys. Lett.* **31,** 24.

Rockley, M. G., and Devlin, J. P. (1980). *Appl. Spectrosc.* **34,** 407.

Rockley, M. G., and Waugh, K. M. (1978). *Chem. Phys. Lett.* **54,** 597.

Rockley, M. G., Davis, D. M., and Richardson, H. H. (1980). *Science* **210,** 918.

Rockley, M. G., Davis, D. M., and Richardson, H. H. (1981). *Appl. Spectrosc.* **35,** 185.

Roessler, D. M., and Faxvog, F. R. (1980). *Appl. Opt.* **19,** 578.

Rohl, R., and Palmer, R. A. (1981). Technical Digest, Second International Conference on Photoacoustic Spectroscopy, Berkeley (unpublished).

Rohlfing, E. A., Rabitz, H., Gelfand, J., Miles, R. B., and DePristo, A. E. (1980). *Chem. Phys.* **51,** 121.

Rohlfing, E. A., Gelfand, J., Miles, R. B., and Rabitz, H. (1981). *J. Chem. Phys.* **75,** 4893.

Rosencwaig, A. (1973). *Opt. Commun.* **7,** 305.

Rosencwaig, A. (1975). *Anal. Chem.* **47,** 592A.

Rosencwaig, A. (1977). *Rev. Sci. Instrum.* **48,** 1133.

Rosencwaig, A. (1978). *Adv. Electron. Electron Phys.* **46,** 207.

Rosencwaig, A. (1979). *Am. Lab.* **11**(4), 39.

Rosencwaig, A. (1980a). *Chem. Anal.* (*N.Y.*) **57.**

Rosencwaig, A. (1980b). *J. Appl. Phys.* **51,** 2210.

Rosencwaig, A. (1981). *Thin Solid Films* **77,** L43.

Rosencwaig, A., and Busse, G. (1980). *Appl. Phys. Lett.* **36,** 725.

Rosencwaig, A., and Gersho, A. (1976). *J. Appl. Phys.* **47,** 64.

Rosencwaig, A., and Hall, S. S. (1975). *Anal. Chem.* **47,** 548.

Rosencwaig, A., and Hildrum, E. A. (1981). *Phys. Rev. B: Condens. Matter* [3] **23,** 3301.

Rosencwaig, A., and White, R. M. (1981). *Appl. Phys. Lett.* **38,** 165.

Rosengren, L. G. (1975). *Appl. Opt.* **14,** 1960.

Rothenberg, M. S. (1979). *Phys. Today,* February, p. 18.

Royce, B. S. H., Teng, Y. C., and Enns, J. (1980). *Ultrason. Symp. Proc.* p. 652.

Sam, C. L., and Shand, M. L. (1979). *Opt. Commun.* **31,** 174.

Sander, U., Strehblow, H. H., and Dohrmann, J. K. (1981). *J. Phys. Chem.* **85,** 447.

Sawada, T., and Oda, S. (1981). *Anal. Chem.* **53,** 471.

Sawada, T., Oda, S., Shimizu, H., and Kamada, H. (1979). *Anal. Chem.* **51,** 688.

Sawada, T., Shimizu, H., and Oda, S. (1981). *Jpn. J. Appl. Phys.* **20,** L25—

Schneider, S., Möller, U., and Coufal, H. (1982a). *In* "Photoacoustics—Principles and Applications" (H. Coufal, ed.) Vieweg Verlag, Braunschweig.

Schneider, S., Möller, U., and Coufal, H. (1982b). *Appl. Opt.* **21,** 44.

Scruby, C. B., Dewhurst, R. J., Hutchins, D. A., and Palmer, S. B. (1980). *J. Appl. Phys.* **51,** 6210.

Shaw, R. W. (1979). *Appl. Phys. Lett.* **35,** 253.

Shtrickman, S., and Slatkine, M. (1977). *Appl. Phys. Lett.* **31,** 830.

Siebert, D. R., West, G. A., and Barrett, J. J. (1980). *Appl. Opt.* **19,** 53.

Sigrist, M. K., and Kneubuhl, F. K. (1978). *J. Acoust. Soc. Am.* **64,** 1652.

Silberg, P. A. (1965). *Can. J. Phys.* **43,** 2078.

Siqueira, M. A. A., Ghizoni, C. C., Vargas, J. I., Menezes, E. A., Vargas, H., and Miranda, L. C. M. (1980). *J. Appl. Phys.* **51,** 1403.

Sladky, P., Danielius, R., Sirutkaitis, V., and Boudys, M. (1977). *Czeck. J. Phys.* **B27,** 1075.

Smith, W. H., and Gelfand, J. (1980). *J. Quant. Spectrosc. Radiat. Transfer* **24,** 15.

Solimini, D. (1966). *Appl. Opt.* **5,** 1931.

Somoano, R. B. (1978). *Angew. Chem., Int. Ed. Engl.* **17,** 238.

Stadler, B., Billie, J., Blatt, P., Frank, M., and Rensch, C. (1981). Technical Digest, Second International Conference on Photoacoustic Spectroscopy, Berkeley (unpublished).

Starobogatov, I. O. (1977). *Opt. Spectrosc.* (*Engl. Transl.*) **42,** 172.

Starobogativ, I. O. (1979). *Opt. Spectrosc. (Engl. Transl.)* **46**, 445.
Stella, G., Gelfand, J., and Smith, W. H. (1976). *Chem. Phys. Lett.* **39**, 146.
Stone, J. (1978). *Appl. Opt.* **17**, 2876.
Suemune, I., Yamanishi, M., Mikoshiba, N., and Kawano, T. (1980). *Jpn. J. Appl. Phys.* **20**, L9.
Sullivan, S. A. (1963). *J. Opt. Soc. Am.* **53**, 962.
Swofford, R. L., Long, M. E., and Albrecht, A. C. (1976). *J. Chem. Phys.* **65**, 179.
Szkarlat, A. C., and Japar, S. M. (1981). *Appl. Opt.* **20**, 1151.
Taine, J., and Lepoutre, F. (1980). *Chem. Phys. Lett.* **75**, 452.
Tam, A. C. (1980a). *Appl. Phys. Lett.* **37**, 978.
Tam, A. C. (1980b). *J. Appl. Phys.* **51**, 4682.
Tam, A. C. (1980c). *J. Opt. Soc. Am.* **70**, 581.
Tam, A. C. (1982). *IBM Tech. Discl. Bull.* (in press).
Tam, A. C., and Coufal, H. (1983). *Appl. Phys. Lett.*, **42** (in press).
Tam, A. C., and Gill, W. D. (1982). *Appl. Opt.* **21**, 1891.
Tam, A. C., and Happer, W. (1977). *Opt. Commun.* **21**, 403.
Tam, A. C., and Patel, C. K. N. (1979a). *Nature (London)* 280, 304.
Tam, A. C., and Patel, C. K. N. (1979b). *Appl. Opt.* **18**, 3348.
Tam, A. C., and Patel, C. K. N. (1979c). *Appl. Phys. Lett.* **35**, 843.
Tam, A. C., and Patel, C. K. N. (1980). *Opt. Lett.* **5**, 27.
Tam, A. C., and Wong, Y. H. (1980). *Appl. Phys. Lett.* **36**, 471.
Tam, A. C., and Zapka, W. (1982). Technical Digest, CLEO'82 Phoenix, p. 106 (unpublished).
Tam, A. C., Patel, C. K. N., and Kerl, R. J. (1979). *Opt. Lett.* **4**, 81.
Tam, A. C., Zapka, W., Chiang, K., and Imaino, W. (1982). *Appl. Opt.* **21**, 69.
Teng, Y. C., and Royce, B. S. H. (1982). *Appl. Opt.* **21**, 77.
Tennal, K., Salamo, G. J., and Gupta, R. (1981). Technical Digest, Second International Conference on Photoacoustic Spectroscopy, Berkeley (unpublished).
Teslenko, V. S. (1977). *Sov. J. Quantum Electron. (Engl. Transl.)* **7**, 981.
Thielemann, W., and Neumann, H. (1980). *Phys. Status Solidi* A **61**, K123.
Thomas, L. J., Kelly, M. J., and Amer, N. M. (1978). *Appl. Phys. Lett.* **32**, 736.
Thomas, R. L., Pouch, J. J., Wong, Y.H., Favro, L. D., Kuo, P. K., and Rosencwaig, A. (1980). *J. Appl. Phys.* **51**, 1152.
Tokumoto, H., Tokumoto, M., and Ishiguro, T. (1981). *J. Phys. Soc. Jpn.* **50**, 602.
Trautmann, M., Rothe, K. W., Wanner, J., and Walther, H. (1981). *Appl. Phys.* **24**, 49.
Twarowski, A. J., and Kliger, D. S. (1977). *Chem. Phys.* **20**, 264.
Vansteenkiste, T. H., Faxvog, F. R., and Roessler, D. M. (1981). *Appl. Spectrosc.* **35**, 194.
Van Stryland, F. W., and Woodall, M. A. (1980). Laser Damage Conference, Boulder, Colorado (unpublished).
Vejux, A. M., and Bae, P. (1980). *J. Opt. Soc. Am.* **70**, 560.
Vidrine, D. W. (1980). *Appl. Spectrosc.* **34**, 314.
Viertl, J. R. M. (1980). *J. Appl. Phys.* **51**, 805.
Vlases, G. C., and Jones, D. L. (1966). *Phys. Fluids* **9**, 478.
Voigtman, E., and Winefordner, J. (1982). *Anal. Chem.* **54**, 1834.
Voigtman, E., Jurgensen, A., and Winefordner, J. (1981). *Anal. Chem.* **53**, 1442.
von Gutfeld, R. J., and Budd, H. F. (1979). *Appl. Phys. Lett.* **34**, 617.
von Gutfeld, R. J., and Melcher, R. L. (1977). *Appl. Phys. Lett.* **30**, 257.
Wake, D. R., and Amer, N. M. (1979). *Appl. Phys. Lett.* **34**, 379.
Wang, K., Burns, V., Wade, G., and Elliott, S. (1977). *Opt. Eng.* **16**, 432.
Wang, T. T., McDavid, J. M., and Yee, S. S. (1979). *Appl. Opt.* **18**, 2354.

Wasa, K., Tsubouchi, K., and Mikoshiba, N. (1980a). *Jpn. J. Appl. Phys.* **19,** L475.
Wasa, K., Tsubouchi, K., and Mikoshiba, N. (1980b). *Jpn. J. Appl. Phys.* **19,** L653.
Webb, J. D., Swift, K. M., and Berstein, E. R. (1980). *J. Chem. Phys.* **73,** 4891.
West, G. A., and Barrett, J. J. (1979). *Opt. Lett.* **4,** 395.
West, G. A., Siebert, D. R., and Barrett, J. J. (1980). *J. Appl. Phys.* **51,** 2823.
Westervelt, P. J., and Larson, R. S. (1973). *J. Acoust. Soc. Am.* **54,** 121.
Wetsel, G. C., Jr., and McDonald, F. A. (1977). *Appl. Phys. Lett.* **30,** 252.
Weulersse, J. M., and Genier, R. (1981). *Appl. Phys.* **24,** 363.
Whinnery, J. R. (1974). *Acc. Chem. Res.* **7,** 225.
White, R. M. (1963). *J. Appl. Phys.* **34,** 3559.
Wickramasinghe, H. K., Bray, R. C., Jipson, V., Quate, C. E., and Salcedo, J. R. (1978). *Appl. Phys. Lett.* **33,** 923.
Wong, J. S., and Moore, C. B. (1981). Technical Digest, Second International Conference on Photoacoustic Spectroscopy, Berkeley (unpublished).
Wong, Y. H., Thomas, R. L., and Hawkins, G. F. (1978). *Appl. Phys. Lett.* **32,** 538.
Wong, Y. H., Thomas, R. L., and Pouch, J. J. (1979). *Appl. Phys. Lett.* **35,** 368.
Wrobel, J., and Vala, M. (1978). *Chem. Phys.* **33,** 93.
Yamagichi, C., Moritani, A., and Nakai, J. (1980). *Jpn. J. Appl. Phys.* **19,** L711.
Yasa, Z. A., Amer, N. M., Rosen, H., Hansen, A. D. A., and Novakov, T. (1979). *Appl. Opt.* **18,** 2528.
Yasa, Z. A., Jackson, W. B., and Amer, N. M. (1982). *Appl. Opt.* **21,** 21.
Yeack, C. E., Melcher, R. L., and Klauser, H. E. (1982). *Appl. Phys. Lett.* **41,** 1043.
Zapka, W., and Tam, A. C. (1982a). *Opt. Lett.* **7,** 86.
Zapka, W., and Tam, A. C. (1982b). *Appl. Phys. Lett* **40,** 310.
Zapka, W., and Tam, A. C. (1982c). *Appl. Phys. Lett.* **40,** 1015.
Zapka, W., Pokrowsky, P., and Tam, A. C. (1982). *Opt. Lett.* **7,** 477.

2 ONE-PHOTON AND TWO-PHOTON EXCITATION SPECTROSCOPY

Robert R. Birge

Department of Chemistry
University of California
Riverside, California

I. INTRODUCTION

The fundamental goal of molecular electronic spectroscopy is to observe and characterize the excited electronic states of molecules. The dominant technique remains absorption spectroscopy, but there are numerous instances when this technique is incapable of directly observing an electronic state. The most common limitation is one in which the absorptivity of the excited state is too small to permit the use of absorption spectroscopy. Provided the molecule emits light with a reasonable quantum efficiency ($\Phi > 10^{-3}$), one-photon or two-photon excitation spectroscopy can often be used to directly observe these states. The purpose of this chapter is to provide the reader with the necessary experimental and theoretical background to obtain optimized one-photon or two-photon excitation spectra. Our emphasis on experimental technique is due in part to the general goals of this volume and in part to this author's recognition that the theoretical aspects of excitation spectroscopy have already been reviewed in great depth (Murrell, 1963; Herzberg, 1966; Dowley *et al.*, 1967; Peticolas, 1967; Suzuki, 1967; Gold, 1970; McClain, 1971; Yariv, 1975; McClain and Harris, 1977; Harris and Bertolucci, 1978; Swofford and Albrecht, 1978; Mortensen and Svendsen, 1981). Nonetheless, a fair amount of theory is required to successfully apply the excitation technique and selected theoretical concepts will be presented.

One-photon and two-photon excitation spectroscopy share many of the same experimental advantages, limitations, and instrumental requirements. It is therefore useful to discuss both techniques in the same chapter. [The reader should be cautioned, however, that portions of the following discussion will apply to only one excitation process (one-photon or two-photon) and therefore skim reading of certain sections may lead to confusion.] An additional advantage of discussing one-photon and two-photon excitation spectroscopy in the same chapter derives from the complementary nature of these two spectroscopic techniques. The electric dipole approximation provides a useful (though not rigorously accurate) approach to the analysis of one-photon and two-photon selection rules. One-photon transitions are allowed when the transition length has a component which transforms under x, y, or z symmetry operations. In contrast, two-photon transitions are allowed for transition-length products which have components that transform under x^2, y^2, z^2, xy, yz, or xz symmetry operations.

A brief example of the use of one-photon and two-photon spectroscopy to study all-*trans* diphenylbutadiene (C_{2h} symmetry) will help to illustrate the complementary nature of these two techniques. This molecule has two low-lying electronic states which are nearly degenerate, a $^1B_u^*$ $\pi\pi^*$ state with a system origin at 28,030 cm^{-1} and a $^1A_g^*$ $\pi\pi^*$ state with a system origin at 27,900 cm^{-1} (EPA, 77 K) (Bennett and Birge, 1980). One-photon selection rules applied to molecules with inversion symmetry lead to the following well-known parity restrictions: u $\longleftrightarrow$ g, u $\longleftrightarrow\!\!\!\!/\!\!\!\!\longrightarrow$ u, and g $\longleftrightarrow\!\!\!\!/\!\!\!\!\longrightarrow$ g. In contrast, two-photon selection rules yield the following parity restrictions: g $\longleftrightarrow$ g, u $\longleftrightarrow$ u, and g $\longleftrightarrow\!\!\!\!/\!\!\!\!\longrightarrow$ u. The complementary nature of one-photon and two-photon spectroscopy is clearly evident in the above selection rules and, in the case of diphenylbutadiene, indicate that the $^1B_u^*$ state is one-photon allowed but two-photon forbidden and the $^1A_g^*$ state is two-photon allowed but one-photon forbidden (the ground state has ^1A_g symmetry). Accordingly, the above two states can be spec-

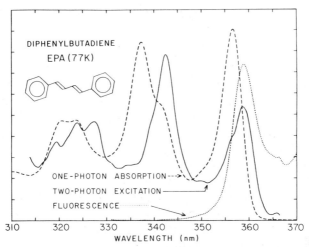

Fig. 1. One-photon absorption, fluorescence, and two-photon excitation spectra of all-*trans* 1,4-diphenyl-1,3-butadiene in EPA at 77 K [from Bennett and Birge (1980)]. The two-photon excitation spectrum is plotted vs $\frac{1}{2}\lambda_{ex}$, where λ_{ex} is the wavelength of the laser excitation.

troscopically observed and characterized despite their near degeneracy by selective use of one- and two-photon spectroscopy (see Fig. 1). The one-photon selection rules for arbitrary point groups have been reviewed by Herzberg (1966) and Harris and Bertolucci (1978). McClain (1971) has reviewed the two-photon selection rules for all 32 crystallographic point groups and the two linear molecule groups.

The allowedness of a one-photon transition is characterized by a dimensionless parameter called the oscillator strength f. This parameter is related to the transition length of the f ← o electronic transition by the expression

$$f_{fo} = (8\pi^2 m_e/3h^2)\Delta\bar{\nu}_{fo}|\langle f|\mathbf{r}|o\rangle|^2 = (1.0847 \times 10^{-5})\Delta\bar{\nu}_{fo}|\langle f|\mathbf{r}|o\rangle|^2 \quad (1)$$

where m_e is the mass of an electron, h is Planck's constant, $\Delta\bar{\nu}_{fo}$ is the transition energy in wavenumbers (cm^{-1}) and $|\langle f|\mathbf{r}|o\rangle|$ is the transition length in angstroms. The oscillator strength is related to the experimentally observed absorption spectrum by the expression

$$f_{fo} = \frac{10^3(\ln 10)m_e c^2}{N_A \pi e^2} \int \varepsilon_{\bar{\nu}}^{fo} \, d\bar{\nu} = (4.3190 \times 10^{-9}) \int \varepsilon_{\bar{\nu}}^{fo} \, d\bar{\nu} \quad (2)$$

where c is the speed of light in a vacuum, N_A is Avogadro's number, e is the charge on an electron, and $\varepsilon_{\bar{\nu}}^{fo}$ is the molar absorptivity at wavenumber

$\tilde{\nu}$ in liters per mole–centimeter. The one-photon cross section $\sigma_{\tilde{\nu}}^{fo}$ is related to the molar absorptivity by the expression

$$\sigma_{\tilde{\nu}}^{fo} = 1000(\ln 10)\,\varepsilon_{\tilde{\nu}}^{fo}/N_A = (3.8235 \times 10^{-21})\,\varepsilon_{\tilde{\nu}}^{fo} \tag{3}$$

where $\sigma_{\tilde{\nu}}^{fo}$ is in units of square centimeters per molecule. Although this parameter is rarely used in the literature, we introduce it here to provide the reader with a better perspective on the experimental parameters used in two-photon spectroscopy (see below).

The one-photon oscillator strength has no direct analogy in two-photon spectroscopy because the allowedness of a two-photon process is dependent upon both the molecular properties and the experimental configuration. This observation does not preclude the definition of an appropriate parameter based on the integrated two-photon cross section for a well-defined experimental condition. No such parameter, however, has gained general acceptance and, consequently, all experimental measurements are reported in terms of the two-photon absorptivity (also called two-photon cross section) $\delta_{\tilde{\nu}}^{fo}$. This parameter is a function both of the properties of the molecule and the laser polarization and energies required to generate a simultaneous two-photon absorption in the molecule (Peticolas, 1967):

$$\delta_{\tilde{\nu}}^{fo} = \frac{(2\pi e)^4}{(ch)^2}\,\tilde{\nu}_\lambda \tilde{\nu}_\mu g(\tilde{\nu}_\lambda + \tilde{\nu}_\mu)|S_{fo}(\lambda, \mu)|^2 \tag{4}$$

where $\tilde{\nu}_\lambda$ and $\tilde{\nu}_\mu$ are the frequencies of the two laser beams, $g(\tilde{\nu}_\lambda + \tilde{\nu}_\mu)$ is the normalized lineshape function (see below) and $S_{fo}(\lambda, \mu)$ is the two-photon tensor:

$$S_{fo}(\lambda, \mu) = \sum_k \left(\frac{(\lambda \cdot \langle k|r|o \rangle)\,(\langle f|r|k \rangle \cdot \mu)}{\tilde{\nu}_k - \tilde{\nu}_\lambda + i\Gamma_k} + \frac{(\mu \cdot \langle k|r|o \rangle)\,(\langle f|r|k \rangle \cdot \lambda)}{\tilde{\nu}_k - \tilde{\nu}_\mu + i\Gamma_k} \right) \tag{5}$$

where λ and μ are the unit vectors defining the polarization of the two photons and $\tilde{\nu}_k$ and Γ_k are the transition frequency and the linewidth of the system origin of state k, respectively. The summation is over all electronic states of the molecule including the ground and final states (Mortensen and Svendsen, 1981).

Each molecule will have a unique lineshape (or bandshape) function which will be very complicated in those instances where vibronic structure can be resolved. It is important to note that this function is solute, solvent, and temperature dependent and therefore a maximum two-photon absorptivity measured under a given set of experimental conditions will not, in general, be meaningful for different solvent or temperature environments. The normalized lineshape function takes on a particularly simple form when the two-photon excitation profile for the final electronic state is Gaussian in shape (Birge and Pierce, 1979):

$$g(\bar{\nu}_\mu + \bar{\nu}_\lambda) = g_{max} \exp\{[-4 \ln 2/(\Delta\bar{\nu})^2](\bar{\nu}_\mu + \bar{\nu}_\lambda - \bar{\nu}_f)^2\}, \tag{6}$$

$$g_{max}(sec) = (3.13363 \times 10^{-11})/\Delta\bar{\nu}, \tag{7}$$

where $\Delta\bar{\nu}$ is the full width at half maximum in wavenumbers. The units of g (sec) derive from the arbitrary convention that this function is normalized in frequency space so that*

$$\int g(\nu)d\nu = 1. \tag{8}$$

The two-photon absorptivity $\delta_{\bar{\nu}}$ has units of (centimeters)^4–second per molecule–photon. This factor has also been called the "two-photon cross section" because of its direct phenomenological relationship to the one-photon cross section $\sigma_{\bar{\nu}}$ [Eq. (3)]. McClain and Harris (1977) have noted that this terminology is dimensionally inappropriate and suggest that the product $\delta_{\bar{\nu}}F$ is a more appropriate "cross section" (F is the photon flux in photons per square centimeter–second).

A rough estimate for a nominal value for the two-photon absorptivity for a two-photon allowed state can be obtained by using a single intermediate-state approximation (Birge and Pierce, 1979):

$$\delta_{max}^{f \leftarrow o} = \frac{8\pi^4 e^4}{15 c^2 h^2} \bar{\nu}_\lambda^2 g_{max} \left(\frac{(a + b) [\langle k|\mathbf{r}|o\rangle \cdot \langle f|\mathbf{r}|k\rangle]^2}{(\bar{\nu}_k - \bar{\nu}_\lambda)^2} + \frac{b|\langle k|\mathbf{r}|o\rangle|^2 |\langle f|\mathbf{r}|k\rangle|^2}{(\bar{\nu}_k - \bar{\nu}_\lambda)^2} \right) \tag{9}$$

where $\bar{\nu}_\lambda$ is the wavenumber of the laser excitation, a and b are photon propagation–polarization variables (Section II.B.8), and k represents a single (one-photon allowed) intermediate state at $\bar{\nu}_k$. The above equation can be simplified further by assuming that the intermediate and final states (k and f) are close in energy [$\bar{\nu}_f \sim \bar{\nu}_k$ and therefore $\bar{\nu}_\lambda = \frac{1}{2}(\bar{\nu}_k)$], the transition lengths $\langle k|\mathbf{r}|o\rangle$ and $\langle f|\mathbf{r}|k\rangle$ are similarly polarized, and the laser excitation is linearly polarized ($a = b = 8$):

$$\delta_{max}^{f \leftarrow o} \cong (1.682 \times 10^{-3})g_{max}|\langle k|\mathbf{r}|o\rangle|^2|\langle f|\mathbf{r}|k\rangle|^2. \tag{10}$$

An allowed one-photon transition will have a transition length on the order of 1 Å. A typical value for g_{max} is 5×10^{-15} sec [$\Delta\bar{\nu} \sim 6000$ cm^{-1}, Eq.

---

* The convention of normalizing the lineshape function by using Eq. (8) is arbitrary and not universal. Some authors choose to normalize $\int g(\omega)d\omega$, which can lead to potential confusion because both integrals yield the same units for g. Since $g(\omega) = 2\pi g(\nu)$, it is important to specify which normalization condition is applied. The constant appearing in brackets on the RHS of Eq. (4) will equal $(2\pi)^5 e^4/(ch)^2$ if $\int g(\omega)d\omega = 1$.

(7)]. Equation (10) yields

$$\delta_{max} = (1.682 \times 10^{-3})(5 \times 10^{-15} \text{ sec})(1 \times 10^{-8} \text{ cm})^4$$

$$= 8 \times 10^{-50} \text{ cm}^4 \text{ sec molecule}^{-1} \text{ photon}^{-1} = 8 \text{ GM.}$$

Accordingly, a broad ($\Delta\bar{\nu} \sim 6000 \text{ cm}^{-1}$) excitation band of a "typical" two-photon allowed state will exhibit a maximum two-photon absorptivity on the order of $10^{-50} \text{ cm}^4 \text{ sec molecule}^{-1} \text{ photon}^{-1}$. This absorptivity value is informally known as 1 Göppert-Mayer (GM) in honor of Maria Göppert-Mayer's pioneering theoretical treatment of the two-photon excitation phenomenon (Göppert-Mayer, 1931).

For the purposes of the present review, we will define excitation spectroscopy as the technique of generating an electronically excited state and observing the light emitted by the solute. The generality of this definition includes the possibility of two substantially different experimental goals: (i) the use of excitation spectroscopy to optimize the experimental observation of an emission spectrum, or (ii) the use of excitation spectroscopy to observe an excitation spectrum. The same apparatus can often be used to accomplish both goals in one-photon spectroscopy. Two-photon excitation techniques are usually used only as a method of observing an excitation spectrum. A requirement inherent in all excitation spectroscopy is that the molecule to be studied either fluoresce or phosphoresce with a reasonable quantum yield (Φ_f or $\Phi_p > 0.001$). Many molecules that do not efficiently emit light in ambient-temperature solution environments do emit in low-temperature solvent environments [liquid nitrogen (77 K) or liquid helium (1.8–4 K)]. The importance of solvent and temperature effects on excitation spectroscopy will therefore be included in the present discussion.

The basic theory and experimental techniques associated with excitation spectroscopy are presented in Section II. Some recent applications of excitation spectroscopy are presented in Section III. Both sections emphasize the use of lasers as excitation sources.

II. EXCITATION SPECTROSCOPY

We begin our discussion of excitation spectroscopy with a brief discussion of typical experimental configurations (Section A). Our primary interest is the use of lasers as excitation sources and Section B reviews the theory and experimental implementation of laser excitation techniques. The goals of optimizing the solvent environment and the excitation process are discussed in Sections C and D. Experimental technique associated with the efficient collection and analysis of the excitation-induced solute emission is described in Section E. A final section reviews basic data reduction techniques which are useful in analyzing two-photon excitation spectra.

A. Experimental Configurations

Various experimental configurations used to generate excitation spectra or laser-selected emission spectra are shown in Figs. 2–5. These experimental arrangements are not presented to provide historical perspective or to represent an overview of all the possibilities. Each apparatus has been selected merely to illustrate general experimental approaches to various types of laser excitation spectroscopy. The apparatus shown in Fig. 2 can be used to generate either one-photon (Hubbard *et al.*, 1981) or two-photon (Birge *et al.*, 1978a) excitation spectra. A unique aspect of this experimental configuration is the use of gated photon counting techniques to monitor the solute emission (Section II.E.4). The apparatus shown in Fig. 3 is designed primarily for two-photon excitation spectroscopy (Birge *et al.*, 1982). This apparatus can be used to generate two-photon excitation spectra of molecules that have relatively low fluorescence quantum yields ($\Phi_f \geq 0.001$) due to the combined use of $f/1$ collection optics and computer-optimized data acquisition (Section II.E.1). The apparatus shown in Fig. 4 is designed for the collection of laser-excited emission spectra of weakly emitting solutes ($\Phi_f \geq 10^{-5}$). The salient feature is the use of an optical multichannel analyzer (OMA) to simultaneously monitor 500 wavelength channels (Section II.E). The apparatus shown in Fig. 5 is designed to generate sharp-line emission spectra of solute molecules using laser selection techniques or Schpol'skii techniques (Section II.C). The key features are the use of narrow-line laser excitation, extremely low sample temperature (liquid helium), and a high-resolution monochromator to monitor the emission spectrum. We will discuss each apparatus in greater detail below.

As previously noted, the above-mentioned configurations are representative of typical experimental approaches to excitation spectroscopy. Numerous other configurations and excitation methods are possible. Christensen and Kohler (1974) describe an excitation spectrometer which uses a standard xenon lamp source and a 1-meter monochromator to scan the excitation spectrum. Their design is optimized for the accurate determination of the wavelength dependence of the fluorescence quantum yield. Drucker and McClain (1974) describe a two-photon excitation spectrometer capable of simultaneously measuring the two-photon excitation

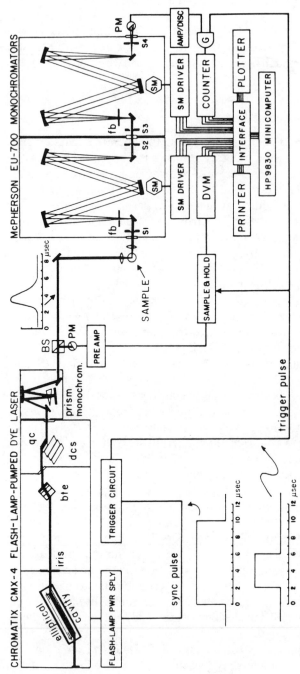

Fig. 2. Flash-lamp-pumped tunable-dye-laser gated photon counting fluorescence excitation spectrometer. Excitation spectra in the wavelength range from 265 to 365 nm are taken using one of the doubling crystals (dcs) and using the prism monochromator to select the second harmonic. Excitation spectra in the wavelength range from 435 to 750 nm are taken by replacing the doubling crystals with a quartz crystal (qc) and using the prism monochromator to select the fundamental wavelength and remove flash-lamp output and trace second-harmonic generation. This spectrometer can be used to generate one-photon or two-photon excitation spectra. The remaining abbreviations denote the following spectrometer components: BS, beam splitter; PM, photomultiplier; SM, stepping motor; fb, field stop baffles (usually removed); G, high-speed TTL gate; bfe, birefringent tuning element (Birge *et al.*, 1978a; Hubbard *et al.*, 1981). Reprinted with permission. Copyright 1981, American Chemical Society.

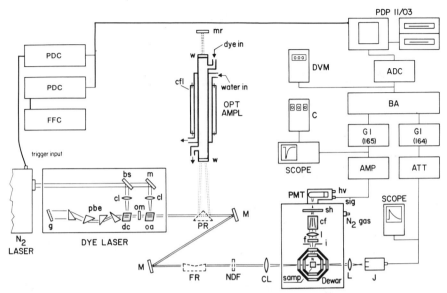

Fig. 3.   Nitrogen-laser-pumped tunable-dye-laser two-photon excitation spectrometer. N_2 LASER, a Molectron UV24 nitrogen laser (firing circuitry provides 200-nsec "sync out" timing pulses for triggering boxcar averager); DYE LASER, a Molectron DL14 UV–visible– near-IR tunable dye laser with optical amplifier (bs, beam splitter; m, mirror; cl, condensing lens; g, diffraction grating; pbe, prism beam expander; dc, dye cuvette; om, output mirror; oa, optical amplifier dye cuvette); OPT AMPL, a double-pass flash-lamp-pumped optical amplifier (modified Phase-R DL 2100C "triax" dye laser; w, antireflection-coated windows; cfl, coaxial flash lamp; mr, broadband maximum reflector); FFC, flash-lamp firing circuit; PR, prism reflector; M, metal-surface mirror; FR, Fresnel rhomb to convert linearly polarized laser light to circular or elliptical polarization; NDF, neutral density filter(s) for attenuation of laser output; CL, achromatic condensing lens; Dewar, optical Dewar containing liquid nitrogen; samp, quartz or Pyrex cell containing sample; i, iris; f, glass filter; l, fluorescence-gathering lens; cf, chemical filter; sh, shutter to protect PMT from ambient light; PMT, RCA 1P28 photomultiplier tube; hv, connection to high-voltage power supply (not shown); sig, output signal from PMT due to sample fluorescence; N_2 gas, connector for dry nitrogen gas passed through Dewar holder to keep Dewar windows from fogging; L, lens to direct laser pulse (slightly defocused) onto pyroelectric detector element of joulemeter; J, joulemeter; AMP, short-pulse amplifier; SCOPE, oscilloscope used to monitor output from amplifier or joulemeter and to adjust gate positions for gated integrators; C, electronic counter to count laser pulses; ATT, variable attenuation for joulemeter output; GI (164), EG&G PAR model 164 gated integrator to obtain average laser peak intensity; GI (165), EG&G PAR model 165 gated integrator, to obtain average sample peak fluorescence intensity; BA, EG&G PAR model 162 boxcar averager main frame (gated integrators are modules which plug into main frame, shown separately for clarity); DVM, digital voltmeter to monitor analog outputs from each channel of boxcar averager; ADC, analog-to-digital converter; PDC, pulse delay circuit (Bennett, 1980; Alameda *et al.*, 1981; Birge *et al.*, 1982). Reprinted with permission. Copyright 1981, American Chemical Society.

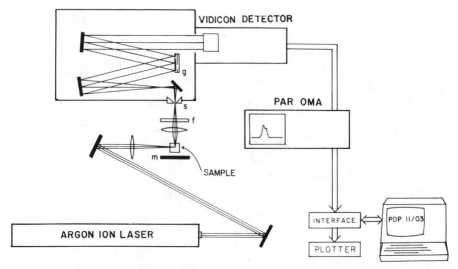

Fig. 4. Optical multichannel analyzer (OMA) emission spectrometer. The UV output from the argon ion laser is focused into the sample cell. The fluorescence (or phosphorescence) is collected using a mirror (m) and lens combination and focused onto the entrance slit (s) of the polychromator. A bandpass filter (f) is used to prevent the scattered laser light from entering the polychromator. The dispersed light falls on a vidicon detector containing a 500-element light-sensitive array. The OMA electronics "read" each of the elements and add the signal for each element in memory. Later, a PDP 11/03 computer scans the OMA memory and plots the spectrum on a digital plotter.

spectrum generated by linearly polarized and circularly polarized laser light. Their apparatus provides for the direct measurement of the wavelength dependence of the two-photon polarization ratio (Section II.B.8). Granville *et al.* (1980) describe a two-photon excitation spectrometer that monitors the laser excitation reference signal and adjusts the nitrogen laser pump energy to maintain constant dye laser power. This approach is recommended for high-resolution vapor-phase studies to minimize power broadening of the vibronic lines. Hermann and Ducuing (1972) describe a two-photon excitation spectrometer designed for the direct measurement of absolute two-photon absorptivities. Their design uses a two-cell arrangement in which the fluorescence intensity produced by the two-photon absorption of a laser beam is compared to that caused by one-photon absorption of the harmonic light generated by the same beam via second-harmonic conversion. The above four examples serve to illustrate the wide range of experimental possibilities. These examples and the more standard configurations depicted in Figs. 2–5 will be discussed in greater detail in subsequent sections.

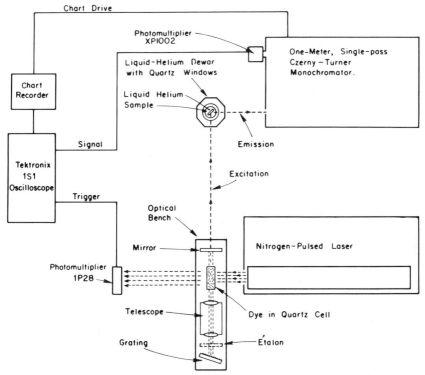

Fig. 5. Nitrogen-laser-pumped tunable-dye-laser fluorescence emission spectrometer. A one-meter Jarrell Ash (model 478-66) with an absolute resolution of 0.4 cm⁻¹ is used to scan the laser-selected emission spectrum. The signal from an Amperex XP1002 photomultiplier is sampled using a Tektronix 1S1 sampling oscilloscope set so that the time window (aperture duration 0.35 nsec) is at the peak of the fluorescence signal. A 1P28 photomultiplier monitors a portion of the nitrogen laser excitation and is driven to saturation to provide reliable triggering of the sampling slope [from Abram *et al.* (1975)].

B. The Laser-Induced Excitation Process

The widespread commercial availability of pulsed and cw tunable dye lasers has encouraged a growing number of researchers to adopt the use of lasers as the excitation source in one-photon spectroscopy. The high peak powers required in two-photon spectroscopy typically mandate the use of pulsed lasers (see, however, Section III). Accordingly, a primary goal of the present review is to discuss the use of lasers as the excitation source in both one- and two-photon excitation spectroscopy. The following sections provide theoretical and experimental background on the interaction of coherent radiation with molecular systems. Although a comprehensive treatment of this subject is beyond the scope of this review, it is our intention to provide the reader with sufficient information and literature sources to gain a reasonable sophistication in the application of laser excitation techniques. Our approach to the presentation of this subject differs from most of the previous literature in that our treatment is primarily numerical rather than analytical. This approach provides generality and notational simplicity. The potential disadvantage of such an approach, however, is that closed-form analytical derivations of many useful relationships are not provided. We partially compensate for this shortcoming by presenting the salient analytical solutions (without derivation), and the literature sources for the analytical derivations.

1. Temporal Behavior of the Laser Intensity

To a first approximation, the temporal behavior of a laser pulse can be represented by a Gaussian profile

$$P(t) = P_0 \exp(-t^2/b^2), \tag{11}$$

where $P(t)$ is the laser power (photons/sec) at time t, P_0 is the laser power at maximum intensity (at time $t = 0$), t is time (sec), and b is related to the pulse width by

$$b = (4 \ln 2)^{-1/2}\Gamma = 0.60056\Gamma \tag{12}$$

where Γ is the full width at half maximum of the laser pulse. Throughout the subsequent discussion, we will refer to Γ as the laser "pulse width."

The intensity of the laser pulse as a function of time is defined in terms of the photon flux,

$$F(t) = P(t)/A \tag{13}$$

where $F(t)$ is the flux (photons cm^{-2} sec^{-1}) and A is the irradiated area (cm^2). The definition of A is somewhat arbitrary and is mode dependent. The most experimentally useful definition for A is $\pi(w_0^2)$, where w_0 is the laser beam waist (see below). Note that the photon flux, represented here by the symbol F, is often represented by the symbol I and called the "intensity." For the purposes of the present discussion these two symbols and their definitions are effectively interchangeable.

The total number of photons in the laser pulse is given by the following expression:

$$N = P_0 \int_{-\infty}^{\infty} \exp\left(\frac{-t^2}{b^2}\right) dt = P_0 b\sqrt{\pi} = 1.06447 P_0\Gamma. \tag{14}$$

The output characteristics of pulsed lasers are usually defined in terms of energy per pulse in millijoules (mJ) or peak power in kilowatts (kW). The following relationships are useful:

$$1 \text{ Watt} = 1 \text{ Joule/sec}, \tag{15a}$$

$$N(\text{photons/pulse}) = (5.03402 \times 10^{12}) \cdot E(\text{mJ/pulse}) \cdot \lambda \text{ (nm)}, \tag{15b}$$

$$P_0(\text{photons/sec}) = (4.72913 \times 10^{21}) \cdot E(\text{mJ/pulse}) \cdot \lambda \text{ (nm)}/\Gamma \text{ (nsec)}, \tag{15c}$$

$$P_0(\text{kW}) = (939.435) \cdot E(\text{mJ/pulse})/\Gamma \text{ (nsec)}, \tag{15d}$$

$$P_{\text{ave}}(\text{kW}) = R(\text{Hz}) \cdot P_0(\text{kW}) \tag{15e}$$

where λ is the laser wavelength (1 nm = 10 Å), E is the pulse energy, P_{ave} is the average laser power, R is the repetition rate, and the remaining symbols were previously defined. The relationships given in Eqs. (15c) and (15d) are f′ r Gaussian pulses only.

The observed temporal behavior of an actual laser pulse usually deviates from a Gaussian profile. In some cases, the deviation is so severe that the assumption of a Gaussian profile will lead to serious error in the analysis of one-photon and two-photon excitation spectra. We will discuss this problem in Section II.B.2.b.

2. Molecular Response to Laser Excitation

The following numerical treatment of the interaction of a laser pulse with an atomic or molecular system serves to illustrate the broader aspects of this phenomenon. Our emphasis is on describing the effect of changes in laser parameters on the efficiency of excitation of the molecular system.

The excitation process is numerically modeled by using small increments in time Δt. This increment must be significantly smaller than the laser pulse width, but from a calculational standpoint, Δt must be large enough to avoid significant roundoff and truncation error. The calculations presented below typically used time increments equal to the laser pulse width or the molecular lifetime τ divided by 2560 (whichever increment is smaller).

The number of laser photons at time t during the time increment Δt is

$$N_{\Delta t}(t) = P(t)\Delta t \tag{16}$$

where $P(t)$ is the laser power at time t [e.g., Eq. (11)]. We assume all $N(t)$ photons are available for molecular excitation and, for the purposes of the immediate discussion, the transverse (spatial) intensity distribution is homogeneous. (The effects of mode structure and "hot spots" on the excitation process will be discussed in Section II.B.4).

The number of photons absorbed at time t during the time increments Δt via one-photon excitation is given by

$$^1N^{abs}_{\Delta t}(t) = N_{\Delta t}(t)A_1, \tag{17}$$

where

$$A_1 = 1 - \exp(-\sigma_\lambda CL), \tag{18}$$

and where A_1 is the wavelength-dependent one-photon absorption parameter, σ_λ is the one-photon cross section (cm^2 molecule^{-1}) [Eq. (3)], C is the concentration (molecule cm^{-3}), and L is the path length (cm).

The number of photons absorbed at time t during the time increment Δt via two-photon excitation is given by

$$^2N^{abs}_{\Delta t}(t) = N_{\Delta t}(t)A_2, \tag{19}$$

where

$$A_2 = 1 - \exp[-F(t)S\delta_\lambda CL] \tag{20a}$$

$$\cong F(t)S\delta_\lambda CL, \tag{20b}$$

and where A_2 is the wavelength-dependent two-photon absorption parameter, $F(t)$ is the photon flux at time t, S is the photon correlation parameter (see below), δ_λ is the two-photon absorptivity (cm^4 sec molecule^{-1} photon^{-1}) [Eq. (4)], and C and L are the concentration and pathlength, respectively. The photon correlation parameter is a dimensionless term equal to 1 for coherent (laser) light and equal to 2 for incoherent (thermal) light (see Section II.B.5). The linearization of Eq. (20a) to produce Eq. (20b) is justified for the vast majority of two-photon excitation processes because the factor $F(t)S\delta_\lambda CL$ is typically orders of magnitude smaller than 1. For example, an approximate upper limit for this term can be calculated by assuming $F(t) \sim 10^{26}$ photons cm^{-2} sec^{-1}, $S \sim 1$, $\delta_\lambda = 10^{-48}$ cm^4 sec molecule^{-1} photon^{-1}, $C \sim 10^{19}$ molecule cm^{-3}, and $L \sim 10$ cm to give $F(t)S\delta_\lambda CL \cong 10^{-2}$. Equations (20a) and (20b) yield $A_2 = 0.00995$ and $A_2 = 0.01$, respectively, which means linearization of A_2 will lead to a nominal worst-case error of only 0.5%.

The total number of molecules in the excited state at time t is given by

$$N^*(t) = N^*(t - \Delta t) + (\xi^{-1})N^{abs}_{\Delta t}(t) - N^{\downarrow}_{\Delta t}(t) \tag{21}$$

where $N^*(t - \Delta t)$ is the total number of molecules in the excited state from the previous evaluation at time $t - \Delta t$, $\xi = 1$ (one-photon) or 2 (two-photon), $N^{abs}_{\Delta t}(t)$ is calculated using Eq. (17) (one-photon) or Eq. (19)

(two-photon), and $N_{\Delta t}^{\downarrow}(t)$ is the number of excited states decaying at time t during the time increment Δt;

$$N_{\Delta t}^{\downarrow}(t) = (\Delta t/\tau)N^*(t) \tag{22}$$

where τ is the excited-state lifetime ($\tau = \Phi\tau_0$ where Φ is the quantum yield of emission and τ_0 is the intrinsic lifetime). Note that $N^*(t)$ appears as a term in Eq. (22), which requires a self-consistent reconciliation of Eqs. (21) and (22) during each time increment cycle of the numerical calculation. The rate of excited-state decay is given by

$$P^{\downarrow}(t) = (1/\tau)N^*(t) = N_{\Delta t}^{\downarrow}(t)/\Delta t \tag{23}$$

where $P^{\downarrow}(t)$ is in units of reciprocal seconds. The number of photons emitted at time t during the time increment Δt is

$$N_{\Delta t}^{\text{em}}(t) = \Phi N_{\Delta t}^{\downarrow}(t), \tag{24}$$

and the rate of photons emitted is given by

$$P^{\text{em}}(t) = \Phi P^{\downarrow}(t). \tag{25}$$

Note that Eqs. (21) and (22) predict an exponential decay of excited-state species after the laser pulse has completely passed through the sample $[N_{\Delta t}^{\text{abs}}(t) = 0$ for all subsequent $t]$:

$$N^*(t) = C_L N^*(0)\exp(-t/\tau) \tag{26}$$

where C_L is a dimensionless constant close to 1 which corrects $N^*(0)$ for asymmetry in the temporal behavior of the laser pulse. Equation (26) provides a convenient check on numerical accuracy provided the lifetime τ of the molecule is larger than the laser pulse width Γ. The program evaluates the lifetime using the following regression equation:

$$\tau_{\text{calc}} = \frac{(\Sigma t_i)^2/n \; \Sigma t_i^2}{\Sigma t_i \ln N^*(t_i) - \{(\Sigma t_i)[\Sigma \ln N^*(t_i)]/n\}} \tag{27}$$

where t_i is the time at iteration i (after the laser pulse has passed), $N^*(t_i)$ is the total number of molecules in the excited state at time t_i [Eq. (21)], and n is the number of samples taken. A second method of checking numerical accuracy is to sum the number of laser photons in the pulse $[\Sigma_i N_{\Delta t}(t_i)]$ and verify equivalence with the analytical number {which for a single Gaussian pulse is $P_0 b\sqrt{\pi}$ [Eq. (14)]}.

We now present the results of various numerical calculations based upon the above procedures to evaluate the effect that changes in laser pulse width and temporal substructure have on the molecular response function of the solute.

a. Laser Pulse-Width Effects

The pulse width of a dye laser is generally not a stable parameter but displays both random fluctuations at a fixed wavelength as well as systematic changes as the dye laser is scanned across the gain curve of the dye [see Fig. 6 and Sorokin *et al.* (1968)]. The effect of changes in the temporal characteristics of a laser pulse on the molecular response function has been treated analytically by a number of investigators (Brody, 1957; Hundley *et al.*, 1967; Carusotto *et al.*, 1970; Swofford and McClain,

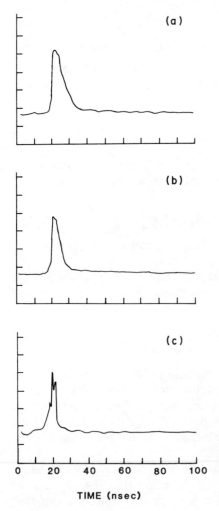

Fig. 6. Single-sweep traces of laser intensity as a function of time for laser pulses generated using a nitrogen-pumped tunable dye laser and rhodamine 6G at 580 nm (a), IR 125 at 905 nm (b), and IR 125 at 885 nm (c). The bottom trace displays "spiking," which is frequently observed in weak dyes near the edges of the gain curve (see text) [from Pierce and Birge (1982)].

1975). The treatment presented by Carusotto *et al.* (1970) for two-photon spectroscopy is highly recommended.

In this section we will examine the effects of laser pulse-width changes on the emission response function of the solute molecule undergoing both one-photon and two-photon excitation. The temporal shape of the laser pulse is assumed to be Gaussian although this restriction will be removed in Section II.B.2.b. Selected examples of laser pulse-width changes on the relative emission response curves for a hypothetical molecule with a lifetime τ of 8 nsec are presented in Fig. 7. The pulse width of the laser is varied from 2 to 24 nsec. Note that the peak intensity of the laser and the one- and two-photon-induced solute emission curves are scaled in each curve to arbitrary values to facilitate comparison. The first observation we can make by reference to Fig. 7 is that the emission maximum invariably occurs after the laser pulse maximum. The delay time for the two-photon-induced emission maximum (t_2 in Fig. 7) is always shorter than the corresponding value for the one-photon-induced emission maximum (t_1 in Fig. 7) for the same molecule and a fixed Γ. These delay times are of considerable experimental utility in that they can be directly related to the emission lifetime and other useful correction factors. A general expression relating t_1 and t_2 to the laser pulse width can be obtained by dividing the observables (t_1, t_2, and Γ) by the molecular lifetime τ. The following polynomial expansions based on 120 numerical evaluations of t_1 and t_2 as a function of Γ and τ are accurate to roughly three significant digits in the range Γ/τ from 0.1 to 10:

$$t_1/\tau = 4.5215X - 18.102X^2 + 56.897X^3 - 111.47X^4 + 125.21X^5$$
$$- 73.797X^6 + 17.665X^7, \tag{28}$$

$$t_2/\tau = 3.4314X - 12.488X^2 + 40.010X^3 - 80.524X^4 + 92.602X^5$$
$$- 55.769X^6 + 13.628X^7 \tag{29}$$

where

$$X = \ln[1 + 0.2(\Gamma/\tau)], \quad 0.1 \leqslant \Gamma/\tau \leqslant 10. \tag{30}$$

The relationship between t_n/τ and Γ/τ is graphed in Fig. 8. In the event that τ is unknown, a measurement of t_1 or t_2 can be used to calculate τ approximately, provided the laser pulse has a Gaussian or nearly Gaussian profile. The following polynomial expansions calculate τ/Γ as a function of t_n/Γ:

$$\tau/\Gamma = 1.0354R - 0.47614R^2 + 9.2208R^3 - 20.200R^4 + 45.116R^5$$
$$- 46.888R^6 + 27.389R^7 \tag{31a}$$

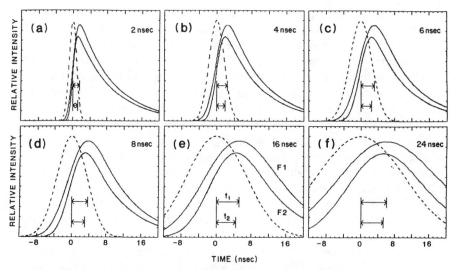

Fig. 7. Effect of changes in laser pulse width on the one-photon- and two-photon-induced fluorescence time profiles for a molecule with an observed fluorescence lifetime τ of 8 nsec. The laser pulse is Gaussian [Eq. (11)] and is represented by a dashed line. The pulse width Γ is indicated in the upper right-hand corner of each graph. The fluorescence time profiles are arbitrarily scaled so that the peak of the one-photon-induced emission curve (F1) is above the two-photon-induced emission curve (F2). The intensity maxima of the fluorescence curves are calculated to be delayed from the peak of the laser pulse by the times t_1 (one-photon) and t_2 (two-photon), which are indicated by horizontal arrows (t_1 always above t_2). The following results are obtained from the numerical simulation of the one-photon excitation process assuming $\sigma CL = 8.4 \times 10^{-3}$ [Eq. (18)] and $P_0 = 1 \times 10^{23}$ photons sec^{-1} (constant): (a) $\Gamma = 2$ nsec, $^1N_{abs} = 1.7808 \times 10^{12}$, $^1I_{em}^{rel} = 1.7670$, $t_1 = 1.49$ nsec; (b) $\Gamma = 4$ nsec, $^1N_{abs} = 3.5616 \times 10^{12}$, $^1I_{em}^{rel} = 2.9874$, $t_1 = 2.44$ nsec; (c) $\Gamma = 6$ nsec, $^1N_{abs} = 5.3424 \times 10^{12}$, $^1I_{em}^{rel} = 3.8907$, $t_1 = 3.15$ nsec; (d) $\Gamma = 8$ nsec, $^1N_{abs} = 7.1218 \times 10^{12}$, $^1I_{em}^{rel} = 4.5841$, $t_1 = 3.73$ nsec; (e) $\Gamma = 16$ nsec, $^1N_{abs} = 1.4223 \times 10^{13}$, $^1I_{em}^{rel} = 6.2235$, $t_1 = 5.23$ nsec; (f) $\Gamma = 24$ nsec, $^1N_{abs} = 2.0840 \times 10^{13}$, $^1I_{em}^{rel} = 7.0077$, $t_1 = 6.05$ nsec, where $^1N_{abs}$ is the number of photons absorbed and $^1I_{em}^{rel}$ is the relative peak fluorescence intensity. The following results are obtained from the numerical simulation of the two-photon excitation process assuming $S\delta CL = 6 \times 10^{-32}$ cm^2 sec photon^{-1} [Eq. (20b)], $A = 0.001$ cm^2 [Eq. (13)], and $P_0 = 1 \times 10^{23}$ photons sec^{-1} (constant): (a) $^2N_{abs} = 9.0323 \times 10^8$, $^2I_{em}^{rel} = 4.745$, $t_2 = 1.15$ nsec; (b) $^2N_{abs} = 1.8065 \times 10^9$, $^2I_{em}^{rel} = 8.324$, $t_2 = 1.93$ nsec; (c) $^2N_{abs} = 2.7097 \times 10^9$, $^2I_{em}^{rel} = 11.158$, $t_2 = 2.54$ nsec; (d) $^2N_{abs} = 3.6129 \times 10^9$, $^2I_{em}^{rel} = 13.459$, $t_2 = 3.04$ nsec; (e) $^2N_{abs} = 7.2257 \times 10^9$, $^2I_{em}^{rel} = 19.475$, $t_2 = 4.46$ nsec; (f) $^2N_{abs} = 1.0809 \times 10^{10}$, $^2I_{em}^{rel} = 22.772$, $t_2 = 5.36$ nsec, where the symbols have definitions analogous to those used in the one-photon simulation. Note that the number of photons per pulse varies as a function of Γ: $N(\Gamma) = 2.1289 \times 10^{14}$ (2 nsec), 4.2579×10^{14} (4 nsec), 6.3868×10^{14} (6 nsec), 8.5140×10^{14} (8 nsec), 1.7004×10^{15} (16 nsec), 2.49×10^{15} (24 nsec).

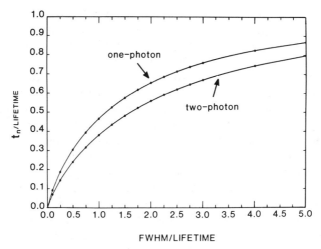

Fig. 8.   Effect of changes in laser pulse width (Γ = FWHM) on the delay times t_1 and t_2 (see Fig. 7), expressed in terms of the molecular lifetime τ (see text) [from Pierce and Birge (1983)].

where

$$R = t_1/\Gamma, \qquad 0.1 \leqslant R \leqslant 0.8, \tag{31b}$$

and

$$\tau/\Gamma = 0.94274S + 1.2416S^2 + 4.2004S^3 - 8.0815S^4 + 101.91S^5$$
$$- 220.26S^6 + 223.23S^7 \tag{32a}$$

where

$$S = t_2/\Gamma, \qquad 0.1 \leqslant S \leqslant 0.6. \tag{32b}$$

The limits on R and S are specified both in terms of the regression accuracy and the experimental constraints inherent in the application of this method. Although the use of the above relationships to calculate the emission lifetime is not recommended if high accuracy is required, the measurement of t_n/Γ is a simple and rapid technique for estimating τ. The following examples illustrate the method. The one-photon-induced fluorescence maximum of dimethyl POPOP is separated from the nearly Gaussian laser pulse ($\Gamma \sim 2.8$ nsec) by ~0.96 nsec [Fig. 7a from Hundley et al. (1967)]. Therefore, $R = 0.34$ and Eq. (31) predicts $\tau/\Gamma = 0.54$ or $\tau = 1.5$ nsec. This value is identical to the value obtained using iterative deconvolution techniques (Hundley et al., 1967). An example for two-photon excitation can be presented using the numerical data from Fig. 7. Here the laser pulse is (by definition) purely Gaussian. Using the data of Fig. 7b, we find that $\Gamma = 4$ nsec and $t^2 = 1.925$ nsec. Therefore, $S = t_2/\Gamma$

= 0.481 and Eq. (32) predicts $\tau/\Gamma = 2.001$ or $\tau = 8.004$ nsec. The correct value for τ is 8 nsec. The accuracy of this method is therefore determined by the accuracy in the measurement of t_n and Γ and the extent to which the laser pulse conforms to a Gaussian temporal profile. We will demonstrate below that the latter constraint is far less critical than one might expect in that lifetime measurements made using the above procedure and laser pulses that deviate significantly from Gaussian are still fairly accurate (about one and a half significant digits). Nonetheless, the above procedures are not intended to replace deconvolution techniques (Hundley et al., 1967), which should be employed where maximum accuracy is desired. Furthermore, if the molecular lifetime is significantly longer than the laser lifetime ($\tau/\Gamma > 5$), one is better off fitting the trailing edge of the emission curve to an exponential to determine the lifetime [replace $N^*(t_1)$ with $I_{em}^{rel}(t_1)$ in Eq. (27)]. We demonstrate below that knowledge of the approximate molecular lifetime is important if one is to obtain accurate one-photon or two-photon excitation spectra using a pulsed laser system.

The excitation spectrometer shown in Fig. 3 measures the excitation spectrum as a function of I_F/N_L^n where I_F is the peak fluorescence intensity, N_L is the pulse energy of the laser (photons/pulse), and $n = 1$ (one-photon) or 2 (two-photon). Many investigators, however, use a photomultiplier or fast photodiode to monitor the laser energy. The excitation spectrum is then obtained as a function of I_F/I_L^n where I_L is the peak laser pulse intensity. Although the latter method requires that one correct I_L for the wavelength dependence of the photomultiplier or photodiode response, it has the advantage that one can directly measure the laser pulse width during the experiment. Whichever method is employed to measure the reference signal, one must correct the excitation spectrum for variations in the laser pulse width whenever the molecular lifetime τ is on the order of, or greater than, the laser pulse width Γ. This important factor has been discussed by a number of investigators (see, for example, Carusotto et al., 1970; Swofford and McClain, 1975). Nonetheless, many investigators appear to ignore this correction factor in the analysis of their excitation spectra.

The following equation describes the relationship between the total number N of photons in the laser pulse and the total number N_{em} of photons emitted by the solute for a one-photon excitation process:

$$^1N_{em} = \Phi N[1 - \exp(-\sigma_\lambda CL)] \tag{33}$$

where the remaining symbols are defined following Eq. (18). Accordingly, $^1N_{em}/N$ is the one-photon experimental invariant in that a direct measurement of this ratio is invariant to changes in laser temporal or spatial properties. No comparable invariant exists in two-photon spectroscopy. If we assume a spatially homogeneous or TEM_{00} laser pulse with a temporal profile that is Gaussian, the following relationship holds for two-photon excitation:

$$^2N_{em} = [1/(2\sqrt{2})]\Phi NF_0\delta_\lambda CL \qquad (34)$$

where the factor of 2 in the denominator corrects for the fact that two photons are absorbed for each photon emitted and the factor of $\sqrt{2}$ corrects F_0, the maximum laser flux (photons cm^{-2} sec^{-1}) to the root-mean-square value:

$$\int_{-\infty}^{\infty} \left[P_0 \exp\left(\frac{-t^2}{b^2}\right)\right]^2 dt = \frac{NP_0}{b\sqrt{\pi}} \, b\sqrt{\frac{\pi}{2}} = \frac{NP_0}{\sqrt{2}}.$$

Accordingly, the ratio N_{em}/NF_0 is invariant to changes in pulse width but not to changes in pulse shape. Unfortunately, the direct experimental measurement of N_{em}/N or N_{em}/NF_0 (or values proportional to these ratios) is often impractical. All of the other experimental ratios are variant to changes in pulse width; for example,

$$^1I_{em} = \eta\Phi I_0[1 - \exp(-\sigma_\lambda CL)]f_1(\Gamma/\tau)_I, \qquad (35)$$

$$^2I_{em} = \eta\Phi I_0^2\delta_\lambda CLf_2(\Gamma/\tau)_I, \qquad (36)$$

$$^1I_{em} = \eta\Phi N[1 - \exp(-\sigma_\lambda CL)]g_1(\Gamma/\tau)_N, \qquad (37)$$

$$^2I_{em} = \eta\Phi N^2\delta_\lambda CLg_2(\Gamma/\tau)_N \qquad (38)$$

where $^nI_{em}$ is the peak intensity of the emitted light, η is a combined instrumentation factor and unit conversion factor assumed to be constant for the entire experiment, and f_n and g_n are correction factors that "normalize" the peak emission intensity values so that they are independent of changes in the laser pulse width. The f_n function corrects one-photon ($n = 1$) or two-photon ($n = 2$) excitation data for experiments which monitor a peak laser intensity (I_0) as a reference signal; the g_n function performs the analogous correction for laser-energy (N) reference signals. Accordingly, these functions display the following limiting behavior:

$$f_n(\Gamma/\tau)_I = 1 \qquad (\Gamma/\tau \to \infty), \qquad (39a)$$

$$g_n(\Gamma/\tau)_N = 0 \qquad (\Gamma/\tau \to \infty) \qquad (39b)$$

where the subscripts I and N are used to indicate constant I_0 and constant N, respectively. Graphs of these correction factors evaluated for Gaussian laser pulses (Pierce and Birge, 1983) are shown in Fig. 9, and their derivatives with respect to the ratio Γ/τ are graphed in Fig. 10. The use of these functions to correct excitation data for changes in laser pulse width Γ is described below. First, however, we note that the derivative curves shown in Fig. 10 can be used to determine when it is necessary to evaluate these correction factors. The magnitude of the derivative of the correction

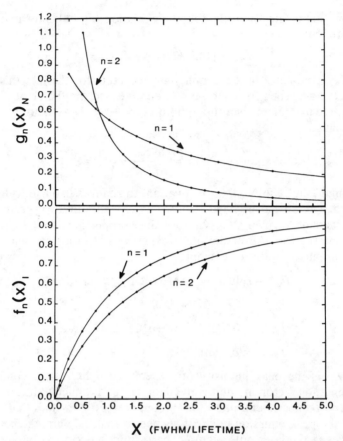

Fig. 9. One-photon ($n = 1$) and two-photon ($n = 2$) pulse-width correction factors [Eqs. (40)–(43)] plotted as a function of Γ/τ (see text) [from Pierce and Birge (1983)].

factor for a given nominal ratio of Γ to τ determines the magnitude of the expected error. If the ratio Γ/τ is above 5, the data of Fig. 10 indicate that neglect of pulse-width changes ($\pm 20\%$) will introduce minor ($<5\%$) error. However, if the ratio Γ/τ is less than 2, the error can become very significant.

The following polynominal expansions can be used to calculate the correction factors (for Gaussian pulses) within the specified limits:

$$f_1\left(\frac{\Gamma}{\tau}\right)_I = 0.9\,\frac{t_1}{\tau} + 1.9112\left(\frac{t_1}{\tau}\right)^2 - 4.5201\left(\frac{t_1}{\tau}\right)^3 + 4.4481\left(\frac{t_1}{\tau}\right)^4$$

$$- 1.7689\left(\frac{t_1}{\tau}\right)^5 \qquad \left(0 \leq \frac{t_1}{\tau} \leq 0.95\right), \tag{40}$$

$$f_2\left(\frac{\Gamma}{\tau}\right)_I = 0.9176\,\frac{t_2}{\tau} + 1.6754\left(\frac{t_2}{\tau}\right)^2 - 3.6007\left(\frac{t_2}{\tau}\right)^3 + 3.0705\left(\frac{t_2}{\tau}\right)^4$$

$$- 1.0667\left(\frac{t_2}{\tau}\right)^5 \qquad \left(0 \leqslant \frac{t_2}{\tau} \leqslant 0.92\right), \tag{41}$$

$$g_1\left(\frac{\Gamma}{\tau}\right)_N = 10.346R - 71.379R^2 + 294.01R^3 - 699.28R^4 + 936.19R^5$$

$$- 652.01R^6 + 183.08R^7, \tag{42a}$$

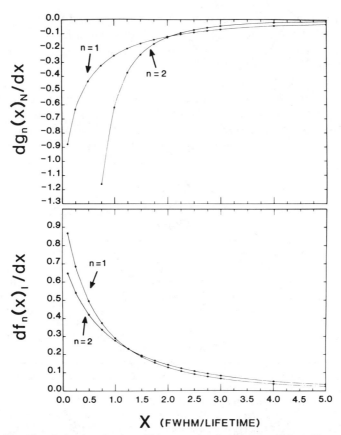

X (FWHM/LIFETIME)

Fig. 10. First derivative of pulse-width correction factors [Eqs. (40)–(43)] as a function of Γ/τ plotted as a function of Γ/τ [from Pierce and Birge (1983)]. If the value of τ and the expected range in Γ are known, these plots can be used to determine the magnitude of the error to be expected if pulse-width effects are ignored (i.e., is the error sufficiently large to warrant the use of pulse-width correction factors?).

where

$$R = 0.1(\tau/\Gamma) \qquad (0.1 \leqslant \tau/\Gamma \leqslant 10), \tag{42b}$$

$$g_2\left(\frac{\Gamma}{\tau}\right)_N = 0.17265S + 17.231S^2 - 41.498S^3 + 47.928S^4 - 24.791S^5$$
$$+ 4.6267S^6 \tag{43a}$$

where

$$S = 0.2(\tau/\Gamma) \qquad (0.1 \leqslant \tau/\Gamma \leqslant 10). \tag{43b}$$

The above four expansions are accurate to two-and-a-half significant digits within the specified ranges. The f_n correction functions have been expressed in terms of t_n/τ because t_n is a more easily measured parameter than Γ during an experiment in which the laser intensity is monitored by a fast photomultiplier or photodiode.

The following example illustrates the use of the g_2 correction factor to correct two-photon excitation data for changes in Γ that occur as one tunes across the gain curve of a dye. We assume the solute has a lifetime of 4 nsec and we are using the dye IR125 (data from Pierce and Birge,

TABLE I

Analysis of Two-Photon Excitation Data Using Pulse-Width Correction Factors[a]

λ (nm)	Γ (nsec)[b]	$^2I_{em}/N^2$[c]	$g_2(\Gamma/\tau)_N$[d]	$^2I_{em}/[N^2g_2(\Gamma/\tau)_N]$[e]
885	3.4	9.57	0.584	16.4
895	5.7	4.49	0.266	16.9
910	5.9	4.26	0.252	16.9
925	5.6	4.60	0.274	16.8
935	5.3	4.99	0.299	16.7
945	4.0	7.65	0.461	16.6

[a] The molecular lifetime τ is equal to 4 nsec and the two-photon absorptivity is defined to be constant over the wavelength range 885–945 nm.

[b] The full width at half maximum pulse widths Γ were measured for the laser dye IR125 ($2.5 \times 10^{-3}M$ in DMSO) using a Molectron UV24 900-kW nitrogen laser ($\Gamma = 10$ nsec) to pump a Molectron DL14 tunable dye laser with optical amplifier (Pierce and Birge, 1982).

[c] Numerical simulation using the procedures outlined in Section II.B.2 assuming a constant δ (885–945 nm). The values are relative.

[d] Two-photon correction factor calculated using Eq. (43).

[e] Corrected two-photon excitation data using Eq. (44a). Note that the data are only accurate to two and a half significant digits owing to error inherent in Eq. (43).

1981). We assume that the solute molecule has a constant two-photon absorptivity from 885 to 945 nm. In other words, the corrected measurement (I_f/N^2) should be invariant to pulse width within this wavelength range.

We want to measure a parameter which is proportional to the two-photon absorptivity. Equation (38) is rearranged to give

$$\delta_\lambda \propto \frac{^2I_{em}}{N^2} \frac{1}{g_2(\Gamma/\tau)_N}. \tag{44a}$$

The experimental procedure therefore requires a measurement of both $^2I_{em}/N^2$ and Γ as a function of wavelength. The results are given in Table I. Note that without any correction, the raw data display the characteristic "U" shape symptomatic of pulse-width changes near the edge of the dye gain curve.

Pulse-width corrections for the other experimental ratios are carried out in an analogous fashion using the relationships derived from Eqs. (35)–(37):

$$\sigma_\lambda \propto -\ln\left(1 - \frac{^1I_{em}}{I_0} \frac{1}{f_1(\Gamma/\tau)_I}\right), \tag{44b}$$

$$\sigma_\lambda \propto -\ln\left(1 - \frac{^1I_{em}}{N} \frac{1}{g_1(\Gamma/\tau)_N}\right), \tag{44c}$$

$$\delta_\lambda \propto \frac{^2I_{em}}{I_0^2} \frac{1}{f_2(\Gamma/\tau)_I}. \tag{44d}$$

Although the correction factors used in Eqs. (44) were derived for laser pulses with Gaussian temporal profiles, they are still useful for correcting data obtained using pulsed lasers that do not exhibit pure Gaussian pulses provided the pulse shape does not vary significantly across the gain curve (see next section).

b. Laser Pulse-Shape Effects

The temporal profile of an actual laser pulse is rarely Gaussian (Fig. 6). The most serious deviations from Gaussian often occur at the edges of the dye gain curve and result in "spiking" (Fig. 6c) (Siegman, 1971). The effect of severe "spiking" on the one-photon and two-photon excitation process is numerically simulated in Fig. 11. The laser pulse ($\Gamma \sim 8$ nsec) is shown in Fig. 11a and its peak intensity and shape are held constant. Figures 11b–11f show the fluorescence response functions for one-photon (top curves) and two-photon (bottom curves) excitation as a function of

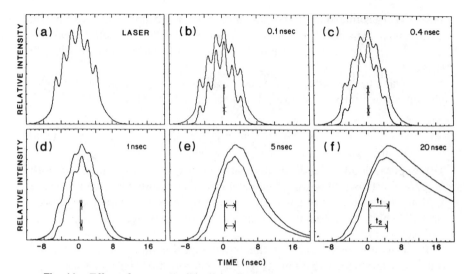

Fig. 11. Effect of severe "spiking" in the laser pulse on the one-photon- and two-photon-induced fluorescence time profiles as a function of molecular lifetime τ. The laser pulse is shown in graph (a) and has a nominal pulse width of ~8 nsec. The fluorescence profiles are shown in graphs (b)–(f) for molecular lifetimes of $\tau = 0.1$ (b), 0.4 (c), 1 (d), 5 (e), and 20 nsec (f) using the same conventions as described in the caption to Fig. 7.

lifetime ($\tau = 0.1, 0.4, 1, 5,$ or 20 nsec). The following observations can be made by reference to Fig. 11 and the associated numerical data:

(i)   The two-photon-induced fluorescence profile is much more sensitive to temporal structure in the laser pulse than the corresponding one-photon-induced fluorescence profile for large ratios of Γ/τ (e.g., Figs. 11b and 11c). This observation follows from the square dependence of the emission profile on the laser intensity.

(ii)   If the ratio Γ/τ is greater than 5, one should use $^1I_{em}/I_0$ (or $^2I_{em}/I_0^2$) to experimentally measure the one-photon (or two-photon) excitation spectrum. If the ratio Γ/τ is less than 5, one should use $^1I_{em}/N$ or $^2I_{em}/N^2$ to measure the corresponding excitation spectra. The use of Gaussian pulse-width correction factors [Eqs. (40)–(44)] will significantly improve the accuracy of the resulting excitation spectrum even though these correction factors do not rigorously apply. However, if the pulse shape changes significantly during the experiment, there are no simple corrections that can be made. The experimentalist is then forced to directly measure N_{em}/N [Eq. (33)] to obtain the one-photon excitation spectrum, or measure $N_{em}/(NF_0)$ [Eq. (34)] or use a two-photon excitation standard (Swofford and McClain, 1975) to obtain the two-photon excitation spectrum.

3. Gaussian Optics and the Propagation of Laser Beams

The number of photons absorbed by a solute molecule undergoing two-photon excitation is proportional to the product of the number N of photons in the laser pulse and the intensity (or flux, F_0) of the laser beam in the sample cell [Eq. (34)]. One might naïvely predict that two-photon absorption can be induced for almost any molecule by focusing the laser beam to a sufficiently small diameter. The problem with this assumption is that lasers operating in a fundamental mode are subject to quantum optical constraints that limit the extent to which the laser beam can be focused (Yariv, 1975). The purpose of this section is to briefly review the theory of Gaussian optics with the goal of accurately predicting the focusing properties of a laser beam in the TEM_{00} mode. The subsequent section discusses the effects of mode structure on the excitation process.

A comprehensive review of the theory of the propagation of laser beams through homogeneous and lenslike media is beyond the scope of this chapter. The interested reader may consult Yariv (1975) for a detailed treatment of this subject. Our presentation is limited to the use of a single thin lens to focus a TEM_{00} laser beam into the sample. The situation is depicted in Fig. 12. The light beam exits the laser with beam radius w_{01} and an angular beam spread θ_1. We assume that the beam divergence is a manifestation of wave diffraction:

$$\theta_1 = \tan^{-1}(\lambda/\pi w_{01}n) \simeq \lambda/\pi w_{01}n \tag{45}$$

where λ is the wavelength, w_{01} is the beam radius (Fig. 12), and n is the refractive index of the medium ($n_{air} = 1.0003$). The angular beam spread θ_1 is defined as the far-field beam divergence half-angle in radians and is equal to $\lambda/\pi w_{01}n$ only for small angles [e.g., $\tan^{-1}(0.1) = 0.0997$ rad]. For example, a 600-nm laser beam with an exit beam diameter of 0.4 mm

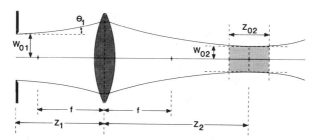

Fig. 12. A single "thin" lens at z_1 of focal length f is used to focus a laser beam with an initial beam radius of w_{01} and divergence half-angle of θ_1 to produce a focused beam radius at z_2 of w_{02} with a confocal distance of z_{02} (see text).

($w_{01} = 0.2$ mm) will exhibit a beam divergence of $\tan^{-1}$ [(600 × 10^{-9} m)/ ($\pi \cdot 0.2 \times 10^{-3}$ m)] ≈ 1 rad.

A "thin lens" of focal length f is placed in the beam a distance z_1 from the laser beam exit (Fig. 12). The following equation can be used to accurately calculate the position of maximum focus z_2:

$$z_2 = f + \frac{(z_1 - f)f^2}{(z_1 - f)^2 + (\pi w_{01}^2 n/\lambda)^2}. \tag{46}$$

The beam radius at maximum focus, w_{02} (Fig. 12), is given by

$$w_{02} = \{w_{01}^{-2}[1 - (z_1/f)]^2 + f^{-2}(\pi w_{01}n/\lambda)^2\}^{-1/2}. \tag{47}$$

This parameter is frequently referred to as the laser beam "spot size" and is defined as the radial distance from the beam axis at which the field amplitude of the TEM_{00} mode has decreased by a factor $1/e$ (~36.8%) compared to its value on axis.

The confocal beam parameter

$$z_{02} = \pi w_{02}^2 n/\lambda \tag{48}$$

is the distance from the beam waist at which the beam spot size increases by $\sqrt{2}$. We will define this distance as the pathlength (centered at z_2) over which the laser beam is in "focus" in the sample cell (Fig. 12). In other words, the two parameters w_{02} and z_{02} define an irradiated-volume element, a cylinder with radius w_{02}, height z_{02}, and volume $\pi z_{02} w_{02}^2$. The number of photons absorbed by the solute within the above volume element via two-photon excitation induced by a TEM_{00} laser pulse with a Gaussian temporal profile is [see Eq. (34)]

$$^2N_{abs} = \frac{1}{\sqrt{2}} NP_0\delta_\lambda C \frac{z_{02}}{\pi w_{02}^2} = \frac{1}{\sqrt{2}} NP_0\delta_\lambda C \frac{n}{\lambda}, \tag{49}$$

providing z_{02} is less than the pathlength of the cell. We therefore want to maximize z_{02} (subject to the pathlength constraints of the cell) and minimize w_{02} (subject to the constraints imposed by optical saturation of the sample, see Section II.B.3.a).

It is of equal importance to focus the laser beam inside the sample cell. While this goal may appear to be a trivial experimental objective, the following example serves to illustrate a potential complication. Assume that a 10-cm-focal-length lens is placed 30 cm from the laser exit port. As shown in Fig. 13, the actual focus point does not coincide with the focal length of the lens. The discrepancy is largest for laser beams of small diameter where beam divergence [Eq. (45)] is responsible for shifting the beam waist a significant distance beyond the focal point. The output of a

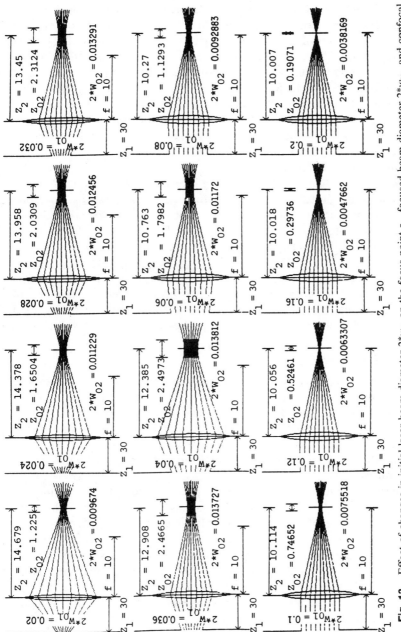

Fig. 13. Effect of changes in initial laser beam diameter $2*w_{01}$ on the focus point z_2, focused beam diameter $2*w_{02}$, and confocal distance z_{02}. A single "thin" lens with a focal length f of 10 cm is placed 30 cm (z_1) from the exit port of the laser. All parameters are given in centimeters for a laser beam of wavelength 600 nm. The vertical scale is linear and is adjusted for each situation to provide visual clarity. The horizontal scale is divided into two linear portions [laser to lens (z_1) and lens to maximum focus (z_2)].

typical nitrogen-pumped dye laser has a beam diameter $2*w_{01}$ of ~ 0.04 cm. The above lens configuration will place the beam waist 2.4 cm beyond the focal length (Fig. 13). If one is using a 1-cm cell, an error of 2.4 cm is quite significant. The experimentalist should use Eq. (46) to determine the actual focus point whenever a laser is being used with a beam diameter that yields a ratio of $f/2*w_{01}$ or $z_1/2*w_{01}$ greater than 50. An interesting exception is the special case when $z_1 = f$ where Eq. (46) reduces to $z_2 = f$ [Table II (row 10)]. Unfortunately, the convenience inherent in this special case is usually obviated by the fact that the confocal beam parameter is larger than the pathlength of the cell (which is inefficient with respect to maximizing two-photon absorption).

The effect of changing the focal length of the lens on the beam "spot size" and confocal parameter is evaluated in Table II (rows 1–12) for one particular configuration. In general, the shorter the lens focal length, the shorter the confocal length z_{02} and the smaller the beam "spot size." Note, however, that an extremely small "spot size" is not necessarily desirable. The salient ratio to maximize is $z_{02}/\pi w_{02}^2$ subject to $z_{02} \leq L$. This parameter equals $\sim 1.7 \times 10^4$ cm^{-1} for the first five configurations listed in Table II. The 8-cm-focal-length lens is to be preferred because it will minimize optical saturation and decrease the probability that laser field effects will alter the photophysical properties of the solute (see below).

a. Optical Saturation

An important consideration in excitation spectroscopy involves the restriction imposed upon the laser intensity to prevent saturation. Holtom et al. (1977) have reported that it is extremely easy to deplete the ground states of polyenes by focused two-photon excitation. The resulting partial population inversion leads to deviations from the power-squared dependence of the two-photon excitation spectrum and to potentially false spectral features.

The following example illustrates the potential for optical saturation in a $10^{-4}M$ solution ($C = 6 \times 10^{16}$ molecules cm^{-3}) of all-*trans* retinol [$\delta_{705} \simeq 2 \times 10^{-49}$ cm^4 sec molecule^{-1} photon^{-1}; Birge et al. (1978a)]. We assume a 1-mJ laser pulse ($\Gamma = 6$ nsec) with an exit beam diameter of 0.08 cm focused into the sample cell by a 5-cm-focal-length lens. If the lens is placed 35 cm from the laser output port, Eqs. (46)–(48) predict the following parameters at $\lambda = 705$ nm: $z_2 = 5.13$ cm, $w_{02} = 2.59 \times 10^{-3}$ cm, and $z_{02} = 0.298$ cm. The irradiated volume $\pi z_{02} w_{02}^2$ equals 6.3×10^{-6} cm^3, and this volume contains 3.8×10^{11} molecules. Equation (49) predicts that 2.4×10^{11} photons will be absorbed during the laser pulse corresponding to 1.2×10^{11} molecules excited. In other words, roughly one-third of the molecules in the irradiated volume will be promoted into the excited state. If the excited-state lifetime of the molecule is comparable to the laser pulse width, the solute molecules within the irradiated volume are approaching optical saturation. A useful (approximate) equation that can be used to estimate the fraction F of solute molecules excited within the irradiated volume during the lifetime of the laser pulse is

$$F \simeq \delta_\lambda N^2 \tau_{ul}^* / (2\sqrt{2}\pi^2 \Gamma^2 w_{02}^4) \tag{50}$$

where τ_{ul}^* is the observed lifetime of the solute excited state or the laser pulse width Γ, whichever is smaller. The experimentalist should adjust conditions to prevent F from exceeding 0.01. Note that a solute molecule with a lifetime much smaller than the laser pulse width will decrease the probability of saturation problems because the same molecule can be optically excited more than once during the laser pulse.

b. Wavelength Effects on Laser Focusing

The focal length of a single-element lens is given by (Hecht and Zazac, 1979)

$$\frac{1}{f_\lambda} = (n_\lambda - 1)\left(\frac{1}{R_1} - \frac{1}{R_2} + \frac{(n_\lambda - 1)d}{n_\lambda R_1 R_2}\right) \tag{51}$$

where f_λ is the wavelength-dependent focal length, n_λ is the refractive index of the lens material at λ, R_1 and R_2 are the radii of the front spherical surface (R_1 is positive if the center is past the lens) and the rear spherical surface (R_2 is negative if the center is before the lens), respectively, and d is the thickness of the lens. Virtually all lens materials exhibit a wavelength-dependent refractive index which increases with decreasing wavelength [e.g., Suprasil quartz: $n = 1.4564$ ($\lambda = 656.3$ nm), $n = 1.4601$ ($\lambda = 546.1$ nm), and $n = 1.4667$ ($\lambda = 435.8$ nm)]. A Suprasil lens such that $R_1 = -R_2 = 5$ cm and $d = 2$ cm will display the following wavelength dependence on focal length: $f = 5.844$ cm ($\lambda = 656.3$ nm), $f = 5.799$ cm ($\lambda = 546.1$ nm), and $f = 5.721$ cm ($\lambda = 435.8$ nm). This effect, known as "chromatic aberration," is a modest source of error and can be avoided by using multielement achromatic focusing lenses (Hecht and Zazac, 1979).

A potentially more serious source of focusing error is encountered when using lasers with small beam diameters ($2*w_{01} < 0.1$ cm). The wavelength dependence of the beam divergence [Eq. (45)] will have an observable effect on the position of the beam waist as shown in Table II. Note that the wavelength dependence on z_2 shown in Table II is a function entirely of laser beam divergence effects; possible chromatic aberration effects

TABLE II

Effects of Lens Focal Length and Laser Wavelength on Various Properties of a Focused TEM_{00} Laser Beam with Initial Beam Radius of 0.04 cm and $z_1 = 30$ cm[a]

Number[b]	λ (nm)[c]	f (cm)[d]	z_2 (cm)[e]	w_{02} (cm)[f]	z_{02} (cm)[g]	Irradiated volume (cm³)[h]	$z_{02}^{eff}/\pi w_{02}^2$ (cm⁻¹)[i]
1	600	1	1.00369	4.5120(−4)	0.010659	6.8173(−9)	16,666
2	600	2	2.01435	9.0568(−4)	0.042949	1.1068(−7)	16,666
3	600	4	4.0541	1.8240(−3)	0.17421	1.8209(−6)	16,666
4	600	6	6.1138	2.7540(−3)	0.39713	9.4626(−6)	16,666
5	600	8	8.1877	3.6945(−3)	0.71466	3.0644(−5)	16,666
6	600	10	10.2696	4.6441(−3)	1.12930	6.7758(−5)	14,758
7	600	15	15.4659	7.0499(−3)	2.60231	1.5614(−4)	6,405
8	600	20	20.5619	9.4820(−3)	4.70757	2.8245(−4)	3,540
9	600	25	25.4437	0.011915	7.43391	4.4603(−4)	2,242
10	600	30	30.0000	0.014324	10.7430	6.4458(−4)	1,551
11	600	35	34.1304	0.016682	14.5705	8.7423(−4)	1,144
12	600	40	37.7523	0.018964	18.8303	1.1298(−3)	885
13	200	8	8.0221	1.2684(−3)	0.25271	1.2773(−6)	50,000
14	300	8	8.0493	1.8936(−3)	0.37550	4.2300(−6)	33,333
15	400	8	8.0865	2.5083(−3)	0.49415	9.7674(−6)	25,000
16	500	8	8.1329	3.1095(−3)	0.60752	1.8454(−5)	20,000
17	700	8	8.2496	4.2609(−3)	0.81478	4.6472(−5)	14,285
18	800	8	8.3177	4.8068(−3)	0.90735	6.5863(−5)	12,500
19	900	8	8.3908	5.3309(−3)	0.99199	8.8565(−5)	11,111
20	1000	8	8.4677	5.8321(−3)	1.06855	1.0685(−4)	9,358

[a] All data calculated assuming a single "thin" lens is used to focus the laser beam as shown in Fig. 12 and the beam divergence of the beam at the laser exit port is calculated using Eq. (45).

[b] Data set number.

[c] Wavelength of the laser beam in nanometers.

[d] Focal length of the lens in centimeters.

[e] Position of maximum focus in centimeters (see Fig. 12).

[f] Beam radius at z_2 in centimeters (see Fig. 12). Exponents to base ten are given in parentheses.

[g] Confocal length in centimeters [see Eq. (48)].

[h] Irradiated volume in cubic centimeters within a 1-cm cell assuming the sample cell is centered at z_2. The irradiated volume is given by $z_{02}^{eff} \pi w_{02}^2$, where z_{02}^{eff} is the effective confocal length (see i). Exponents to the base ten are given in parentheses.

[i] Two-photon excitation efficiency for a 1-cm cell (in cm^{-1}) [see Eq. (49)] where z_{02}^{eff} is the effective confocal length defined as z_{02}, or 1 cm (the cell path length), whichever is smaller. Note that tighter focusing of the laser beam within the sample cell (rows 1–4) does not increase the efficiency of two-photon excitation provided z_{02} remains smaller than the cell path length L. However, if $z_{02} > L$, the efficiency of two-photon excitation decreases (rows 6–12). Increasing the laser wavelength (rows 13–20) decreases the two-photon excitation efficiency by a factor proportional to $1/\lambda$ provided $z_{02} < L$ and the laser beam divergence is governed by diffraction [Eq. (48)].

[Eq. (51)] have been ignored. The experimentalist should use Eq. (46) to calculate z_2 as a function of wavelength and adjust the lens position accordingly. Furthermore, the efficiency of the two-photon excitation process is proportional to the factor $z_{02}^{\text{eff}}/\pi w_{02}^2$ where z_{02}^{eff} is the confocal beam length z_{02} or the cell path length L, whichever is smaller [see Eq. (49)]. The effect of wavelength on this factor is analyzed in rows 13–20 of Table II for a 1-cm cell and a particular optical configuration. If the confocal beam length z_{02} is less than the cell path length at all wavelengths used in a particular investigation, and the lens position is adjusted so that the TEM_{00} excitation is focused at the center of the cell, this wavelength-dependent "focusing effect" can be corrected for by multiplying the observed two-photon excitation signal by λ. This correction factor derives from the fact that $z_{02}^{\text{eff}}/\pi w_{02}^2 = n/\lambda$ for $z_{02} < L$ [Eqs. (48) and (49)].

4. Mode Structure and Transverse Field Distribution

The discussion presented in the previous sections has evaluated the effect of changes in the temporal structure and the beam diameter on the excitation process. The transverse field distribution has either been neglected or assumed to be TEM_{00}. We now discuss the effect of mode structure on the efficiency of the two-photon excitation process. (In the absence of saturation effects, the one-photon excitation process is invariant to mode structure.)

Most commercially available dye lasers operate in the TEM_{00} mode with a small contribution from a few higher-order modes. This is fortunate, because the principal conclusion of this section is that the TEM_{00} mode is the most efficient mode for the generation of two-photon excitation.

The intensity of light at a specific location (x, y, z) in the laser beam is related to the electric field amplitude by the relation

$$I(x, y, z) = (c\varepsilon/2n)|E(x, y, z)|^2 \tag{52}$$

where c is the speed of light in a vacuum, ε and n are the dielectric constant and refractive index of the medium, respectively. The real part of the electric field amplitude is given by (Kogelnik and Li, 1966; Yariv, 1975)

$$\text{Re}[E_{\text{l,m}}(x, y, z)]$$

$$= E_0 \frac{w_0}{w(z)} H_1\left(\frac{x\sqrt{2}}{w(z)}\right) H_{\text{m}}\left(\frac{y\sqrt{2}}{w(z)}\right) \exp\left(\frac{-(x^2 + y^2)}{w^2(z)}\right) \tag{53}$$

where E_0 is the maximum electric field amplitude, w_0 is the spot size at the focus point ($z = 0$), $H_{\text{m}}(\xi)$ is the Hermite polynomial of order m, and x and

y are the transverse coordinates. The beam radius at position z is

$$w(z) = \{w_0^2[1 - (z^2/z_0^2)]\}^{1/2} \tag{54}$$

where z_0 is the confocal beam parameter [Eq. (48)]. The quantum numbers l and m define the transverse mode (l = m = 0 for TEM_{00}). Equation (53) is derived for rectangular-geometry modes, which are the most common. Transverse modes of circular geometry are rarely observed because most resonators are not truly symmetric about their axis. Intensity distribution contours of various transverse excitation modes are shown in Fig. 14. If a laser is not operating in the TEM_{00} mode, the transverse field distribution is typically a complex and often unsymmetric combination of many modes. Hypothetical multimode transverse intensity distributions are shown in Fig. 15 based on weighting factors for the higher modes given by $(1/n)^q$, where n is the number of intensity maxima ($n = 6$ for TEM_{21}, Fig. 14) and q is an exponent ranging from 0 (all modes given equal weight) to ∞ (TEM_{00}). The intensity distributions shown in Fig. 15 correspond to exponents ranging from 3 (Fig. 15a) to 0.5 (Fig. 15f).

In order to compare the two-photon excitation efficiency of the fundamental (Fig. 14) and complex (Fig. 15) transverse intensity distributions, a basis for comparison is required. We arbitrarily assume that each mode contains the same number of photons (N is constant) and that the beam

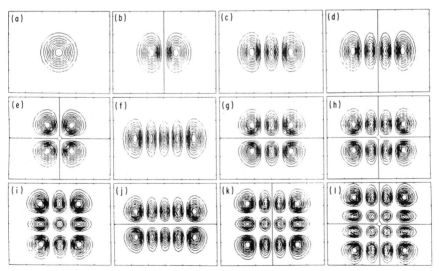

Fig. 14. Intensity contour maps $[I(x, y, 0)$, Eq. (52)] of various transverse excitation modes plotted for identical beam radii [w_0 = const, Eq. (53)]. The individual modes are defined in Table III.

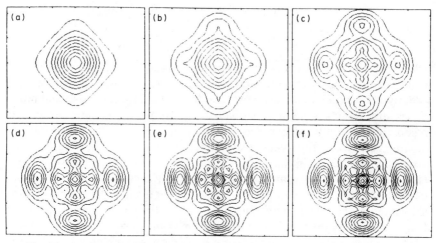

Fig. 15. Intensity contour maps $[I(x, y, 0)$, Eq. (52)] of hypothetical multiple (complex) transverse excitation modes plotted for identical beam radii (w_0 = const). The TEM_{00} mode is given unity weight and all higher modes are weighted by the factor $(1/n)^q$ where n is the number of intensity maxima ($n = 1$ for TEM_{00}, $n = 2$ for TEM_{10}, etc.) and q equals 3 (a), 2.5 (b), 2 (c), 1.5 (d), 1 (e), and 0.5 (f) (see also Table III).

radius and beam waist are equal and constant $[w_0 = w(z) = \text{const}]$. Accordingly, E_0 is varied such that the integral

$$\int_{-\infty}^{\infty} \int_{-\infty}^{\infty} |E(x, y, z)|^p \, dy \, dx$$

is normalized to unity for $p = 2$. The two-photon excitation efficiency is then obtained by (numerically) evaluating the above integral for $p = 4$ (i.e., intensity squared). The results are given in Table III and indicate that the TEM_{00} mode is significantly more efficient at generating two-photon excitation than higher-order modes (assuming w_0 is constant). An equally important observation is that changes in mode structure during the course of an experiment can introduce significant error in the observed two-photon excitation spectrum. Flash-lamp-pumped dye lasers are particularly troublesome in this respect because higher-order modes are usually prevalent and their intensity relative to the TEM_{00} mode is flash-lamp intensity, dye concentration, and dye temperature dependent (Bennett and Birge, 1980).

A useful formula for estimating the two-photon excitation efficiency (²Eff) of a given higher-order mode is

$$^2\text{Eff} \sim n^{-0.4} \qquad [w_0 = w(z) = \text{const}] \qquad (55)$$

where n is the number of maxima in the transverse intensity distribution.

TABLE III

Effects of Transverse Excitation Mode on Peak Intensities and Two-Photon
Excitation Efficiencies of Laser Beams[a]

Mode (q)[b]	Figure[c]	Relative peak intensity[d]	^{2}Eff[e]	$n^{-0.4}$[f]
0,0	14a	(1.000)	(1.000)	1
1,0	14b	0.731	0.750	0.758
2,0	14c	0.641	0.641	0.644
3,0	14d	0.600	0.574	0.574
1,1	14e	0.534	0.563	0.574
4,0	14f	0.569	0.528	0.525
2,1	14g	0.468	0.481	0.488
3,1	14h	0.438	0.431	0.435
2,2	14i	0.406	0.411	0.415
4,1	14j	0.416	0.396	0.398
3,2	14k	0.385	0.368	0.370
3,3	14l	0.362	0.333	0.330
Cl(3)	15a	0.530	0.477	—
C2(2.5)	15b	0.403	0.346	—
C3(2)	15c	0.318	0.272	—
C4(1.5)	15d	0.288	0.249	—
C5(1)	15e	0.292	0.249	—
C6(0.5)	15f	0.306	0.256	—

[a] All data are numerically simulated using a grid of 256×192 (49,152) data points assuming $w_0 = w(z) = $ const [Eq. (53)].

[b] Transverse excitation mode. The value of the exponent q in the weighting factor of the multimode distributions is given in parentheses for complex modes C1–C6 (see text).

[c] Figure number followed by the letter designating the contour plot of the intensity distribution of the corresponding mode.

[d] Peak intensity of the transverse distribution relative to that calculated for the 0,0 mode.

[e] Relative two-photon excitation efficiency of the mode relative to that calculated for the 0,0 mode (see text). (The relative one-photon excitation efficiency is mode independent.)

[f] Relative two-photon excitation efficiency of the mode approximated using Eq. (55).

5. Photon Statistics

The effect of photon statistics on the probability of the two-photon absorption process has been the subject of extensive theoretical study (Lambropoulos *et al.*, 1966; Carusotta *et al.*, 1967; Guccione-Gush *et al.*, 1967; Shen, 1967; Mollow, 1968; Mandel and Wolf, 1970; Chang and Ancker-Johnson, 1971; Weber, 1971; Barashev, 1973; Carusotto and Strati, 1973; Mahr, 1975; Salomaa and Stenholm, 1978). The salient conclusion of these studies is that incoherent (thermal) light is twice as efficient as coherent (laser) light in producing a two-photon absorption. The

Hamiltonian for the radiation field can be written in the form (Shen, 1967; Yariv, 1975)

$$\mathcal{H} = \sum_k \hbar\omega_k(a_k^\dagger a_k + \tfrac{1}{2}) \tag{56}$$

where $a_k^\dagger$ and a_k are the creation and annihilation operators, respectively, for the kth mode. Provided the photon distribution is not appreciably modified by absorption processes, the two-photon correlation parameter [see Eq. (20)] is

$$S = \langle a^\dagger a^\dagger aa\rangle/(\langle a^\dagger a\rangle)^2 \tag{57}$$

where $\langle a^\dagger a\rangle$ and $\langle a^\dagger a^\dagger aa\rangle$ are the first-order and second-order correlation functions of the radiation field (Lambropoulos *et al.*, 1966). A lengthy analytical treatment (e.g., Mahr, 1975) leads to the following values:

$$S_{\text{chaotic}} = 2, \tag{58a}$$

$$S_{\text{coherent}} = 1. \tag{58b}$$

This result has been confirmed by experiment (Shiga and Imamura, 1967).

Numerical simulations of one coherent and three incoherent (chaotic) distributions are shown in Fig. 16. The incoherent distributions are simulated using a pseudo-random-number generator to assign the temporal domain ("channel") into which the photon is placed. The ratio $\langle a^\dagger a^\dagger aa\rangle/(\langle a^\dagger a\rangle)^2$ is equal to $(1/n)(\sum_{k=1}^n N_k^2)$ where N_k is the number of photons in channel k and n is the total number of channels (provided there are the same number of channels available as there are total photons in the field). A coherent source will yield $(\sum_{k=1}^n N_k^2)/n = 1$ because each channel will have unit photon occupation. The photon occupation irregularities inherent in a chaotic distribution will lead to "bunching" of two or more photons in the same channel which enhances the probability of a two-photon absorption because of the intensity-squared dependence of the molecular response function. The simulations shown in Fig. 16 used 256 photons and 256 channels. The small sample leads to potentially large error in the numerically determined correlation parameter. A more elaborate simulation which evaluated the average of 1000 determinations of S_{chaotic} each involving 2560 photons and 2560 channels gave 1.9988 $\pm$ 0.0017 (95% confidence limit). A numerical simulation is unlikely to yield exactly 2 because of error inherent in the use of a pseudo-random-number generator.

The above discussion is not intended to persuade the experimentalist to use a thermal source instead of a coherent laser source to generate two-photon absorption. Although short-pulse, high-intensity flash lamps and

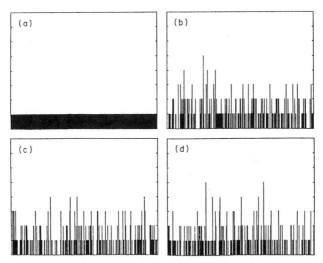

Fig. 16. Numerical simulation of the effect of coherent (a) vs incoherent light (b)–(d) on the photon statistics of the radiation field (see text). Photon occupation (vertical axis) as a function of temporal domain (horizontal axis) is assigned for the three incoherent fields using a pseudo-random-number generator. A total of 256 photons and 256 channels are included and the numerical values of the two-photon correlation parameter $\langle \ddot{a}\ddot{a}\dot{a}\dot{a}\rangle/(\langle \ddot{a}\dot{a}\rangle)^2$ (Eq. 57) are 1 for (a), 1.9766 (b), 2 (c), and 2.0156 (d).

arc sources are available (Hundley *et al.*, 1967; Hartig *et al.*, 1976; Leskovar *et al.*, 1976), their use in two-photon spectroscopy has not been fully explored and is probably not worth the effort. There are, however, techniques that can be used to alter the photon statistics of a laser pulse to enhance nonlinear absorption processes. The following section describes one such technique.

6. Excitation via Mode-Locked Lasers

The output of a laser can be modified to produce a train of extremely short pulses separated in time (e.g., Fig. 17). The technique is known as mode locking and the principles and methodology have been extensively reviewed (Shank and Ippen, 1973; Smith *et al.*, 1975). The purpose of the present section is to evaluate the advantages and disadvantages of using mode-locked lasers in excitation spectroscopy.

The emission response function of molecules with varying τ excited via a mode-locked laser pulse is numerically simulated in Fig. 17. We have arbitrarily assigned the mode structure of the excitation pulse to consist of 16 peaks separated in time by 0.5 nsec with pulse widths of $\Gamma = 100$ psec. The intensity profile was defined to follow a log-normal distribution which

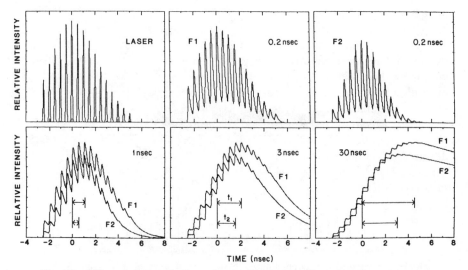

Fig. 17. Effect of changes in the molecular lifetime τ on the one-photon (F1) and two-photon (F2) induced fluorescence time profiles for excitation by a simulated mode-locked laser pulse (upper left) (see text). Molecular lifetimes of $\tau = 0.2$ (upper middle and upper right), 1 (lower left), 3 (lower middle), and 30 nsec (lower right) are depicted.

is somewhat characteristic of pulsed dye lasers mode locked using a saturable absorber.

The most important observation to be made concerns the significant enhancement in the two-photon excitation efficiency using mode-locked pulse excitation versus single (Gaussian) pulse excitation. If the laser pulse shown in Fig. 17 has a peak power (at $t = 0$) of $P_0 = 1 \times 10^{23}$ photons sec^{-1} and a total energy of $N = 1.02 \times 10^{14}$ photons/pulse, the following number of photons are absorbed via one-photon [assuming $1 - \exp(-\sigma CL) = 0.00836$] and two-photon [assuming $\delta CL = 6 \times 10^{-32}$ cm^2 sec photon^{-1} and $\pi w_{02}^2 = 0.001$ cm^2] excitation

$$^1N_{abs}^{ML} = 8.54 \times 10^{11}, \qquad ^2N_{abs}^{ML} = 3.25 \times 10^8.$$

The analogous values for single laser pulse excitation with comparable energy ($N = 1.02 \times 10^{14}$ photons/pulse) and total pulse width ($\Gamma = 5$ nsec) are

$$^1N_{abs}^{G} = 8.54 \times 10^{11}, \qquad ^2N_{abs}^{G} = 8.30 \times 10^7$$

where the superscript G refers to a Gaussian pulse [rather than ML (mode locked)] and all other molecular and experimental parameters are the same. As expected, the one-photon excitation efficiency is invariant with the pulse structure. The two-photon excitation efficiency, however, has increased by ~4 because the second-order correlation function is significantly enhanced by the "bunching" of photons into the individual, shorter pulses. The above example using 100-psec mode-locked pulses is by no means representative of the potential of this technique. Pulse widths on the order of 1 psec and trains with thousands of pulses can be generated by using mode locking (Shank and Ippen, 1973; Smith *et al.*, 1975). The enhancement in S [Eq. (57)] can exceed 10^3.

The use of synchronously pumped mode-locked lasers to generate two-photon excitation spectra is likely to become routine during the next decade. These lasers are capable of providing highly reproducible pulse trains and the significant enhancement in the two-photon correlation parameter makes these systems of considerable utility for two-photon excitation studies.

Perhaps the greatest utility of mode-locked lasers in excitation spectroscopy is in fluorescence lifetime measurements of molecules with lifetimes of less than 1 nsec. The combination of mode locking and pulse extraction techniques provides a convenient source of high-intensity light with pulse widths as short as 1 psec. The analysis of the fluorescence using streak camera techniques can provide lifetime measurements in the picosecond domain.

7. Field Effects

A focused pulsed laser beam is capable of generating large electric fields within the irradiated volume. The calculations of Birge and Pierce (1979) indicate that the one-photon and two-photon properties of polyene excited states are dramatically affected by fields at or above 10^7 V cm^{-1}. For example, the low-lying $^1A_g^{*-}$ and $^1B_u^{*+}$ states of octatetraene are theoretically predicted to invert in the presence of a field of 5×10^7 V cm^{-1}.

The average magnitude E_{ave} of the transient electric field induced by a laser beam with beam radius w_0 and peak power P_0 is given by (Birge and Pierce, 1979)

$$E_{ave} = (1/w_0)(P_0\mu_0 C/2\pi n)^{1/2} \qquad (59a)$$

where μ_0 is the permeability constant ($4\pi \times 10^{-5}$ H cm^{-1}), n is the refractive index of the solution, and C is the speed of light in a vacuum. This equation can be most conveniently evaluated in the following form (1 H = V sec A^{-1}, 1 W = V A):

$$E_{ave}(\text{V cm}^{-1}) = [245/w_0(\text{cm})][P_0(\text{kW})/n]^{1/2}. \qquad (59b)$$

A laser with a peak power of 940 kW (10 mJ, $\Gamma = 10$ nsec) focused into a sample with a refractive index of $n = 1.2$ using a 1-cm-focal-length lens will generate an electric field of $\sim 2 \times 10^7$ V cm^{-1} within the irradiated volume. This field is sufficient to modify the photophysical properties of the solute molecule (see above). Spectroscopists should always consider the possibility of field-induced changes in the electronic properties of any solute probed by focused laser beams. This is a particularly important factor when the solute molecule contains a conjugated π-electron system (see Birge and Pierce, 1979).

8. Polarization Effects

One-photon and two-photon spectroscopy differ markedly with respect to polarization phenomena. The use of one-photon spectroscopy to assign the symmetry of an excited state requires the use of samples in which the solute molecule is oriented. Although this restriction can be partially circumvented by using photoselection techniques to determine the relative polarization of an absorbing versus an emitting state (see, for example, Murrell, 1963; Suzuki, 1967; Harris and Bertolucci, 1978), the symmetry information is relative and not absolute. Two-photon spectroscopy can provide absolute symmetry assignments for many electronic states using randomly oriented samples (Monson and McClain, 1970; McClain, 1971, 1974; McClain and Harris, 1977; Birge and Pierce, 1979; Friedrich and McClain, 1980). This unique characteristic of two-photon spectroscopy derives from the requirement that two photons must be absorbed simultaneously by the molecule. Hence, the relative polarization of the two photons will affect the two-photon absorption probability.

The diagnostic utility of two-photon polarization spectroscopy can be demonstrated using diphenylbutadiene (Fig. 1). Molecular orbital calculations predict that two low-lying $^1A_g^*$ $\pi\pi^*$ excited states may be present. Both states are two-photon allowed, but one is polyene localized and the other is phenyl-group localized (Bennett and Birge, 1980). Although both states have the same symmetry classification ($^1A_g^{*-}$), the calculations predict that the polyene-like state will exhibit a higher two-photon absorptivity for linearly rather than circularly polarized light; the opposite is predicted for the phenyl-like state. The ratio of the two-photon absorptivity for circularly polarized light versus linearly polarized light can be easily measured by using a Fresnel Rhomb (Fig. 3) or a quarter-wave plate to convert linearly polarized to circularly polarized light (Drucker and McClain, 1974). The experimental accessibility combined with the diagnostic utility of this parameter has led to its definition as the "two-photon polarization ratio"

$$\Omega = \delta(\text{circ})/\delta(\text{lin}). \qquad (60)$$

The single-intermediate-state approximation can be used to examine the range to be expected for this parameter [Eq. (9)]. The photon propagation–polarization variables, a and b in Eq. (9), are (Birge and Pierce, 1979)

$$a = b = 8 \qquad \text{(linearly polarized)},$$

and

$$a = -8, \quad b = 12 \qquad \text{(circularly polarized)}$$

where parallel propagation is assumed in both cases. Note that Eq. (9) assumes both photons have the same energy. If the transition length vectors $\langle k|r|o \rangle$ and $\langle f|r|k \rangle$ are identically polarized, then

$$\Omega = (a + 2b)_{circ}/(a' + 2b')_{lin} = \tfrac{16}{24} = 0.667.$$

If the transition length vectors $\langle k|r|o \rangle$ and $\langle f|r|k \rangle$ are orthogonal, then

$$\Omega = b_{circ}/b'_{lin} = \tfrac{12}{8} = 1.5.$$

The polarization ratios for molecules probed using a single laser source (parallel propagation, $E_\lambda = E_\mu$) therefore fall in the range $\tfrac{2}{3} \leqslant \Omega \leqslant 1.5$. The two low-lying $^1A_g^* \, \pi\pi^*$ states of diphenylbutadiene are calculated to have polarization ratios of 0.7 (polyene-like state) and 1.3 (phenyl-like state) (Bennett and Birge, 1980). The observed value for the lowest-energy two-photon allowed state is 0.8 ± 0.1 (Swofford and McClain, 1973). The lowest-lying $^1A_g^*$ state is therefore assigned to be the polyene-like $\pi\pi^*$ state.

The above example serves to illustrate the diagnostic utility of the polarization ratio. This parameter, however, is only one of many experimentally measurable polarization variables. The review by McClain and Harris (1977) and the symmetry tables of McClain (1971) should be consulted for a more extensive treatment of the use of two-photon polarization measurements to assign excited-state electronic symmetries.

C. Solvent and Temperature Effects

The effect of solvent environment and temperature on the photophysical properties of the solute is an important consideration in optimizing the excitation spectrum. The use of low temperatures is required in many instances in order to increase the quantum yield of emission. Many molecules that do not fluoresce or phosphoresce at room temperature exhibit relatively high quantum yields of emission at liquid-nitrogen (77 K) or liquid-helium (1.8–4 K) temperatures. For example, the quantum yields of fluorescence of the visual chromophores increase by as much as two orders of magnitude at 77 K relative to room temperature (Birge, 1981). This consideration has mandated the use of low-temperature (77 K) solvent glasses (e.g., EPA) to obtain two-photon excitation spectra of these compounds (Birge *et al.,* 1978a,b, 1981; Birge, 1982). A further advantage of low-temperature solvent glasses is that reduced temperature usually enhances the vibronic structure that can be observed in the excitation spectra of molecules broadened by environmental inhomogeneity (Fang *et al.,* 1977, 1978; Bennett and Birge, 1980). For example, the room-temperature two-photon excitation spectrum of diphenylbutadiene is broad and shows virtually no discernible vibronic structure (Swofford and McClain, 1973). In contrast, the low-temperature (77 K) two-photon excitation spectrum of this molecule in EPA displays distinct vibronic structure (see Fig. 1). High-resolution excitation spectra in which the individual vibronic lines are fully resolved are possible by using liquid-helium temperature (1.8–4 K) and single-crystal (Hochstrasser *et al.,* 1973a,b, 1974; Bree *et al.,* 1979), Shpol'skii or mixed crystal environments (Hudson and Kohler, 1972, 1973; Abram *et al.,* 1975; Andrews and Hudson, 1978; Granville *et al.,* 1979, 1980). Naturally, those molecules which have adequate vapor pressures and fluoresce with reasonable quantum efficiency in the gas phase can be studied by using one-photon or two-photon vapor-phase excitation spectroscopy (Bray *et al.,* 1975; Friedrich and McClain, 1975; Robey and Schlag, 1978; Vasudev and Brand, 1979). The advantages of vapor-phase excitation spectra include the potential of observing both the individual vibronic lines as well as the discrete rotational structure (McClain and Harris, 1977).

1. Inhomogeneous Broadening and Laser Selection Techniques

When a molecule is dissolved in a solvent, the interaction between the solute and the solvent will produce changes in the energies of the ground state and the excited states. Frequently, the solvation energy of the ground state is different from that of the excited state because of differences in the ground- versus excited-state dipole moments or polarizabilities. The solvent will then produce a shift in the transition energy (Basu, 1964, and references cited therein). If the molecule occupies a nonspherical cavity, the solvent environment can also alter the one-photon oscillator strength of the electronic transition (Myers and Birge, 1980, and references cited therein). In the majority of cases, the microscopic orientation of the solvent surrounding the solute will have an appreciable effect on the magnitude of the solvent-induced shift. Statistical fluctuations of the solvent can therefore introduce a range of spectral shifts leading to inhomogeneous broadening. Provided the solute does not have an intrinsically broadened excitation spectrum due to intramolecular mechanisms, laser selection techniques can often be used to generate a sharp-line excitation or emission spectrum.

The fluorescence spectrum of the polynuclear aromatic hydrocarbon perylene in a 4.2-K n-octane Shpol'skii matrix displays distinct vibronic structure with linewidths on the order of 5 cm^{-1} (Abram *et al.*, 1974, 1975). This same molecule displays a relatively broad, poorly developed vibronic fluorescence spectrum at 4.2 K in ethanol using broadband excitation (Fig. 18). However, when the emission spectrum is obtained by using sharp-line laser excitation, using the apparatus shown in Fig. 5, dramatic sharpening of the emission spectrum occurs (Fig. 18) (Abram *et al.*, 1975). The sharp-line laser excitation "selects" the small fraction of the inhomogeneously broadened perylene molecules that have vibronic absorption bands that coincide with the laser excitation wavelength. This small fraction of the total population then emits with the distinct sharp-line vibronic development characteristic of a quasi-single-molecule spec-

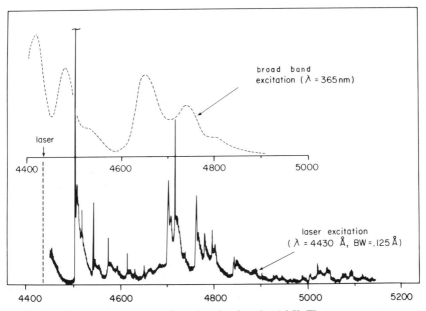

Fig. 18. Fluorescence spectra of perylene in ethanol at 4.2 K. The upper, spectrum was obtained by using a pressure-broadened mercury arc and a broadband filter as the excitation source ($\lambda_{ex} \cong 365$ nm). The lower spectrum was obtained by using the apparatus shown in Fig. 5 and 443-nm laser excitation ($\Delta\lambda = 0.125$ Å) [from Abram *et al.* (1975)].

trum. A more complete discussion of this phenomenon may be found in Abram *et al.* (1975) and references cited therein.

Sharp-line excitation spectra of an inhomogeneously broadened solute can also be obtained by using the apparatus shown in Fig. 5. The monochromator is fixed at a wavelength corresponding to a sharp-line vibronic peak in the laser-selected emission spectrum (e.g., the ~450-nm band in the laser-induced fluorescence spectrum of Fig. 18). The laser excitation wavelength is then scanned to produce a sharp-line excitation spectrum. This excitation technique, however, is usually not as successful as the corresponding emission technique at generating a sharp-line spectrum because excitation into the higher-energy regions of the absorption spectrum will typically result in the excitation of multiple vibronic levels. Additional peaks appear in the excitation spectrum corresponding to molecules with different solvent shifts.

D. Optimizing the Excitation Process

We now briefly review some of the salient conclusions of the previous sections with the goal of outlining some of the experimental variables that can be adjusted to optimize the signal-to-noise ratio of the excitation spectrum.

1. Choosing the Right Laser

It is generally accepted that high average power rather than high peak power is the desired parameter in one-photon laser excitation spectroscopy. It is frequently assumed that the principal criterion for choosing a laser for use in two-photon spectroscopy is peak power. As noted by Salomaa and Stenholm (1978) and Birge and Bennett (1979), however, very high peak power is rarely desirable. A more meaningful parameter is the product of peak power and pulse energy, or P_0N [see Eq. (34)]. Optimum results are obtained for a majority of cases when $P_0N \geqslant 1 \times 10^{38}$ photons^2 sec^{-1} and P_0 is greater than 10^{22} photons sec^{-1} but less than 10^{24} photons sec^{-1}.

High peak powers ($P_0 > 10^{24}$ photons sec^{-1}) can cause numerous problems which include optical saturation (Section II.B.3.a), field-induced changes in solute electronic properties (Section II.B.7), dielectric breakdown in the sample, thermal lensing effects which can produce self-focusing of the beam (Twarowski and Kliger, 1977) and enhance the above deleterious effects, local heating of the sample, and, in low-temperature (77 K) solvent glasses, sample cracking. One is better off using modest laser powers and sophisticated data collection techniques rather than high laser powers to generate a two-photon excitation spectrum.

The wavelength range and the repetition rate of the laser are also important variables to consider. Recent advances in the design of stimulated-Raman cells (e.g., Trutna and Byer, 1980) have extended the available wavelength range of two-photon spectrometers into the 1–3 μm region. The stimulated-Raman process, however, does require fairly high peak laser powers to be efficient.

2. Choosing the Correct Optical Configuration

The optimum optical configuration for observing the solute emission will be discussed in Section II.E.1. We limit the present discussion to optimization of the "excitation optics." The first, and perhaps the most obvious, fact to keep in mind is that every optical surface that the laser beam comes in contact with will produce loss. The total fractional light loss due to optical surfaces can be estimated by using the following formula:

$$\text{Loss} \cong 1 - (E_r)^m [4n/(1 + n)^2]^{2l} \tag{61}$$

where E_r is the mirror surface efficiency, m is the number of mirrors, n is the refractive index of the lens(es), and l is the number of lenses. Two beam-steering mirrors (98% efficiency; $E_r = 0.98$) and a single quartz lens ($n \cong 1.46$) will lead to a fractional loss of $1 - (0.98)^2 (0.965)^2 = 0.11$ or 11%. The use of inefficient mirrors ($E_r = 0.93$) will increase the loss to 20%. A complicated optical arrangement involving many lenses and mirrors is therefore to be avoided if laser energy is a critical variable.

The methodology and importance of properly focusing the laser beam in two-photon spectroscopy was examined in detail in Section II.B.3. One should choose a lens with a focal length such that the confocal beam parameter (z_{02} in Fig. 12) is as close as possible to the path length of the cell. This will generate the maximum two-photon excitation efficiency and minimize optical saturation.

3. Choosing the Best Solvent Environment

The advantage of using low-temperature solvent glasses in excitation spectroscopy has been discussed in Section II.C. Two disadvantages of these solvent glasses, however, are their tendency to scatter light to a greater extent than ambient-temperature solvent environments and their tendency to "crack" when probed by high-intensity laser excitation. The latter problem appears to derive from local heating effects creating a small shock wave that "fractures" the glass. The end result is a translucent rather than transparent sample which dramatically increases scattered light. One of the best solvent glasses to use is EPA, a mixture of ethyl ether, isopentane, and ethyl alcohol in a 5 : 5 : 2 volume ratio. This solvent mixture forms an excellent glass at 77 K and does not significantly enhance the amount of scattered laser light above levels encountered using ambient-temperature solvents. Other useful 77-K solvent glasses include 3MP (3-methylpentane), PMH (isopentane–methyl cyclohexane; 5 : 1, v/v) and EIP (ethyl ether–isopropyl ether; 1 : 1 v/v).

If one is working with an ambient-temperature solvent environment and one's goal is to choose a solvent that will perturb the electronic spectrum as little as possible (relative to the vapor-phase spectrum), the best solvent may be isopentane. This solvent has a small refractive index ($n = 1.3509$ at 25°C) which will minimize dispersive solvent shifts, and a negligible dipole moment, which will minimize electrostatic solvent shifts. Unfortunately, many molecules do not dissolve in sufficient concentration in isopentane.

4. Choosing the Best Sample Cell

The use of quartz cells in two-photon excitation spectroscopy should be avoided whenever possible. Crystalline quartz or annealed quartz with microscopic polycrystalline regions can produce in situ second-harmonic generation within the walls of the sample cell. The conversion efficiency of this process is usually quite low ($\sim 10^{-8}$; Franken et al., 1961). Nonetheless, the second-harmonic light can generate one-photon excitation of the solute and the subsequent emission can be on the same order of magnitude as the two-photon-induced emission. If one is not careful to circumvent this problem, a false two-photon excitation spectrum can result.

There are numerous situations, however, when the improved UV transmission characteristics of quartz mandate its use in order to monitor the two-photon-induced solute emission. The problem of in situ second-harmonic generation can be minimized by (i) masking the sample cell so that emission near the cell walls is blocked, (ii) using a shorter focal length lens and carefully focusing at the center of the cell, and (iii) using thin-walled, carefully annealed quartz cells (Birge et al., 1978a).

E. Detecting the Emission

The previous discussion has centered on an analysis and the experimental optimization of the excitation process. Of equal importance in excitation spectroscopy is the efficient collection and analysis of the solute emission. The theory and experimental technique of detecting the solute emission are, for the most part, independent of whether one is using one-photon or two-photon excitation to prepare the excited state. Consequently, the following discussion applies to both techniques unless stated otherwise.

1. Collection Optics

We assume for the purposes of the present discussion that the ensemble of excited-state solute molecules emits light with equal probability in all directions. This assumption is valid for the majority of ambient-temperature and low-temperature "glassy" environments because the solute maintains sufficient rotational freedom to permit random reorientation prior to emission. Most excitation configurations monitor the solute emission perpendicular to the excitation beam to minimize the amount of scattered excitation light (cf. Figs. 2–5). A lens system is typically used to collect a fraction of the scattered light and focus the emitted light onto the entrance slit of a monochromator (Figs. 2, 4, and 5) or direct it through a series of filters onto a photomultiplier (Fig. 3). The f *number* ($f/\#$) of a lens is defined as the ratio of the focal length f of the lens divided by the diameter d of the lens ($f/\# = f/d$). The inverse of the $f/\#$ gives the relative aperture (RA) of the lens (RA $= d/f$). A single-element lens is physically limited to an f number of roughly unity or greater. The collection efficiency of a lens is proportional to the square of the relative aperture; consequently, the smaller the f number, the higher the collection efficiency.

The optical configuration shown in Fig. 19 has been optimized to provide maximum collection efficiency subject to a minimum collection of scattered laser light. All three collection lenses (CL1, CL2, and CL3 in Fig. 19) are $f/1$ lenses. The first two lenses (CL1 and CL2) have focal lengths equal to their distance from the center of the sample cell. The third lens (CL3) has a focal length approximately equal to the distance separating the lens (CL3) from filter 2 (F2) so that the emitted light is defocused slightly prior to imaging on the photocathode of the detector (PM). The lens combination CL2 plus CL3 collects roughly $\frac{1}{16}$ of the emitted light. Similarly, the lens–dielectric mirror combination CL1 plus DM collects $\sim\frac{1}{16}$ of the emitted light and returns this light back through the sample such that it is gathered by CL2 and CL3. The total optical system therefore collects $\sim\frac{1}{8}$ of the emitted light (ignoring light loss due to reflections at the lens and window surfaces and mirror inefficiency) prior to filtering (Section II.E.2). A more rigorous analysis of the collection efficiency of the optical system shown in Fig. 19 using a ray tracing program including Fresnel losses at all the surfaces and assuming a Pyrex sample cell in liquid nitrogen, quartz lenses and Dewar windows, and a dielectric

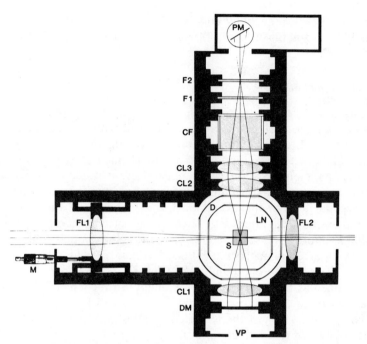

Fig. 19. Lens, filter, and low-temperature Dewar assembly for use in laser excitation studies of samples at liquid-nitrogen temperature. The laser excitation beam enters through the aperture on the left, is focused into the sample (S) by an achromatic focusing lens FL1, and exits through the aperture on the right. Lens FL2 is used to refocus the beam onto an external joulemeter or photomultiplier (not shown) to provide a reference signal for two-photon excitation studies (cf. Fig. 3). The emitted light is collected by lenses CL1 and CL2 and focused via lens CL3 onto the photomultiplier (see text). All of the lenses (except FL1), the dielectric mirror (DM), and the filters (CF, F1, and F2) are mounted in square black anodized aluminum blocks which can be dropped into any of the identical slots equally spaced along the excitation or emission axes of the assembly. This approach allows flexibility in changing the optical configuration or changing the chemical (CF) or broadband (F1 and F2) filters. The dielectric mirror (DM) should be chosen to provide maximum reflection of the emitted light and maximum transmission of the scattered laser light. The micrometer (M) is used to adjust the position of the primary focusing lens (FL1) in relation to the sample (S), which should remain stationary during the course of the experiment and be positioned as accurately as possible at the center of the optical Dewar (D) containing liquid nitrogen (LN). [Sample positioning is easily accomplished by removing the dielectric mirror and lens FL2 and viewing through the "view port" (VP) and the exit aperture (laser off!).] The Dewar can be removed for room-temperature studies.

mirror with 98% reflectivity and a confocal length of $0.1 f_{CL1}$ ($f_{CL1} = f_{CL2}$ are the focal lengths of lenses CL1 and CL2) yields $\sim\frac{1}{12}$ instead of $\frac{1}{8}$ (prior to filtering).

The basic optical configuration shown in Fig. 19 can also be used to image the solute emission onto the entrance slit of a monochromator. The photomultiplier housing is replaced with a monochromator positioned so that the entrance slit is aligned with the optical axis. Lens CL3 is moved two or three slots further away from the sample so that the emission light is focused onto the monochromator entrance slit [the chemical filter (CF) must be removed]. An important consideration in using a monochromator instead of broadband filters is that the f number of the monochromator will be the limiting factor in determining the effective collection aperture and the spectral bandwidth of the monochromator will probably be much narrower than that of the broadband filter (see below). Consequently, the improved wavelength discrimination of the monochromator will invariably result in a significant loss in signal.

2. Wavelength Selection Methods

The optimum wavelength selection method is determined by numerous criteria which include (i) the absolute intensity of the solute emission, (ii) the intensity of emitted light relative to scattered excitation light, (iii) the difference between the excitation and emission wavelengths, and (iv) the ultimate objectives of the excitation technique (is the emission spectrum to be scanned or not?). Criterion (ii) is usually the primary determinant in two-photon excitation spectroscopy because the number of photons emitted by the solute is likely to be 4–10 orders of magnitude smaller than the number of photons in the excitation pulse. Consequently, extremely effective filtering is required to prevent scattered laser light from reaching the detector. A double monochromator (Fig. 2), or a series of carefully chosen filters (Figs. 3 and 19), is required. The principal advantage of using a double monochromator is the ability to scan the two-photon-induced emission spectrum to verify that one is observing the excitation spectrum of the solute and not some luminescent impurity (Birge *et al.*, 1978a). However, a double monochromator will typically transfer only a small fraction of the total solute emission ($\sim 10^{-5}$) to the detector, thereby precluding its use in studies on solute molecules with low two-photon absorptivities ($\delta < 10^{-51}$ cm^4 sec molecule^{-1} photon^{-1}) or low quantum yields of emission ($\phi_{em} < 10^{-2}$). Fixed-wavelength filter combinations have the potential of providing both greater throughput and wavelength discrimination, but they must be optimized for the solute and excitation wavelength range to be studied. We have found that the combination of a chemical filter and two (usually different) broadband glass filters (Fig. 19) is usually optimum. The literature on chemical bandpass and cutoff filters has recently been reviewed by Klein (1979) and Laporta and Zaraga (1981). Other useful references include Kasha (1948), Withrow and Price (1953), Wladimiroff (1966), and Calvert and Pitts (1966). One should be cautious in using the transmission spectra published by commercial manufacturers of bandpass filters. Frequently, there are secondary bandpass regions that are not shown. For example, Corning filter #5-57 is reported to have a transmission maximum at 420 nm and a bandpass ($T > 1\%$) from ~320 to ~580 nm. Unfortunately, this filter also has a secondary bandpass region from ~730 to ~850 nm ($\lambda_{max}^{T} = 780$ nm), which is somewhat inconvenient if one wants to use this filter (by itself) to remove scattered laser light in the 700–900 nm region. Consequently, one should independently verify the transmission characteristics of each filter prior to use. It is a simple matter to check the efficiency of a filter combination by replacing the sample cell with a cell containing small glass beads and measuring the signal due to scattered light.

3. Photomultipliers

The filtered light emitted by the solute molecule during an excitation experiment is typically of such low intensity that multistage photomultiplier tubes providing current amplification of 10^6–10^7 must be used to observe the signal. Excellent discussions of the theory and use of photomultiplier tubes can be found in the literature available from the manufacturers (e.g., the EMI GENCOM, Hamamatsu, and RCA photomultiplier catalogs). The following brief discussion presents some of the salient parameters that should be considered prior to selecting a photomultiplier tube and associated housing.

a. Quantum Efficiency and Spectral Response

The quantum efficiency $\Phi_{PM}(\lambda)$ of a photocathode at a given wavelength λ is defined as the average number of photoelectrons emitted from the photocathode per incident photon. The wavelength dependence of this parameter defines the spectral response of the photocathode. The peak quantum efficiency $\Phi_{PM}(\lambda_{peak})$ typically ranges from 0.1 to 0.3. Many manufacturers graph spectral response in terms of the photocathode radiant sensitivity $S(\lambda)$. The following formula can be used to convert this parameter to the quantum efficiency:

$$\Phi_{PM}(\lambda) = 1.2395 S(\lambda)/\lambda, \tag{62}$$

where $S(\lambda)$ is in milliamperes per watt and λ is the wavelength in nanometers. Naturally, one wants to use a photomultiplier that has a high quantum efficiency at the wavelength maximum of the solute emission. However, it is also desirable to select a photomultiplier which has a poor quantum efficiency in the excitation-wavelength region. This technique is particularly useful in two-photon excitation spectroscopy to diminish the contribution of scattered laser light to the observed signal. For example, the RCA 1P28 photomultiplier tube has a nominal spectral response range of 185–650 nm with a peak quantum efficiency of 0.18 at 320 nm. Two-photon excitation spectroscopy using laser excitation above 650 nm can be optimized by using a 1P28 photomultiplier because the quantum efficiency at wavelengths beyond 650 nm is extremely low: $\Phi_{1P28}(650 \text{ nm}) \cong 0.003$; $\Phi_{1P28}(>700 \text{ nm}) > 1.8 \times 10^{-4}$.

b. Response Time

The temporal response of a photomultiplier tube is usually measured in terms of the *rise time,* the time required for the output pulse to rise from 10 to 90% of the peak output when the entire photocathode is illuminated by a short-duration pulse of light. Photomultiplier rise times usually range from 1 to 20 nsec. If the photomultiplier response is to be used to measure the lifetime of the excited state, a photomultiplier with a rise time less than or equal to the lifetime should be used if possible. However, time-delay measurements (Section II.B.2) or deconvolution techniques (Hundley *et al.,* 1967) can be used to relax this restriction. Small, side-on-type photomultiplier tubes such as the RCA 1P21, RCA 1P28, and the Hamamatsu R955 display rise times of ~2.6 nsec. This rise time is adequate for monitoring the fluorescence response of any molecule excited by laser pulses with $\Gamma > 6$ nsec.

c. Dark Noise

Dark noise is a phenomenon associated with the tendency of all photocathodes and dynodes to randomly eject electrons at a low rate in the absence of photon excitation. The dark current associated with this phenomenon is usually very small (~10 nA) and can be ignored for most pulsed laser experiments. Continuous-wave excitation spectroscopy, however, often requires that dark noise be at a minimum to provide good signal-to-noise ratios. The dark noise can be significantly reduced by cooling the photomultiplier to $-20°C$ (or lower), reducing the applied voltage, or using photon counting or gated photon counting techniques (Section II.E.4). Reducing the applied voltage, however, is rarely a useful method because it also reduces the signal amplification. Photomultiplier tubes vary significantly in their dark noise, and many manufacturers sell tubes specially selected for low dark noise. In general, the more red the response curve of the photomultiplier, the larger the dark noise. Accordingly, one should avoid using red-sensitive photomultiplier tubes unless the extended response is required for the experiment.

4. Photon Counting and Gated Photon-Counting Techniques

A single photon with wavelength λ striking the photocathode will cause an electron to be emitted from the photocathode with a probability equal to $\Phi_{PM}(\lambda)$. The charge Q_a at the anode for each electron emitted is given by

$$Q_a = A_{PM}e = (1.602 \times 10^{-19} \, C)A_{PM} \tag{63a}$$

where A_{PM} is the photomultiplier gain. Each electron will follow a slightly different path in passing through the multiplier structure. The anode current pulse dQ_a/dt will therefore display a finite temporal width with a full width at half maximum Γ_{PM} on the order of 4–10 nsec for photomultiplier tubes normally used for photon counting. The peak anode current I_a is given by

$$I_a(A) \cong Q_a \, (C)/\Gamma_{PM} \, (sec). \tag{63b}$$

For example, a 1P28 photomultiplier ($\Gamma_{PM} = 6$ nsec) operated at a gain of $A_{PM} = 2.6 \times 10^6$ will generate a Q_a/photon of 4.2×10^{-13} C and a peak anode current of 69 μA. The peak voltage into 50 Ω will therefore equal 3.5 mV. Unfortunately, spurious "dark" electrons emitted from the photocathode will also produce 3.5-mV signals, but spurious electrons emitted from the dynodes will be amplified to a lesser extent and will therefore produce signals of ~1 mV or less. Photon counting reduces dark noise by using an amplifier-discrimination circuit to reject pulses below a certain threshold (i.e., 3 mV referenced to input for the above example). Therefore, the only source of dark noise (i.e., dark counts) is thermal emission from the photocathode. The end result is an improvement in signal-to-noise by roughly one order of magnitude.

If one is using a laser with a sufficiently long pulse width or studying a molecule with a sufficiently long lifetime (~1 μsec or longer), gated photon-counting (GPC) techniques should be considered (see Fig. 2). One of the principal advantages of GPC derives from the ability to monitor the discriminator output only during the lifetime of the laser pulse. Consequently, the dark noise (dark count) that normally accompanies a photon counting detector system is negligible. The trigger circuit used in the spectrometer shown in Fig. 2 gates the counter for a 4-μsec period during each laser pulse, thus reducing dark noise from 500 to 0.1 cps for a 30-pps laser repetition rate. Accordingly, the signal-to-dark-noise ratio is significantly increased. An additional advantage of GPC over analog boxcar integration is the inherently smaller baseline drift of GPC, which permits the use of a computer to average scans over extended time periods.

The maximum (random pulse) counting rate in megahertz required of the photon counting system is approximately given by

$$R \text{ (MHz)} \cong 2\Phi_{PM}N_{det}/\Gamma \text{ (μsec)} \tag{64}$$

where Φ_{PM} is the quantum efficiency of the photomultiplier (at the emission maximum), Γ is the pulse width (FWHM) of the laser in microseconds, and N_{det} is the number of photons per pulse reaching the detector. This number is a function of the light-harvesting efficiency of the focusing lenses, slit assemblies, and folding mirrors, and the dispersion and slit width of the monochromator. An accurate calculation requires considerable computation and is dependent upon the emission profile of the solute and the optical configuration of the monochromator and the external optics. The following formula provides a very rough estimate of N_{det} (Birge and Bennett, 1979):

$$N_{det} \cong 0.06N_{em}A^2(E_g)^n(E_r)^m(\text{SBW/FWHM}). \tag{65}$$

Equation (65) assumes Gaussian optics, negligible stray light, an external collecting-lens system with an aperture ratio larger than the aperture ratio of the monochromator, and a Gaussian profile for the solute emission. The monochromator aperture ratio A is expressed as the reciprocal of the f-stop number (e.g., $A = 0.2$ for $f/5$). E_g is the grating efficiency (unity corresponds to 100% efficiency), n is the number of gratings, E_r is the mirror surface efficiency, m is the number of mirror surfaces, SBW is the spectral bandwidth of the monochromator, and FWHM is the full width at half maximum of the emission band of the solute. Application of Eq. (65) requires that the monochromator be set at the emission maximum and that the values for A, E_g, and E_r be appropriate for this wavelength.

The experimental arrangement shown in Fig. 2 passes a very small fraction ($\sim 10^{-5}$) of the emitted photons N_{em} to the detector. A typical two-photon experiment using this apparatus will require a photon-counting system with a maximum random pulse counting rate of ~ 40 MHz, which is well within the capabilities of commercially available photon-counting systems. One should note, however, that the individual pulse counts must be added up for a sufficient number of laser pulses to produce a statistically valid signal (100–600 laser pulses are usually required). A laser with a high repetition rate ($\geqslant 5$ pps) is usually required for the successful implementation of GPC techniques.

The observed signal must be corrected for the laser intensity by dividing by the square of the observed reference signal. If the laser has sufficient pulse-to-pulse stability ($\sim 5\%$) it is not necessary to perform the correction on each data point. That is, the average signal can be divided by the average reference squared at the end of the sampling interval. It is very important to use average values of the signal and reference and not arbitrarily summed values since the square of the sum of the signals will not equal the sum of the squares of the signals:

$$\delta = (\text{const}) \frac{(\Sigma S_i)/N}{[(\Sigma R_i)/N]^2} \neq (\text{const}) \frac{\Sigma S_i}{(\Sigma R_i)^2}$$

where S_i and R_i are the individual sample and reference signals for pulse i and N is the total number of laser pulses. Consequently, GPC techniques require that the number of laser pulses included in a given sampling interval be divided into the total signal and total reference counts.

5. Analog Detection and Boxcar-Averaging Techniques

Most pulsed lasers have pulse widths which are too short to permit the use of gated photon-counting techniques. Analog detection of the sample and reference signal is then required. The most straightforward approach is to display the sample and reference signals on a pair of fast storage oscilliscopes in single-shot mode. This approach lacks a certain elegance, but it provides for the simultaneous measurement of I_{em}, I_{ref}, and Γ (or t_n) at each wavelength. Pulse-width corrections [Eqs. (35)–(43)] can then be used, if necessary, to calculate the corrected excitation spectrum. Alternatively, one can use boxcar-averaging techniques to monitor the sample and reference signals (see Fig. 3). Boxcar averaging is the analog signal analogy of gated photon counting in that the signal is only sampled over a short period of time; for the remaining time (between triggers) the signal is held. During sampling, the signal can be averaged with the previous signal so that the output is equal to an exponentially weighted average of all the samples. The output is then proportional to the average peak intensity of the signal (provided the aperture duration is less than $\frac{1}{2}\Gamma$ and the aperture delay is adjusted so that the pulse maximum occurs within the sampling window). Alternatively, the signal can be linearly added to the previous signal. The output is then equal to the sum of the integrals of the signals. If the aperture duration is greater than 2Γ and the pulse is roughly centered in the sampling window, the output is proportional to the average number of photons in each pulse times the number of pulses. Accordingly, boxcar averaging can be used to monitor either peak intensity (I_{em} or I_o) or photons per pulse (N_{em} or N). In practice, exponential averaging is easier to use than linear summation because the latter is more subject to drift and biasing problems and the number of samples must be held relatively low to prevent overflow. Consequently, boxcar averaging is usually used to monitor peak intensity.

One of the important advantages of boxcar averaging is that noise (any signal that is not synchronous to the trigger rate) is reduced by the averaging process. The actual signal-to-noise improvement ratio (SNIR) is dependent upon the type of averaging performed. In the case of linear summation, the SNIR is proportional to $\sqrt{N}$ where N is the number of signal repetitions. In the case of exponential averaging, the SNIR is proportional to $\sqrt{2N}$ where N is the number of triggers required to reach 63% of steady state and at least $5N$ triggers occur.

The key to the successful application of boxcar averaging is an accurate (jitter-free) trigger signal. Most commercial lasers provide a "sync" pulse output which precedes the laser pulse. However, this trigger signal is only useful if it displays a jitter less than the aperture duration divided by four.

An excellent discussion of boxcar-averaging techniques can be found in the "Model 162 Boxcar Averager System" brochure available from EG&G Princeton Applied Research, Princeton, NJ.

F. Data-Analysis Techniques in Two-Photon Spectroscopy

The number of solute photons emitted following two-photon excitation is typically four or more orders of magnitude smaller than the number of photons in the laser excitation pulse. If the solute molecule has a very low quantum yield of emission or two-photon absorptivity, the filtering techniques described in Section II.E.2 may be incapable of completely removing laser scattering contributions to the observed signal. The following data analysis approach offers a potential solution to the problem of laser scattering interference.

The total signal S_i can be divided into two portions:

$$S_i = a_1 R_i + a_2 R_i^2 \tag{66a}$$

where R_i is the observed reference signal, a_1 is the scattering component of the signal, and a_2 is the two-photon component of the signal and equals $^2S_{em}/R^2$, the desired two-photon excitation signal. This equation can be rearranged to yield a closed-form linear least-squares regression problem by dividing both sides by R_i:

$$y_i = S_i/R_i = a_1 + a_2 x_i \tag{66b}$$

where $x_i = R_i$. The following equations can be used to evaluate the regression coefficients:

$$a_2 = [\Sigma x_i y_i - (\Sigma x_i \Sigma y_i)/n]/[\Sigma x_i^2 - (\Sigma x_i)^2/n], \tag{67a}$$

$$a_1 = [(\Sigma y_i)/n] - a_2[(\Sigma x_i)/n], \tag{67b}$$

and their standard errors,

$$s_{yx} = [(\Sigma y_i^2 - a_1 \Sigma y_i - a_2 \Sigma x_i y_i)/(n - 2)]^{1/2}, \tag{68a}$$

$$s(a_1) = s_{yx}\{\Sigma x_i^2/[n\Sigma x_i^2 - (\Sigma x_i)^2]\}^{1/2}, \tag{68b}$$

$$s(a_2) = s_{yx}/\{\Sigma x_i^2 - [(\Sigma x_i)^2/n]\}^{1/2} \tag{68c}$$

where $s(a_1)$ and $s(a_2)$ are the standard errors in the a_1 and a_2 regression coefficients, respectively.

An example of the use of Eqs. (66)–(68) is presented in Table IV using actual two-photon data obtained for diphenylbutadiene (Fig. 1) at 707 nm. Both the laser dye and the two-photon signal were relatively weak at this wavelength, and this combination frequently results in scattering problems. The most useful indicator of potential scattering interference is the slope of $\ln(S_i)$ versus $\ln(R_i)$, which should equal two for a two-photon excitation process. Values significantly less than two indicate the presence of a one-photon component in the signal provided the difference is statistically meaningful. (The calculated slope will invariably differ from two because of experimental error. A calculation of the standard error in the slope will indicate the significance of the deviation.) The data from Table IV yield a slope Q_2 of 1.59 ± 0.13, which suggests that laser scattering is a component of the signal. The average value of S_i/R_i^2 will therefore be overestimated. This is confirmed by carrying out a linear least-squares regression using Eqs. (66b) and (67a) to give $a_2 = 2.35 \pm 0.52$, a value significantly lower than the raw average of S_i/R_i^2 of 3.94 (see Table IV).

A measure of the reliability of the analysis can be obtained by evaluating the equation

$$r^2 = \frac{[\Sigma x_i y_i - (\Sigma x_i \Sigma y_i/n)]^2}{\{\Sigma x_i^2 - [(\Sigma x_i)^2/n]\}\{\Sigma y_i^2 - [(\Sigma y_i)^2/n]\}} \tag{69}$$

where r^2 is the coefficient of determination. A value of $r^2 = 1$ indicates a perfect fit. In practice, r^2 is invariably less than 1, but if the calculated value is less than 0.85, the values calculated for a_1 and a_2 [Eq. (67)] should be considered very approximate and potentially less meaningful physically than values obtained by using the raw data to calculated S/R^2. A value of r^2 below 0.85 is usually due to a sampling problem (either too few samples or the range in the reference signal was too small relative to the noise component).

TABLE IV

Analysis of Two-Photon Excitation Data to Remove the Laser Scattering Component[a]

S_i^b	$R_i \ (=x_i)^c$	$S_i/R_i \ (=y_i)^d$	$S_i/R_i^2 \ ^e$
0.0085	0.044	0.193	4.39
0.0195	0.070	0.279	3.98
0.023	0.076	0.303	3.98
0.012	0.058	0.207	3.57
0.019	0.072	0.264	3.67
0.018	0.072	0.250	3.47
0.013	0.056	0.232	4.15
0.021	0.074	0.284	3.83
0.023	0.082	0.280	3.42
0.0115	0.048	0.240	4.99
Average		(0.253)	(3.94)

Regression analysis[f]	Parameter	Value	Standard error
$\ln S_i = Q_1 + Q_2 \ln R_i$	Q_2 (slope)	1.59	±0.13
$y_i = a_1 + a_2 x_i$ [Eq. (66b)]	a_1 (scattering)	0.10	±0.03
	a_2 (two-photon)	2.35	±0.52

[a] Data from Bennett (1980) (see text).
[b] Peak fluorescence signal [proportional to I_{em}, Eq. (36)].
[c] Peak laser intensity [proportional to I_0, Eq. (36)].
[d] S_i/R_i is proportional to I_{em}/I_0.
[e] S_i/R_i^2 is proportional to I_{em}/I_0^2.
[f] See text.

III. RECENT APPLICATIONS

The presentation in Section II has concentrated on the theory and basic experimental techniques of one-photon and two-photon excitation spectroscopy. We conclude this review with a brief discussion of some relatively new applications of excitation spectroscopy with an emphasis on techniques which yield potential improvement in sensitivity or resolution.

A. Analytical Excitation Spectroscopy Using Pulsed Lasers and Time-Filtered Detection

Excitation spectroscopy is recognized as one of the most sensitive analytical techniques for both the qualitative identification and quantitative analysis of luminescent atoms and molecules (Winefordner, 1971).

The sensitivity of excitation spectroscopy derives from the excellent signal-to-noise ratio potential of a technique which measures a signal which is directly proportional (rather than inversely proportional) to the solute concentration. Most techniques (such as absorption spectroscopy) measure the presence of small quantities of a given species by measuring small differences in a large signal. The sensitivity of the latter approach is limited by the stability of the interrogating beam and the sophistication of the optics and electronics required to measure an accurate, noise-free reference signal.

Haugen and Lytle (1981) have recently proposed an analytical one-photon fluorescence excitation technique that has the potential of detecting fluorescing species in solution at concentrations on the order of $10^{-14}M$. Their apparatus uses single-photon counting, time-filtered detection, and mode-locked laser excitation. The combination of pulsed laser excitation and time resolution of the detected signal effectively eliminates three sources of potential noise in excitation-based spectrochemical analyses: (i) competitive emission from the instrumental components, (ii) laser scattering, and (iii) impurity luminescence in multicomponent mixtures. The key feature of this technique is the selection of a temporal "acceptance window" to monitor the fluorescence of the desired species. The combination of pulsed laser excitation and time-filtered detection, optimization of the wavelength of the laser excitation, and optimization of the detection wavelength provides for both sensitivity and selectivity. The major limitation, however, remains the requirement that the solute under investigation have a reasonable quantum yield of emission.

B. Double-Resonance Fluorescence Excitation Vibrational Spectroscopy

Seilmeier and co-workers (1978a,b) have recently demonstrated the feasibility of observing double-resonance fluorescence excitation in condensed phases. Wright (1980) has analyzed this technique in detail and concludes that this approach has the potential of generating a sharp-line vibrational spectrum characteristic of Raman or infrared spectroscopy while simultaneously providing the increased sensitivity characteristic of fluorescence excitation spectroscopy. The spectroscopic and analytical potentials of this technique are therefore considerable.

Double-resonance fluorescence excitation vibrational spectroscopy (DFEVS) involves the use of two (or more) laser excitation sources. The first laser prepares a ground-state vibrational level via infrared absorption or (in conjunction with a second laser) stimulated Raman scattering. A second (or third) laser is used to subsequently excite the molecule from the vibrationally excited level in the ground state to an electronically excited state from which fluorescence occurs. The fluorescence is monitored perpendicular to the laser excitation beams, thereby providing for the inherent sensitivity of the technique. The principal experimental difficulty of this method is the necessity of using relatively high laser intensities to overcome the rapid vibrational relaxation rates that are characteristic of molecules in condensed phases. As noted by Wright (1980), interference due to coherent anti-Stokes Raman scattering (CARS) and two-photon-induced fluorescence has the potential of obscuring the double-resonance-induced fluorescence. Although the CARS signal can be attenuated by choosing an optical geometry that minimizes phase matching, if the lowest excited state is strongly two-photon allowed ($\delta \lesssim 10^{-49}$ cm^4 sec molecule^{-1} photon^{-1}), the two-photon-induced signal is likely to be comparable to the DFEVS signal. Nonetheless, DFEVS has some unique characteristics which under certain circumstances can be exploited to yield information unobtainable by any other technique. The discussion by Wright (1980) is highly recommended.

C. Supersonic-Molecular-Jet Excitation Spectroscopy

One of the principal advantages of gas-phase excitation spectroscopy is the ability to observe the vibronic, and for small molecules the rovibronic, structure of isolated (environmentally unperturbed) molecules. Unfortunately, many polyatomic molecules have ambient-temperature vapor pressures that preclude vapor-phase spectroscopy or exhibit such complex vibrational and/or rovibrational structure that their vapor-phase spectra are hopelessly complex or completely unresolved. The use of low-temperature matrix environments can frequently generate sharp-line vibronic spectra, but the intermolecular perturbations associated with the condensed phase are considerable and an analysis of the excited-state geometry based on rovibronic structure is clearly impossible. The use of supersonic molecular jets to prepare internally cold, isolated, vapor-phase molecules retains the advantage of low-temperature matrix isolation spectroscopy without the disadvantage of environmental perturbations caused by intermolecular interactions. The use of excitation and emission spectroscopy to probe molecules in supersonic jets represents one of the more important recent advances in spectroscopic technique (for an excellent review, see Levy, 1980).

The molecule to be studied via this technique is expanded through a small orifice, either neat or in a carrier gas. The expansion lowers the temperature of the gas (or gas mixture) and under typical experimental conditions produces a translational temperature of 0.05–1 K, a rotational temperature of 0.1–50 K, and a vibrational temperature of 10–180 K. The spectral simplification associated with cooling of the internal degrees of freedom results in a significant improvement in intrinsic resolution and allows one to analyze spectra that would otherwise be hopelessly complicated had they been taken in ambient-temperature vapor-phase environments.

The above-mentioned spectral simplification is most apparent for large molecules [recent examples include the phenylalkynes (Powers *et al.*, 1981), all-*trans* octatetraene (Heimbrook *et al.*, 1981), naphthalene (Beck *et al.*, 1981), tetracene (Amirav *et al.*, 1981), and phthalocyanine (Fitch *et al.*, 1981)]. The elegance and power of this technique has prompted a growing number of researchers to apply supersonic-jet excitation spectroscopy to study polyatomic and van der Waals molecules. The recent review by Levy (1980) is highly recommended.

D. Doppler-Free Two-Photon Excitation Spectroscopy

The supersonic-molecular-jet technique described in the previous section achieves "spectral sharpening" by cooling the internal degrees of freedom in the molecule. Although this technique has significant advantages (see above), one notable disadvantage is the potential loss of structural information inherent in the rotational structure associated with high-J transitions. The precise determination of the moments of inertia is usually impossible if only transitions involving low J are observed [in the case of benzene, the levels above $J = 8$ are of significant structural importance (Riedle *et al.*, 1981)]. Unfortunately, the rotational structure in the electronic spectra of large polyatomic molecules is usually so dense that individual rotational lines are obscured by the Doppler profiles of the individual transitions. Riedle *et al.* (1981) have recently obtained a Doppler-free cw two-photon excitation spectrum of benzene with an observed resolution of ~0.0027 cm^{-1}, significantly less than the ambient-temperature Doppler width of ~0.06 cm^{-1}. Over 400 individual rotational transitions involving J levels up to ~30 were observed within a 4-cm^{-1} scan from 39,653 to 39,657 cm^{-1}. This remarkable accomplishment represents the first Doppler-free cw two-photon excitation spectrum reported for a polyatomic molecule and amply illustrates the intrinsic capabilities of this ultra-high-resolution technique.

The Doppler-free two-photon technique takes advantage of the requirement that two photons must be absorbed simultaneously (Salour, 1978). The use of two counterpropagating beams to generate the excitation removes the contribution of the molecule's velocity to the frequency of the light generating the excitation. The technique is therefore the two-photon analog of one-photon laser saturation spectroscopy (Hänsch, 1973). Riedle *et al.* (1981) used two relatively low-powered (100 and 300 mW) cw lasers operating at different wavelengths. This approach has the added advantage that the Doppler-broadened background due to simultaneous absorption of two photons from a single laser beam can be effectively eliminated provided the more powerful of the two lasers is fixed at a nonresonant wavelength and the less powerful laser is used to scan the two-photon excitation spectrum.

ACKNOWLEDGMENTS

This work was supported in part by grants from the National Institutes of Health (EY-02202), the National Science Foundation (CME-7916336), and a Grant-in-Aid from the Standard Oil Company of Ohio (SOHIO). I am grateful to Professors David Kliger, George Leroi, and Martin McClain; Drs. James Bennett, Howard Fang, and Robert Swofford; and Thomas Haw and Brian Justus for interesting and helpful discussions. I would especially like to thank Brian Pierce for preparing Figs. 6, 8, 9, and 10 and his contributions to the section on pulse-width effects (Section II.B.1.a).

REFERENCES

Abram, I. I., Auerbach, R. A., Birge, R. R., Kohler, B. E., and Stevenson, J. M. (1974). *J. Chem. Phys.* **61,** 3857–3858.

Abram, I. I., Auerbach, R. A., Birge, R. R., Kohler, B. E., and Stevenson, J. M. (1975). *J. Chem. Phys.* **63,** 2473–2478.

Alameda, G. K., Bennett, J. A., and Birge, R. R. (1981). *Rev. Sci. Instrum.* **52,** 1664–1670.

Amirav, A., Even, U., and Jortner, J. (1981). *J. Chem. Phys.* **75,** 3770–3793.

Andrews, J. R., and Hudson, B. S. (1978). *Chem. Phys. Lett.* **57,** 600–604.

Barashev, P. P. (1973). *Chem. Phys. Lett.* **19,** 143–147.

Basu, S. (1964). *Adv. Quantum Chem.* **1,** 145–169.

Beck, S. M., Hopkins, J. B., Powers, D. E., and Smalley, R. E. (1981). *J. Chem. Phys.* **74,** 43–52.

Bennett, J. A. (1980). Ph.D. Thesis, University of California, Riverside.

Bennett, J. A., and Birge, R. R. (1980). *J. Chem. Phys.* **73,** 4234–4246.

Birge, R. R. (1981). *Annu. Rev. Biophys. Bioeng.* **10,** 315–354.

Birge, R. R. (1982). *In* "Methods in Enzymology," Vol. 88 (L. Packer, ed.), pp. 522–532. Academic Press, New York.

Birge, R. R., and Bennett, J. A. (1979). "Experimental Techniques in Two-Photon Spectroscopy," Chromatix Appl. Note No. 8. Chromatix, Sunnyvale, California.

Birge, R. R., and Pierce, B. M. (1979). *J. Chem. Phys.* **70,** 165–178.

Birge, R. R., Bennett, J. A., Pierce, B. M., and Thomas, T. M. (1978a). *J. Am. Chem. Soc.* **100,** 1533–1539.

Birge, R. R., Bennett, J. A., Fang, H. L., and Leroi, G. E. (1978b). *Springer Ser. Chem. Phys.* **3,** 347–354.

Birge, R. R., Bennett, J. A., Hubbard, L. M., Fang, H. L., Pierce, B. M., Kliger, D. S., and Leroi, G. E. (1982). *J. Am. Chem. Soc.* **104,** 2519–2525.

Bray, R. G., Hochstrasser, R. M., and Sung, H. N. (1975). *Chem. Phys. Lett.* **33**, 1–4.
Bree, A., Pai, Y. H. F., and Taliani, C. (1979). *Chem. Phys. Lett.* **63**, 190–192.
Brody, S. S. (1957). *Rev. Sci. Instrum.* **28**, 1021–1026.
Calvert, J. G., and Pitts, J. N., Jr. (1966). "Photochemistry." Wiley, New York.
Carusotto, S., and Strati, C. (1973). *Nuovo Cimento B* **15**, 159–179.
Carusotto, S., Fornaca, G., and Polacco, E. (1967). *Phys. Rev.* **157**, 1207–1213.
Carusotto, S., Giulietti, A., and Vaselli, M. (1970). *Lett. Nuovo Cimento* **4**, 1243–1248.
Chang, D. B., and Ancker-Johnson, B. (1971). *Phys. Lett. A* **35A**, 113–115.
Christensen, R. L., and Kohler, B. E. (1974). *Photochem. Photobiol.* **19**, 401–410.
Dowley, M. W., Eisenthal, K. B., and Peticolas, W. L. (1967). *J. Chem. Phys.* **47**, 1609–1619.
Drucker, R. P., and McClain, W. M. (1974). *J. Chem. Phys.* **61**, 2609–2615, 2616–2619.
Fang, H. L., Thrash, R. J., and Leroi, G. E. (1977). *J. Chem. Phys.* **67**, 3389–3390.
Fang, H. L., Thrash, R. J., and Leroi, G. E. (1978). *Chem. Phys. Lett.* **57**, 59–63.
Fitch, P. S. H., Haynam, C. A., and Levy, D. H. (1981). *J. Chem. Phys.* **74**, 6612–6620.
Franken, P. A., Hill, A. E., Peters, C. W., and Weinreich, G. (1961). *Phys. Rev. Lett.* **7**, 118–124.
Friedrich, D. M., and McClain, W. M. (1975). *Chem. Phys. Lett.* **32**, 541–549.
Friedrich, D. M., and McClain, W. M. (1980). *Annu. Rev. Phys. Chem.* **31**, 559–577.
Gold, A. (1970). *Proc. Scott. Univ. Summer Sch. Phys.* **10**, 397–420.
Göppert-Mayer, M. (1931). *Ann. Phys. (Leipzig)* [5] **9**, 273–285.
Granville, M. F., Holtom, G. R., Kohler, B. E., Christensen, R. L., and D'Amico, K. L. (1979). *J. Chem. Phys.* **70**, 593–594.
Granville, M. F., Holtom, G. R., and Kohler, B. E. (1980). *J. Chem. Phys.* **72**, 4671–4675.
Guccione-Gush, R., Gush, H. P., and Van Kranedonk, J. (1967). *Can. J. Phys.* **45**, 2513–2524.
Hänsch, T. W. (1973). *In* "Dye Lasers" (F. P. Schäfer, ed.), pp. 194–258. Springer-Verlag, Berlin and New York.
Harris, D. C., and Bertolucci, M. D. (1978). "Symmetry and Spectroscopy." Oxford Univ. Press, London and New York.
Hartig, P. R., Sauer, K., Lo, C. C., and Leskovar, B. (1976). *Rev. Sci. Instrum.* **47**, 1122–1129.
Haugen, G. R., and Lytle, F. E. (1981). *Anal. Chem.* **53**, 1554–1559.
Hecht, E., and Zazac, A. (1979). "Optics," pp. 99–194. Addison-Wesley, Reading, Massachusetts.
Heimbrook, L. A., Kenny, J. E., Kohler, B. E., and Scott, G. W. (1981). *J. Chem. Phys.* **75**, 4338–4342.
Hermann, J. P., and Ducuing, J. (1972). *Phys. Rev. A* **5**, 2557–2568.
Herzberg, G. (1966). "Molecular Spectra and Molecular Structure," Vols. I and III. Van Nostrand-Reinhold, Princeton, New Jersey.
Hochstrasser, R. M., Sung, H. N., and Wessel, J. E. (1973a). *J. Am. Chem. Soc.* **95**, 8179–8180.
Hochstrasser, R. M., Sung, H. N., and Wessel, J. E. (1973b). *J. Chem. Phys.* **58**, 4694–4695.
Hochstrasser, R. M., Sung, H. N., and Wessel, J. E. (1974). *Chem. Phys. Lett.* **24**, 168–171.
Hohlneicher, G., and Dick, B. (1979). *J. Chem. Phys.* **70**, 5427–5437.
Holtom, G., Anderson, R. J. M., and McClain, W. M. (1977). *Abstracts of 32nd Symposium on Molecular Spectroscopy,* Ohio State University, Columbus, Ohio.
Hopfield, J. J., and Worlock, J. M. (1965). *Phys. Rev.* **137**, A1455–A1458.
Hubbard, L. M., Bocian, D. F., and Birge, R. R. (1981). *J. Am. Chem. Soc.* **103**, 3313–3320.
Hudson, B. S., and Kohler, B. E. (1972). *Chem. Phys. Lett.* **14**, 299–304.

Hudson, B. S., and Kohler, B. E. (1973). *J. Chem. Phys.* **59**, 4984–5002.
Hudson, B. S., and Kohler, B. E. (1974). *Annu. Rev. Phys. Chem.* **25**, 437–460.
Hundley, L., Coburn, T., Garwin, E., and Stryer, L. (1967). *Rev. Sci. Instrum.* **38**, 488–492.
Kasha, M. (1948). *J. Opt. Soc. Am.* **38**, 929–934.
Klein, R. M. (1979). *Photochem. Photobiol.* **29**, 1053–1054.
Kogelnik, H., and Li, T. (1966). *Proc. IEEE* **54**, 1312–1329.
Lambropoulos, P., Kikuchi, C., and Osborn, R. K. (1966). *Phys. Rev.* **144**, 1081–1086.
Laporta, P., and Zaraga, F. (1981). *Appl. Opt.* **20**, 2946–2950.
Leskovar, B., Lo, C. C., Hartig, P. R., and Sauer, K. (1976). *Rev. Sci. Instrum.* **47**, 1113–1121.
Levy, D. H. (1980). *Annu. Rev. Phys. Chem.* **31**, 197–225.
McClain, W. M. (1971). *J. Chem. Phys.* **55**, 2789–2796.
McClain, W. M. (1974). *Acc. Chem. Res.* **7**, 129–135.
McClain, W. M., and Harris, R. A. (1977). *Excited States* **3**, 2–56.
Mahr, H. (1975). *In* "Quantum Electronics: A Treatise" (H. Rabin and C. L. Tang, eds.), Vol. 1, Part A, pp. 285–361. Academic Press, New York.
Mandel, L., and Wolf, E., eds. (1970). "Coherence and Fluctuations of Light," Vols. I and II. Dover, New York.
Mollow, B. R. (1968). *Phys. Rev.* **175**, 1555–1563.
Monson, P. R., and McClain, W. M. (1970). *J. Chem. Phys.* **53**, 29–37.
Mortensen, O. S., and Svendsen, E. N. (1981). *J. Chem. Phys.* **74**, 3185–3189.
Murrell, J. N. (1963). "The Theory of the Electronic Spectra of Organic Molecules," Chapman and Hall, London.
Myers, A. B., and Birge, R. R. (1980). *J. Chem. Phys.* **73**, 5314–5320.
Peticolas, W. L. (1967). *Annu. Rev. Phys. Chem.* **18**, 233–260.
Pierce, B. M., and Birge, R. R. (1982). *IEEE J. Quantum Electron.* **QE-18**, 1164–1170.
Pierce, B. M., and Birge, R. R. (1983). *IEEE J. Quantum Electron.* (in press).
Powers, D. E., Hopkins, J. B., and Smalley, R. E. (1981). *J. Chem. Phys.* **74**, 5971–5976.
Riedle, E., Neusser, H. J., and Schlag, E. W. (1981). *J. Chem. Phys.* **75**, 4231–4240.
Robey, M. J., and Schlag, E. W. (1978). *Chem. Phys.* **30**, 9–17.
Salomaa, R., and Stenholm, S. (1978). *Appl. Phys.* **17**, 309–316.
Salour, M. M. (1978). *Rev. Mod. Phys.* **50**, 667–681.
Seilmeier, A., Kaiser, W., and Laubereau, A. (1978a). *Opt. Commun.* **26**, 441–448.
Seilmeier, A., Kaiser, W., Laubereau, A., and Fischer, S. F. (1978b). *Chem. Phys. Lett.* **58**, 225–235.
Shank, C. V., and Ippen, E. P. (1973). *In* "Dye Lasers" (F. P. Schäfer, ed.), pp. 121–143. Springer-Verlag, Berlin and New York.
Shen, Y. R. (1967). *Phys. Rev.* **155**, 921–931.
Shiga, F., and Imamura, S. (1967). *Phys. Lett. A* **25A**, 706–707.
Siegman, A. E. (1971). "An Introduction to Lasers and Masers," pp. 443–449. McGraw-Hill, New York.
Smith, P. W., Duguay, M. A., and Ippen, E. P. (1975). *Prog. Quantum Electron.* **3**, 107–229.
Sorokin, P. P., Lankard, J. R., Moruzzi, V. L., and Hammond, E. C. (1968). *J. Chem. Phys.* **48**, 4726–4741.
Suzuki, H. (1967). "Electronic Absorption Spectra and Geometry of Molecules." Academic Press, New York.
Swofford, R. L., and Albrecht, A. C. (1978). *Annu. Rev. Phys. Chem.* **29**, 421–440.
Swofford, R. L., and McClain, W. M. (1973). *J. Chem. Phys.* **59**, 5740–5742.
Swofford, R. L., and McClain, W. M. (1975). *Chem. Phys. Lett.* **34**, 455–460.
Trutna, W. R., and Byer, R. L. (1980). *Appl. Opt.* **19**, 301–312.

Twarowski, A. J., and Kliger, D. S. (1977). *Chem. Phys.* **20,** 253–258, 259–265.

Vasudev, R., and Brand, J. C. D. (1979). *Chem. Phys.* **37,** 211–217.

Weber, H. P. (1971). *IEEE J. Quantum Electron.* **QE-7,** 189–195.

Winefordner, J. D., ed. (1971). "Spectrochemical Methods of Analysis," Adv. Anal. Chem. Instrum. Ser., Vol. 9, Wiley-Interscience, New York.

Withrow, R. B., and Price, L. (1953). *Plant Physiol.* **28,** 105–115.

Wright, J. C. (1980). *Appl. Spectrosc.* **34,** 151–157.

Wladimiroff, W. W. (1966). *Photochem. Photobiol.* **5,** 234–250.

Yariv, A. (1975). "Quantum Electronics," 2nd ed. Wiley, New York.

3 THE THERMAL LENS IN ABSORPTION SPECTROSCOPY

Howard L. Fang and Robert L. Swofford

The Standard Oil Company (Ohio)
Corporate Research Center
Warrensville Heights, Ohio

175

I. INTRODUCTION

The present chapter describes the development of measurement techniques based on the warming effect of the incident radiation on a sample. The goal of this work is primarily the development of these techniques for sensitive absorption measurements. It should be noted, however, that the phenomena discussed here are intrinsic in the use of lasers to probe the optical properties of materials. This chapter may therefore also be of interest to experimenters wishing to *avoid* the influence of these thermal phenomena on a particular measurement technique being used.

The treatment begins with a brief discussion of the thermal effects of radiation and some order-of-magnitude calculations of the effects. Next are reviewed the early accounts of the experimental observations of the phenomena and the model developed to describe the effects. Refinements in the model proposed by several authors are discussed. Early experimental techniques are described as well as more recent refinements and innovations. The attention of the present chapter is on the most extensively developed of the techniques for monitoring the thermal effects of radiation, i.e., the measurement of the heat distribution in the sample which is manifested as a lenslike element (the "thermal lens").

A. Thermal Effects of Radiation

The warming effect of electromagnetic radiation is perhaps one of nature's most pleasurable phenomena. Early man no doubt quickly learned that a colored object was warmed when placed in sunlight. We now understand such a process to be the result of the dissipation of the incident energy within the material. The observed heat is a measure of the equilibrium distribution of that energy among the many degrees of freedom of the material. While processes such as phase change, photochemical transformation, or reemission of radiation may compete for a share of the excitation energy, in almost all cases a portion of the incident energy appears as heat in the illuminated substance.

Observation of the thermal effects of electromagnetic radiation has played an important role in the discovery and understanding of many natural phenomena. Accounts of William Herschel's discovery of infrared radiation in 1800 describe how Herschel observed the warming of a thermometer bulb placed adjacent to the red color band in a prism-dispersed beam of sunlight. Until the invention of reliable photoelectric sensors for infrared radiation, the direct thermal detection of the IR transmitted by a sample was commonly used for the tedious point-by-point measurement of a sample's infrared absorption spectrum. The now-famous discovery of the photoacoustic (or optoacoustic) effect by Alexander Graham Bell in 1880 during his experiments in optical communications was the result of Bell's interest in the thermal effects of radiation (Bell, 1880).

Modern interest in the thermal effects of radiation developed with the invention of the laser, which can provide both a means for extremely high energy delivery to the sample and a means for extremely sensitive detection of the resultant heating of the sample. As will be seen, both of these features are used in the measurement technique to be described here.

We propose that measurement schemes which monitor the thermal effects of incident radiation be collectively referred to as *photothermal* techniques, a name which is already being used in reference to certain types of measurements. We include in our definition all techniques that employ some thermal phenomenon as a monitor of the residual energy content of a substance. The attention here is focused exclusively on optical techniques for monitoring the sample heating, although it should be noted that other sensitive techniques are available for detecting sample heating, for example, calorimetry (Fujishima *et al.*, 1980; Severin and van Esveld, 1981; White, 1976) or bulk physical measurements (e.g., extensometry; Helander *et al.*, 1981).

B. Development of Photothermal Measurement Techniques

There are several factors which have fostered the development of photothermal techniques. The foremost reason is the widespread availability and use of the laser. Scientists quickly learned to recognize a variety of phenomena which could be generated in experimental measurements involving lasers. The energy delivery capability and, in some cases, the high peak power capability of the laser can result in significant photothermal effects in the samples. With the understanding of these phenomena has come an appreciation of the sensitivity of the various effects to the absorptivity of the sample. This appreciation has resulted in numerous studies which use photothermal techniques for very sensitive absorption measurements. Reports of these studies have received widespread attention, just at the time when trace analysis has become a topic of widespread interest to the analytical chemist. Furthermore, the continuing improvements and greater commercial availability of well-behaved tunable lasers have brought these sensitive absorption techniques out of the physics laboratory and into the hands of the analytical chemist and spectroscopist. Finally, detection schemes such as the one to be described here are applicable to a wide range of excitation energies, and the techniques approach closely the "ideal" universal detectors in a number of situations. Unlike detection techniques which are linked to the spectral range of the incident exciting radiation, photothermal detection can be thought of as the "greatest common denominator," working satisfactorily over virtually the entire electromagnetic spectrum.

C. Magnitude of Photothermal Effects—Simple Calculations

Before the photothermal measurement techniques are described, it is valuable to have an appreciation for the magnitude of the effects being considered. At the same time, these simple calculations will serve to introduce the physical parameters which will be important in the more detailed calculations presented later.

The starting point of the calculations is Beer's law, which gives the power P_{trans} (W) transmitted through a thickness l (cm) of sample with absorptivity α (cm^{-1}), as

$$P_{trans} = Pe^{-\alpha l} \tag{1}$$

where P (W) is the incident power. For a weakly absorbing sample, for example, $\alpha \approx 10^{-3}$ cm^{-1}, the absorbed power is

$$P_{abs} = P - P_{trans} \approx P\alpha l \tag{2}$$

since $e^{-x} = 1 - x$ for small x. Consider the situation of an isolated cylindrical sample with radius $w = 0.1$ cm and length $l = 1$ cm illuminated by a $P = 10^{-2}$-W laser beam. The rate of energy delivery to the sample, $\dot{Q}_{in}$, is $P\alpha l/J$, where $J = 4.18$ J cal^{-1} is Joule's constant, so $\dot{Q}_{in} = 2.39 \times 10^{-6}$ cal sec^{-1}. We assume complete dissipation of the incident energy as heat. The resulting temperature rise in the sample depends on the sample volume $\pi w^2 l$ (cm^3), the specific heat C_p (cal g^{-1} K^{-1}), and the sample density ρ (g cm^{-3}). With liquid benzene as a typical organic sample, where $\rho = 0.88$ g cm^{-3} and $C_p = 0.41$ cal g^{-1} K^{-1}, the rate of temperature increase, $\dot{T}$, under the above conditions is $\dot{T} = P\alpha l/J\pi w^2 l\rho C_p = 2.1 \times 10^{-4}$ K sec^{-1}. If the illuminated sample is surrounded by ambient-temperature sample, then heat is conducted radially out of the illuminated region. The thermal conductivity k (cal sec^{-1} cm^{-1} K^{-1}) relates $\dot{Q}_{out}$, the time rate of heat transfer by conduction, to the surface area, in the present case $2\pi w l$, the material thickness b, and the difference in temperature ΔT; hence $\dot{Q}_{out} = (k)(2\pi w l) (\Delta T)/b$. The thermal conductivity for liquid benzene is 3.41×10^{-4} cal sec^{-1} cm^{-1} K^{-1}.

It should be noted that thermal conductivities of organic liquids are notoriously difficult to determine with reasonably high accuracy (Calmettes and Laj, 1972; Gupta et al., 1980; Tada et al., 1978). One should be very cautious therefore in extracting values from the scientific literature. The value of the thermal conductivity of liquid benzene used here comes from the critically evaluated data compiled by Jamieson et al. (1975). Other values quoted in the literature differ by as much as 40%. Because of the fundamental dependence of the photothermal effects on the thermal conductivity, values of this parameter must be carefully evaluated before use.

It is easy to see how, under illumination, the sample temperature reaches an equilibrium value when the rate of heat input from the laser is just balanced by the rate of heat conduction out of the illuminated sample, $\dot{Q}_{in} = \dot{Q}_{out}$. This requires in the present case of a cylinder of surface area $2\pi w l = 0.63$ cm^2 a temperature gradient across the material of $\Delta T/b = \dot{Q}_{in}/k2\pi w l = 1.1 \times 10^{-2}$ K cm^{-1}. If the laser beam has a "bell-shaped" or Gaussian power profile, $P \exp(-2r^2/w^2)$ (see below), then a major portion of the temperature gradient is developed across the beam radius w, and the temperature difference between the illuminated and nonilluminated sample is approximately $\Delta T = [(P\alpha l/J)/k2\pi w l]w = 1.1 \times 10^{-3}$ K. At the rate of temperature increase $\dot{T}$ calculated above, equilibrium would be achieved in a time $\Delta t = \Delta T/\dot{T} = [(P\alpha l/J)w/k2\pi w l]/(P\alpha l/J\pi w^2 l\rho C_p) = \frac{1}{2}w^2\rho C_p/k$, or approximately 5.35 sec. Of course, these calculations cannot be taken too literally and are only meant to provide an appreciation for the order of magnitude of the effects being considered here.

The expression for Δt above represents a fundamental response time of the photothermal effect. Thus, it is seen that the time required to establish an equilibrium temperature distribution in the sample is proportional to the area of the illuminated cross section of the sample. Under the condition of an unfocused cw laser beam incident on the sample, the thermal response is quite slow.

A complete derivation of the heat conduction equation, which is beyond the scope of the present work (see, for example, Kern, 1950), gives $\dot{T}$ $= (k/\rho C_p)\text{div}\, T$. For weak absorption of the Gaussian beam above, we have $\dot{T}_{r=0} \propto (k/\rho C_p)4/w^2 \equiv t_c^{-1}$, where the critical time t_c is defined by analogy with the expression for Δt derived earlier. The critical time is essentially the time response of the molecule to the heat input. In the example of benzene above, for $w = 0.1$ cm, $t_c = 2.65$ sec.

The collection of molecular parameters $k/\rho C_p$ is called the thermal diffusivity D (cm^2 sec^{-1}), since it contains all of the molecular properties involved in the conduction of heat. The thermal diffusivity specifies the change in temperature which would be produced in unit volume of the substance by the quantity of heat which flows in unit time across unit area through a layer of the substance of unit thickness with unit difference in temperature between its faces. In the example of benzene above, $D = 9.45 \times 10^{-4}$ cm^2 sec^{-1}. In a later section, we will return to the calculation of heat conduction, since it is a crucial factor in the understanding of photothermal effects.

D. Thermal Effect on Index of Refraction

It is the combination of performing these precision temperature measurements and doing so over such a relatively long time scale that presents the challenge of photothermal spectroscopy. How are such measurements performed? As will be seen, the photothermal techniques measure the influence of the temperature change on the index of refraction of the sample directly or on the index of some material in thermal contact with the sample. By appropriate arrangement of sample illumination, one can create an index-of-refraction distribution in the medium which can then be probed optically. From the measured index-of-refraction distribution and the thermal coefficient of the index of refraction, one can calculate the temperature distribution in the sample. Using the known thermal properties of the material, one can then calculate the amount of energy deposited in the sample and eventually the sample absorbance.

In regions of the optical spectrum far removed from sample absorption features and at sample temperatures away from phase changes or critical points, the temperature coefficient of the index of refraction, dn/dT (K^{-1}),

$$dn/dT = (\partial n/\partial T)_\rho + (\partial n/\partial \rho)(\partial \rho/\partial T) \tag{3}$$

is determined primarily by changes in the sample density (second term). Most liquids expand when heated, resulting in a negative value of dn/dT. In the example of benzene liquid above, where $dn/dT = -3.9 \times 10^{-4}$ K^{-1}, the change in the index of refraction of the illuminated liquid is given by

$$\Delta n = (dn/dT)\Delta T = -4.3 \times 10^{-7}. \tag{4}$$

The magnitude of Δn calculated in Eq. (4) is at first appearance a hopelessly small number to measure accurately. By comparison, however, we see that the spectral dependence of the index of refraction, $dn/d\lambda$, of a conventional prism is approximately 3×10^{-5} nm^{-1} at 600 nm. Furthermore, a lens with 1 m focal length has an index-of-refraction curvature at its center of approximately 7.5×10^{-4} cm^{-1}. Thus a laser beam of 0.1 cm radius incident on such a lens would experience a change in the index of refraction of 7.5×10^{-5} from the center to the edge of the beam. Finally, the optical path through a 1-cm sample changes by one-half wave at 600 nm for an index-of-refraction change of 3×10^{-5}. Thus we see that measurement of such small changes in the index of refraction of a sample is not totally beyond our reach.

Techniques based on the photothermal creation of different optical elements have been described in recent years. By far the most common technique is based on the photothermal lens (or simply "thermal lens"), in part because it was the first of the modern photothermal effects to be observed and understood. Other photothermal optical elements which have been employed include the prism (Boccara *et al.*, 1980; Murphy and Aamodt, 1980; Jackson *et al.*, 1981), the interferometer (Stone, 1972, 1973; Aung and Katayama, 1975a,b; Miller *et al.*, 1975; Beysens and Calmettes, 1977; Zitter *et al.*, 1980; Abbate *et al.*, 1981), the transmission diffraction grating (Pelletier *et al.*, 1982), and the curved mirror (Da Costa and Calatroni, 1978). These latter devices will not be discussed in the present chapter, which concentrates on the photothermal lens.

II. THE PHOTOTHERMAL LENS

The thermal lens was the first of the modern photothermal phenomena to be described (Gordon *et al.*, 1964). The effect was observed quite by accident when Porto and associates at Bell Laboratories attempted to enhance the intensity of laser Raman scattering from benzene by placing a 1-cm Brewster-angle sample cell of the liquid inside the resonator of a He–Ne laser in order to take advantage of the higher circulating intracavity laser power (0.8 W). With the sample cell in place, the laser exhibits power buildup and decay transients with a time constant of a few seconds. When steady state is achieved, the beam spot size at the laser mirrors is changed, as if a diverging lens of approximately 1 m focal length has been formed in the sample. The scientists concluded that a "thermal lens" is produced by the heating action of the Gaussian laser beam, and their analysis (Gordon *et al.*, 1965) forms the basis of the present discussion. The creation of the thermal lens by a laser beam with a Gaussian intensity profile is easy to understand. Near the center of the laser beam, the power profile of the laser and hence the temperature profile of the heated sample are nearly parabolic. With a negative value of dn/dT, the refractive index profile very closely approximates that of a high-quality diverging lens.

A. Focal Length of the Thermal Lens

Born and Wolf (1970) show how to calculate the effective focal length of such an index-of-refraction distribution. We begin by writing the index of refraction of the cylindrically symmetric distribution as a Maclaurin series in the radial displacement from the center of the distribution,

$$n(r) = n(0) + r(\partial n/\partial r)_{r=0} + (\tfrac{1}{2})r^2(\partial^2 n/\partial r^2)_{r=0} + \cdots. \tag{5}$$

We note that such a distribution has a vanishing first derivative at $r = 0$. Referring to Fig. 1, consider the path of ray $\overrightarrow{AB}$ incident parallel to the central axis $\overline{HM}$ (i.e., the paraxial approximation) but displaced a distance $|KB| = r$. The ray follows the circular path BC with radius of curvature R, given by Born and Wolf (1970) as

$$1/R = \mathbf{v} \cdot \text{grad } \ln(n) \tag{6}$$

where $\mathbf{v}$ is the unit principal normal at a typical point along $\overparen{BC}$. Since we assume that the gradient is only in the radial direction and that the sample cell is thin (so that point O is fixed), then Eq. (6) reduces to

$$1/R = (1/n)(\partial n/\partial r) = (r/n)(\partial^2 n/\partial r^2)_{r=0}. \tag{7}$$

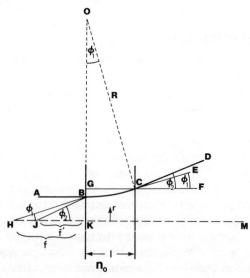

Fig. 1.   Ray-tracing diagram for calculation of the focal length of the thermal lens.

The ray would normally depart along $\overrightarrow{CE}$, but refraction at the glass–air interface sends it along $\overrightarrow{CD}$ instead. By construction, $\langle BJK = \langle DCF$ and $\langle BHK = \langle ECF$. Without refraction, the ray departing along $\overrightarrow{CE}$ would have a virtual image at H, resulting in a focal length of $-|HK| = -f$. Instead, the ray departs along CD and has a virtual image at J, or a focal length $-|KJ| = -f'$. Snell's law of refraction gives

$$\sin(\phi_2)/\sin(\phi_1) = n \approx \phi_2/\phi_1 \tag{8}$$

since $n_{air} = 1$, and both ϕ_1 and ϕ_2 are small. We also have from geometric construction,

$$\tan\phi_1 = r/f \approx \phi_1, \tag{9}$$

$$\tan\phi_2 = r/f' \approx \phi_2, \tag{10}$$

and hence

$$\phi_2/\phi_1 = f/f' = n. \tag{11}$$

From $\triangle COG$,

$$\sin\phi_1 = l/R \approx \phi_1, \tag{12}$$

and thus,

$$\phi_1 = r/f = l/R. \tag{13}$$

Finally,

$$ln/rR = l(\partial^2 n/\partial r^2)_{r=0}, \tag{14}$$

giving the focal length of the distribution in Eq. (5) as

$$F = -f' = -[l(\partial^2 n/\partial r^2)_{r=0}]^{-1}. \tag{15}$$

We can write the radial distribution of the index of refraction in terms of the radial temperature distribution by repeated use of the chain rule for derivatives and recalling that $(\partial T/\partial r)_{r=0} = 0$, so that Eq. (15) becomes

$$F = -[l(dn/dT)(\partial^2 T/\partial r^2)_{r=0}]^{-1}. \tag{16}$$

Thus we see in Eq. (16) how the strength of the thermal lens depends on the temperature distribution in the sample. Note that both dn/dT and $(\partial^2 T/\partial r^2)_{r=0}$ are negative in the situation being considered here, so that the thermal lens is divergent ($F < 0$).

An estimate of the strength of the thermal lens in the above example can be made by recalling that the total temperature gradient from the center to the edge of the beam, $\Delta T/w$, has been calculated to be 1.1×10^{-2} K cm^{-1}. Since the temperature gradient at beam center is zero by symmetry and the temperature gradient in the wings of the Gaussian profile approaches zero beyond a few times w, then the maximum rate of change in the gradient with radius occurs within approximately $1w$ of beam center. Thus $\partial^2 T/\partial r^2$ near beam center should be approximately $\Delta T/w^2 = \alpha P/2\pi Jkw^2 = 1.1 \times 10^{-1}$ K cm^{-2}, and from Eq. (16), $F = -2.3 \times 10^4$ cm. For the experiment of Gordon *et al.* (1965), where the intracavity laser power is 0.8 W, the focal length of the intracavity thermal lens is estimated to be $F = -2.88 \times 10^2$ cm, which compares favorably with the observed focal length of approximately 1 m.

B. Restrictions Imposed by the Thermal Lens Model

There are several important restrictions implicit in the development of this "thin-lens" approximation, the most crucial being on the path length of the sample cell. The model assumes an index-of-refraction distribution which remains constant along the beam path through the cell, requiring that the laser beam remain reasonably well collimated over this distance. A rigorous discussion of the propagation of Gaussian laser beams is beyond the scope of this chapter. The interested reader is referred to the excellent treatment of the subject of Yariv (1975) or the compact summary by Harris and Dovichi (1980). For the present discussion, the most important result is that the depth of focus, or Rayleigh length, of a laser beam is related to the focal spot size w_0 and the wavelength λ by $b = \pi w_0^2/\lambda$. The parameter b (cm), also called the confocal length, is the distance in which the beam expands from its minimum size to a radius $\sqrt{2}w_0$. Thus it is seen that the analysis of Gordon *et al.* (1965) assumes a sample length which is a fraction of the confocal length of the incident beam. For the present example of a 0.1-cm-radius laser beam at 600 nm, the confocal length is calculated to be $b = 500$ cm. An additional restriction is that the cell length be short compared with the focal length of the thermal lens. Both of these restrictions are easily met in the experiment of Gordon *et al.*

A third restriction is that the absorbed power must be limited to avoid full wave shifts in the phase front of the laser beam which would otherwise lead to interference rings in the transmitted beam, a phenomenon known as spherical aberration. Such a situation is demonstrated in a dramatic fashion in Fig. 2, where (c) shows serious aberration purposely created in a sample by the absorption of approximately 10^{-2} W of laser power. Such strong absorption must be avoided under normal circumstances, either by use of lower laser power, shorter pathlengths, or more dilute samples. The total change in optical pathlength in the sample is $l\Delta n/\lambda = l\Delta T(dn/dT)/\lambda = P\alpha l(dn/dT)/2\pi Jk\lambda$. The maximum allowable phase shift is approximately one-quarter wave, which limits total absorbed

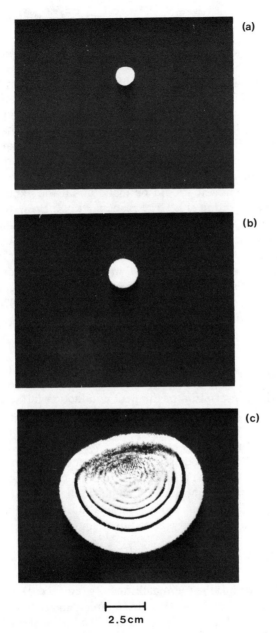

(a)

(b)

(c)

|———————|
2.5 cm

Fig. 2. Time-resolved photographs of the far-field laser beam profile: (a) beam profile produced by a CCl_4 blank, which approximates within 2% the unperturbed, $t = 0$, profile; (b) beam profile produced at $t = 1.5$ msec using a 2.7×10^{-5} M solution of iodine in CCl_4, where $I/I(t) \approx 3$; (c) beam profile at steady state in the solution of (b), where $I/I(\infty) \approx 840$. [Reprinted with permission from Dovichi and Harris (1981a). Copyright 1981, American Chemical Society.]

power to $P\alpha l < \pi J k \lambda / 2(dn/dT)$. For the example of benzene, this gives $P\alpha l < 0.5$ mW.

Finally, it is implicitly assumed that heat travels out of the illuminated region of the sample only by conduction. Any convection which occurs would seriously distort the index of refraction profile and hence the transmitted beam (Vest, 1974; Boyd and Vest, 1975). This restriction has been discussed by Hu and Whinnery (1973), who demonstrate that it is generally a much less serious restriction than that imposed by the spherical aberration limit above.

One additional comment should be made about the experimental arrangement used by Gordon et al. The placement of a Brewster-angle cell inside a laser cavity can produce a significant amount of astigmatism in the laser beam since the liquid sample behaves as a thin lens tilted at Brewster's angle. A lens of focal length f_0 tilted at Brewster's angle will have an effective focal length $f_\perp = 0.23 f_0$ in the plane perpendicular to the axis around which the lens is rotated, and an effective focal length $f_\| = 0.72 f_0$ in the plane containing the rotation axis (Sinclair, 1970). For the thermal lens, this will lead to a stronger effect than expected from the model of Gordon et al.

C. Heat Conduction

The appropriate description of the heat conduction in the illuminated medium is essential to the development of a proper understanding of the thermal lens and other photothermal effects. The description given here follows closely that of Gordon et al. (1965), Whinnery (1974), and Swofford and Morrell (1978). The classic work on heat conduction by Carslaw and Jaeger (1963) derives the propagation function describing the temperature rise in an infinite cylindrically symmetric medium at radius r and time t due to an instantaneous cylindrical heat pulse of strength per unit length Q' applied to the medium at radius r' and time 0 as

$$G(r, r', t) = \frac{Q'}{4\pi Dt} \exp\left(\frac{-(r^2 + r'^2)}{4Dt}\right) I_0 \left(\frac{rr'}{2Dt}\right), \qquad (17a)$$

where $D = k/\rho C_p$, with D the thermal diffusivity (cm^2 sec^{-1}), k the thermal conductivity (cal sec^{-1} cm^{-1} K^{-1}), ρ the sample density (g cm^{-3}), C_p the heat capacity (cal g^{-1} K^{-1}), and I_0 a modified Bessel function. The strength (cm^3 K) of a heat pulse is defined by Carslaw and Jaeger as the temperature to which the amount of heat liberated would raise unit volume of the sample. Thus the quantity of heat instantaneously generated per unit length of the cylinder is $\rho C_p Q'$ (cal cm^{-1}). The problem is formulated by Carslaw and Jaeger in terms of individual line sources parallel to the z (cylinder) axis of the sample, each line of strength per unit length $Qr'\,d\theta$, and the line sources distributed uniformly around the circle. Integration over $d\theta$ gives $Q' = 2\pi r'Q$. For a continuous heat source, the temperature rise at radius r and time t is calculated by integrating Eq. (17a) over all time t', $0 \leqslant t' \leqslant t$, where the integration can be considered as a backward look in time, with the individual heat pulses each propagating from their instant of generation to the present time according to Eq. (17a). It is necessary to insert the *rate* of continuous heat generation per unit length of the cylinder $\dot{Q}$ (cal sec^{-1} cm^{-1}) for the instantaneous heat pulse per unit length given by $\rho C_p Q'$. Thus Eq. (17a) can be rewritten as

$$\dot{G}(r, r', t')dt' = \frac{2\pi r' \dot{Q}/\rho C_p}{4\pi Dt'}\exp\left(\frac{-(r^2 + r'^2)}{4Dt'}\right)I_0\left(\frac{rr'}{2Dt'}\right). \quad (17b)$$

For a distributed heat source $\dot{Q}(r')dr'$ (cal sec^{-1} cm^{-2}), which specifies the rate of heat generation per unit length of the sample, per unit thickness of cylinder between radius r' and $r' + dr'$, the temperature rise at r is calculated by integrating over all r' in the cylindrically symmetric medium. Thus, for a distributed and continuous heat source, the temperature rise is given by

$$T(r, t) = \int_0^t dt' \int_0^\infty dr'\, \dot{G}(r, r', t'). \quad (18)$$

In the limit of low absorptivity, the rate of heat generation between r' and $r' + dr'$ per unit length of the sample as a result of absorption of a Gaussian laser beam is given by Beer's law,

$$\dot{Q}(r') = \frac{2P\alpha}{\pi J w_1^2}\exp\left(\frac{-2r'^2}{w_1^2}\right)dr', \quad (19)$$

where $J = 4.18$ J cal^{-1}, α (cm^{-1}) is the absorptivity of the sample, P (W) is the incident laser power, and w_1 is the "$1/e^2$" beam radius of the laser in the sample cell, i.e., the radius which encompasses $1 - e^{-2} = 0.86$ of the total beam power,

$$\int_0^{w_1} I(r')2\pi r'\, dr' = \int_0^{w_1}\frac{2P}{\pi w_1^2}\exp\left(\frac{-2r'^2}{w_1^2}\right)2\pi r'\, dr' = P(1 - e^{-2}). \quad (20)$$

It is important to note that in Eq. (19) it has been implicitly assumed that the laser beam size w_1 is constant along the cell length. For Gaussian laser beams, this assumption must be treated with some care.

We can now proceed to calculate the temperature rise in the medium, evaluate its second derivative at $r = 0$, and compute the focal length of the resulting lens by the use of Eq. (16). The integral over r' in Eq. (18) is evaluated (Abramowitz and Stegun, 1965) to give

$$\int_0^\infty \dot{G}(r, r', t')\, dr' = \int_0^\infty \frac{2P\alpha}{\pi J w_1^2} \exp\left(\frac{-2r'^2}{w_1^2}\right) \frac{1}{4\pi kt'}$$

$$\times \exp\left(\frac{-(r^2 + r'^2)}{4Dt'}\right) I_0\left(\frac{rr'}{2Dt'}\right) 2\pi r'\, dr'$$

$$= \frac{2\alpha P}{\pi J k} \frac{D}{w_1^2 + 8Dt'} \exp\left(\frac{-2r^2}{w_1^2 + 8Dt'}\right). \qquad (21)$$

It is simpler to evaluate the second derivative with respect to r at $r = 0$ before performing the time integral in Eq. (18). Thus,

$$\left[\frac{\partial^2}{\partial r^2}\left(\int_0^\infty \dot{G}(r, r', t')\, dr'\right)\right]_{r=0} = \frac{-8\alpha PD}{\pi J k(w_1^2 + 8Dt')^2}. \qquad (22)$$

The integral over t' in Eq. (18) is evaluated to give

$$\left(\frac{\partial^2 T}{\partial r^2}\right)_{r=0} = \frac{\alpha P}{\pi J k} \int_0^t \frac{-8D\, dt'}{(w_1^2 + 8Dt')^2}$$

$$= \frac{\alpha P}{\pi J k w_1^2} \int_0^t \frac{-8D\, dt'/w_1^2}{(1 + 8Dt'/w_1^2)^2}$$

$$= \frac{\alpha P}{\pi J k w_1^2} \left[\frac{1}{1 + 8Dt'/w_1^2}\right]_0^t$$

$$= \frac{-\alpha P}{\pi J k w_1^2} \frac{1}{1 + t_c/2t}. \qquad (23)$$

The characteristic time $t_c = w_1^2/4D$ has been shown earlier to be the response time of the medium to the heat input, with an effective "half-life" of $\frac{1}{2}t_c$. At equilibrium, the curvature of the center of temperature profile in the example case $(\partial^2 T/\partial r^2)_{r=0} = -0.23$ K cm^{-2}, which compares favorably with the earlier estimate of -0.11 K cm^{-2}. The focal length F of the resulting thermal lens formed can be calculated by using Eq. (16);

$$F(t) = \frac{\pi J k w_1^2}{\alpha P l(dn/dT)}\left(1 + \frac{t_c}{2t}\right). \qquad (24)$$

The equilibrium focal length for the thermal lens in the example of benzene, with a 10^{-2}-W laser, is $F = -1.14 \times 10^4$ cm. In the experiment of Gordon et al., which uses a 0.8-W laser, the focal length is calculated to be -1.42×10^2 cm. It is seen that the focal lengths calculated here are a factor of two smaller than those given by the simple estimates earlier.

There are several important points to be made regarding the treatment of heat conduction reviewed briefly here. First is the difficulty of the initial assumption of the radially infinite medium. As noted by Gordon *et al.* (1965), in the limit of $t \rightarrow \infty$, the temperature rise $T \rightarrow \infty$ for all finite radii. For realistic times, however, Eq. (24) has been shown to describe the time behavior of the strength of the thermal lens quite well. The steady-state limit of Eq. (24) is derived by Wesfreid *et al.* (1977) without resort to integration of the heat conduction equation [Eq. (18)]. Thus we can use the results derived here with reasonable confidence in their validity.

We see again how the development of the model imposes a restriction on the length of the sample cell since the model assumes a constant beam size in the sample. It is not clear from this development what behavior to expect with longer sample lengths. Finally, it is important to note that the series expansion of the index-of-refraction profile [Eq. (5)] is only valid near the center of the beam. Thus the measurement of the change in beam diameter caused by the thermal lens will lead to erroneously small values for the strength of the thermal lens, since the wings of the Gaussian intensity profile result in the formation of a smaller gradient in the index of refraction (and hence a weaker lens) than predicted for the parabolic approximation of Eq. (5). Part of the difficulty may be avoided by fitting the intensity profile of the transmitted laser beam to a Gaussian and then measuring the "$1/e^2$" radius of the fitted Gaussian. Ideally, however, the problem can be more correctly resolved by measuring the intensity behavior in the center of the laser beam. As will be seen, just such a technique has been developed and is enjoying widespread use.

D. Early Measurements Based on the Thermal Lens

The first use of the thermal lens to measure the absorptivity of liquids was reported by Leite *et al.* (1964) of Bell Laboratories. The workers used the original intracavity sample cell and He–Ne laser setup of Gordon *et al.* (1965) and measured the equilibrium laser beam size at the cavity mirrors. From the beam size was calculated the focal length of the thermal lens and eventually the sample absorptivities, which range from 2.3 × 10^{-4} cm^{-1} for CCl$_4$ to 5.9 × 10^{-4} cm^{-1} for CS$_2$. An additional check on the accuracy of the method was performed (at the suggestion of C. H. Townes) by measuring the absorptivity of sequential dilutions of aqueous copper sulfate. It should be noted that water is a poor solvent for use in photothermal measurements owing to its high thermal conductivity, which is nearly an order of magnitude larger than that of common organic liquids. Thus only a relatively small temperature gradient can be created in water; and since the magnitude of the temperature coefficient of the index of refraction of water is only 10–20% that of common organic liquids, only a very weak thermal lens is formed in water. Leite *et al.*

remarked that for the aqueous copper sulfate, an absorptivity of 2.4×10^{-2} cm^{-1} is too low for accurate measurement by the technique.

Shortly after the work of Gordon *et al.*, other scientists working with the He–Ne laser reported observation of the thermal lens effect. Rieckhoff (1966) described the "self-induced divergence" of a He–Ne laser beam as it passes through a 98-cm liquid sample. This is the first report of the extracavity thermal lens; however, Rieckhoff incorrectly interpreted the results in terms of nonlinear sample absorption, overlooking the possibility of weak linear absorption in "transparent" liquids.

Solimini (1966a) measured the absorptivity of a number of organic liquids at 632.8 nm using the technique of Leite *et al.* (1964), and later reported an extensive study of the accuracy and sensitivity of the intracavity thermal lens for measuring sample absorptivity (Solimini, 1966b). That author was also the first to speculate that the absorptivity in organic liquids might arise from transitions between vibrational energy states in the molecule (Solimini, 1966a).

Callen *et al.* (1967) reported the observation of a pattern of concentric annuli that form with 0.5–7.5 mW of He–Ne laser 10-cm sample of 0.6 cm^{-1} absorptivity. The effect is now recognized as spherical aberration of the thermal lens. Similar observations were reported by Akhmanov *et al.* (1967), Craddock and Jackson (1968), Dabby *et al.* (1970), and Aleshkevich *et al.* (1972). A comprehensive discussion of these aberration effects is given by Akhmanov *et al.* (1972).

Carman and Kelley (1968) studied the time evolution of the thermal lens created in CCl$_4$ solutions of iodine by 0.2 W of the 488-nm line of the argon laser. Those authors recorded the growth of the thermal lens by use of a movie camera and measured the size of the beam image frame by frame. This method has the advantage that it can follow the time behavior of the thermal lens formation even in cases where the equilibrium lens is strongly distorted by convection or aberration.

Other experimental techniques have been devised to monitor thermal lens formation. One of the most interesting is the method of Kohanzadeh and Auston (1970), which measures the shift in the beat frequency between the two lowest axial modes (TEM$_{00}$ and TEM$_{01}$) of a He–Ne laser cavity in which a thermal lens is formed. The laser acts as a "self-heterodyning" device to produce a beat-frequenty rf signal, which in the laser used occurs at approximately 85 MHz. Because the two modes are influenced to differing degrees by the thermal lens, a shift in the beat frequency of a few megahertz is observed. Absorptivity measurements as low as 5×10^{-5} cm^{-1} (for CCl$_4$) are reported. The technique was also used by Banilis *et al.* (1973), who later presented a thorough theoretical development of the method (Burgos *et al.*, 1977). Although the technique is not significantly more sensitive than other techniques, its use of a heterodyne detection scheme suggests that additional improvements may be possible through modifications in the method (see, e.g., Davis, 1979; Davis and Petuchowski, 1981).

III. THE EXTRACAVITY THERMAL LENS TECHNIQUE OF HU AND WHINNERY

The most important advance in the thermal lens technique was introduced by Hu and Whinnery (1973). Those authors recognized that the largest fractional change in the radius of curvature of a laser beam wave front (and hence in the divergence of the beam) can be achieved for a given lens by placing that lens at the point where the radius of curvature of the beam is minimum, i.e., at a distance of one confocal length $b = \pi w_0^2/\lambda$ from the minimum beam waist w_0 of the laser (see, for example, Yariv, 1975). Most importantly, Hu and Whinnery demonstrated that a sensitive detection of thermal lens formation can be achieved for samples placed *outside* the laser cavity by use of an auxiliary lens to form a beam waist at a distance of one confocal length before the sample cell. The change in the laser beam size is monitored in the far field. Hu and Whinnery noted that the *fractional* change of the far-field beam size is independent of the beam width in the sample because an increase in the strength of the thermal lens obtained by reducing the beam width in the cell also increases the initial far-field beam size, and the fractional change of the beam size remains constant (see below).

The experimental setup used by Hu and Whinnery for absorptivity measurements is quite simple (see Fig. 3). A lens of focal length 20–60 cm is used to create a beam waist in the laser. The sample cell is centered one confocal length past the lens focus, where the laser beam size is $w_1 = \sqrt{2}\, w_0$. An aperture and detector located on-axis in the far field sample the laser beam intensity (i.e., the detected power passing through the aperture). The aperture size is chosen so that a relatively small fraction (e.g., 10^{-4}–10^{-6}) of the total beam area is sampled. The beam path to the detector may be folded to allow the entire apparatus to be mounted on a single rigid support base of manageable size. The mirrors also provide for convenient beam positioning to center the beam on the detector aperture. A shutter placed at the input lens focus allows rapid unblocking of the laser beam. The measurement is conveniently performed by observing the transient behavior of the detector signal on an oscilloscope in order to determine the change in the far-field beam size, w_2, as a change in the beam intensity, which is proportional to w_2^2, as shown below.

In the example of benzene above, where an incident laser beam size of $w_1 = 0.1$ cm is used, it is shown how a thermal lens of focal length -1.14

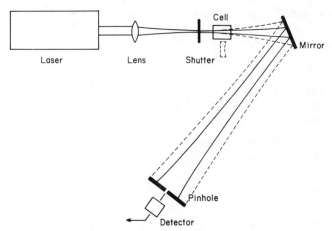

Fig. 3. Single-beam thermal lens experiment. The dotted lines are meant to indicate the increased beam size in the presence of the thermal lens.

$\times 10^4$ cm is formed. In the present situation, for example, a laser beam *waist* of $w_0 = 0.01$ cm may be produced by a 60-cm-focal-length lens. From Eq. (24) one can calculate that the focal length of the thermal lens for $w_1 = 0.014$ cm to be $F = -2.27 \times 10^2$ cm. Note that the confocal

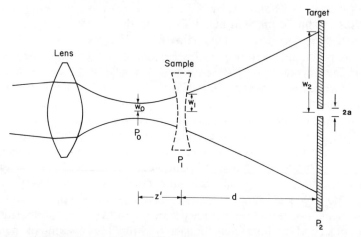

Fig. 4. Optical arrangement for measuring the thermal lens signal. The drawing is not to scale, and in reality $d \gg z'$. There are three important planes perpendicular to the optical axis. The lens focuses the incoming laser beam to a minimum waist w_0 in the plane P_0. The beam expands to a radius w_1 as it propagates a distance z' to the sample cell at P_1. Then the beam expands further to a radius w_2 as it propagates a distance d to the target at P_2. [From Swofford and Morrell (1978).]

distance for the 0.01-cm beam is 5.2 cm, so the paraxial approximation of Gordon *et al.* (1965) is valid for the present case.

A. Detection of the Thermal Lens

The magnitude of the thermal lens signal depends on the laser power hitting the detector located behind the target pinhole (see Fig. 4). It is easy to see that the detected laser beam power is inversely proportional to the beam area, defined as πw_2^2 where w_2 is the laser beam "$1/e^2$" spot size at the target. The Gaussian intensity profile of the laser beam at the target is

$$I(r) = \frac{2P}{\pi w_2^2} \exp\left(\frac{-2r^2}{w_2^2}\right). \tag{25}$$

If the target pinhole has radius a ($a \ll w_2$), then the detected laser beam power is

$$P_{det} = \int_0^a I(r)2\pi r \, dr \approx 2P \frac{\pi a^2}{\pi w_2^2}. \tag{26}$$

The thermal lens signal can be conveniently expressed as the fractional change in the detected laser beam power relative to its equilibrium value in the presence of the thermal lens as

$$\frac{P_{det}(t = 0) - P_{det}(t = \infty)}{P_{det}(t = \infty)} = \frac{w_2^2(t = \infty)}{w_2^2(t = 0)} - 1. \tag{27}$$

B. Mathematics of the Hu and Whinnery Method

The mathematical description presented by Hu and Whinnery of the behavior of the thermal lens begins with the equation for the time-dependent focal length of the thermal lens derived by Gordon *et al.* (1965) and given previously by us as Eq. (24). Hu and Whinnery demonstrated analytically by straightforward, tedious calculation that the increase in the far-field beam size of the laser is maximized when the thermal lens is located at the confocal distance beyond the focus of the input lens. Under that condition, the fractional increase in beam size as a function of time is

$$\frac{\Delta w_2(t)}{w_2(0)} = \frac{w_2(t) - w_2(0)}{w_2(0)} = \frac{-P\alpha l(dn/dT)}{2\lambda Jk(1 + t_c/2t)}. \tag{28}$$

Note that in Eq. (28), the *explicit* dependence of the ratio on the incident beam size w_1 has canceled; however, the time behavior of Eq. (28) still depends on the incident beam size since $t_c = w_1^2/4D$. The far-field laser beam intensity depends inversely on the far-field beam area, which is proportional to the beam radius squared. We first rewrite Eq. (28) as

$$\frac{w_2(t)}{w_2(0)} = 1 - \frac{P\alpha l(dn/dT)}{2\lambda Jk(1 + t_c/2t)} .$$

(29)

Since the relative expansion of the beam by the thermal lens is small, the second term on the right-hand side of Eq. (29) is small, so that we can write

$$\frac{P_{det}(t = 0)}{P_{det}(t)} = \frac{w_2^2(t)}{w_2^2(0)} \approx 1 - \frac{P\alpha l(dn/dT)}{\lambda Jk(1 + t_c/2t)} .$$

(30)

In the limit as $t \to \infty$,

$$\frac{P_{det}(t = 0)}{P_{det}(t = \infty)} = 1 - \frac{P\alpha l(dn/dT)}{\lambda Jk}$$

(31)

and finally

$$\alpha = \frac{-[P_{det}(t = 0) - P_{det}(t = \infty)]}{P_{det}(t = \infty)} \frac{\lambda Jk}{Pl(dn/dT)} .$$

(32)

Thus, from a measurement of the relative change in the detected power of the transmitted laser beam at equilibrium, one can calculate the sample absorptivity. In the continuing example of benzene above, the change in the detected beam power relative to its equilibrium value produced by the thermal lens in a 1-cm sample cell by a 10^{-2}-W laser beam at 600 nm is calculated to be 0.045.

The derivation by Hu and Whinnery of the behavior of the laser beam propagating under the influence of the thermal lens relies heavily on a proper mathematical description of the laser beam properties. A remarkably convenient technique for calculating the propagation of Gaussian laser beams has been developed (see, for example, Yariv, 1975). The method is introduced in the next section and will subsequently be used to provide a convenient derivation of Hu and Whinnery's result shown in Eq. (32).

C. Propagation of Gaussian Laser Beams—The *ABCD* law

The fundamental Gaussian beam solution $q(z)$ is a complex function which describes the radius of curvature $R(z)$ at any point z along the direction of the laser propagation, once the minimum spot size w_0 and its location (i.e., the plane $z = 0$) are specified. The function $q(z)$ which thus defines the beam is given by

$$\frac{1}{q(z)} = \frac{1}{R(z)} - i\,\frac{\lambda}{\pi w^2(z)} \tag{33}$$

where λ is the laser wavelength. The propagation of the laser beam from its minimum beam waist at $z = 0$ to any point is conveniently analyzed with the $ABCD$ law for Gaussian beams. At any point along the beam path, the Gaussian beam is described by

$$q = (Aq_0 + B)/(Cq_0 + D) \tag{34}$$

where $\begin{bmatrix} A & B \\ C & D \end{bmatrix}$ is the matrix of transformation which describes the propagation of the laser through the optical elements in the beam. Tabulations of $ABCD$ matrices for common optical elements are given by several authors (Yariv, 1975; Kogelnik and Li, 1966). With the laws of matrix multiplication it can be demonstrated that the resultant matrix for the propagation of the beam through a series of optical elements is simply the ordered product of the $ABCD$ matrices for the individual optical elements.

The Gaussian beam solution can be used to calculate the propagation of the laser beam through the thermal lens with focal length given by Eq. (16). Such a method has been used successfully by Twarowski and Kliger (1977a) and Swofford and Morrell (1978). The transformation matrix for the thermal lens is

$$\begin{bmatrix} 1 & 0 \\ -1/F(t) & 1 \end{bmatrix} \tag{35}$$

where $F(t)$ is the (time-dependent) focal length given by Eq. (16). It remains now to calculate the resultant beam size at the detector. The use of the Gaussian $ABCD$ law makes this a simple task. The most convenient starting point for the calculation of the beam's propagation is at the location of the minimum waist formed by the input lens. At the minimum beam waist, the radius of curvature of the wave front is infinite, and the Gaussian beam solution [Eq. (33)] is $q = i\pi w_0^2/\lambda = ib$. The $ABCD$ matrix for propagation through distance d of material with index n is

$$\begin{bmatrix} 1 & d/n \\ 0 & 1 \end{bmatrix} \tag{36}$$

and the propagation through an interface from index n_1 into index n_2 is

$$\begin{bmatrix} 1 & 0 \\ 0 & n_1/n_2 \end{bmatrix}. \tag{37}$$

At any point along the beam path, the beam radius càn be calculated as

$$w = \{-\lambda[\pi \, \mathrm{Im}(1/q)]^{-1}\}^{1/2}, \qquad (38)$$

where $\mathrm{Im}(1/q)$ is the imaginary part of $1/q$.

Figure 4 shows the propagation distances for the calculation of the laser beam size at the detector. At the plane of the target the $ABCD$ matrix is

$$\begin{bmatrix} A & B \\ C & D \end{bmatrix} = \begin{bmatrix} 1 - d/F(t) & z' + d[1 - z'/F(t)] \\ -1/F(t) & 1 - z'/F(t) \end{bmatrix}, \qquad (39)$$

which represents the three-step propagation of the laser from $z = 0$ to $z = z'$, through the thermal lens of focal length $F(t)$, and finally from $z = z'$ to $z = z' + d$. Noting that the inverted Eq. (34) can be directly compared to Eq. (33), we substitute Eq. (39) into Eq. (34) and rationalize the imaginary expression of Eq. (34) using $(x + iy)^{-1} = (x - iy)(x^2 + y^2)^{-1}$. Comparing the imaginary part of the resulting expression with that of Eq. (33), we can write

$$-\frac{\lambda}{\pi w_2^2(t)} = \frac{-b}{b^2[1 - d/F(t)]^2 + [z' + d(1 - z'/F(t))]^2}; \qquad (40)$$

Eq. (40) can be rearranged to give w_2^2, the square of the radius of the beam at the detector,

$$w_2^2(t) = (\lambda/\pi b)[(b^2 + z'^2)(1 - d/F(t))^2 + 2dz' + d^2(1 - 2z'/F(t))]. \qquad (41)$$

At the beginning of the experiment, no thermal lens has formed ($F = \infty$), and the square of the beam size at the target is given by

$$w_2^2(0) = (\lambda/\pi b)(b^2 + z'^2 + 2dz' + d^2). \qquad (42)$$

Following Eq. (27), we can write

$$\frac{P_{\mathrm{det}}(t = 0) - P_{\mathrm{det}}(t)}{P_{\mathrm{det}}(t)} = \frac{w_2^2(t)}{w_2^2(t = 0)} - 1$$

$$= \frac{(b^2 + z'^2)[1 - d/F(t)]^2 + 2dz' + d^2[1 - 2z'/F(t)]}{b^2 + z'^2 + 2dz' + d^2} - 1. \qquad (43)$$

Equation (43) is derived by Twarowski and Kliger (1977a). The experimental setup assures that $d \gg z' \approx b$, so that the denominator reduces to d^2, and Eq. (43) can be written

$$\frac{P_{\mathrm{det}}(t = 0) - P_{\mathrm{det}}(t)}{P_{\mathrm{det}}(t)} = \frac{b^2 + z'^2}{F^2(t)} - \frac{2(b^2 + z'^2)}{dF(t)} - \frac{2z'}{F(t)}. \qquad (44)$$

The intermediate result, Eq. (44), is displayed to indicate that no assumption need be made concerning the relative absolute magnitudes of d and

$F(t)$; indeed, as has been shown in the example above, a lens of focal length -2.27×10^2 cm is produced under quite normal conditions so that $d \approx |F(t)|$ in many cases. This is in contrast to the restriction $d < |F(t)|$ implied by Twarowski and Kliger (1977a). Finally, Eq. (44) is dominated by the third term and reduces to

$$\frac{P_{\text{det}}(t = 0) - P_{\text{det}}(t)}{P_{\text{det}}(t)} = \frac{-2z'}{F(t)}. \tag{45}$$

In the example above, where $F(t = \infty) = -2.27 \times 10^2$ cm and $z' = b = 5.2$ cm, the fractional change in detected power relative to the equilibrium value is 0.0458, in agreement with the calculation following Eq. (32). Since Hu and Whinnery have shown that $z' = b$ maximizes the thermal lens signal, and using Eq. (24), we can write Eq. (45)

$$\begin{aligned}
\frac{P_{\text{det}}(t = 0) - P_{\text{det}}(t)}{P_{\text{det}}(t)} &= \frac{-2\alpha Pl(dn/dT)}{\pi Jkw_1^2(1 + t_c/2t)} \frac{\pi w_0^2}{\lambda} \\
&= \frac{-\alpha Pl(dn/dT)}{Jk\lambda(1 + t_c/2t)},
\end{aligned} \tag{46}$$

where we have used the fact that at $z' = b$, $w_1 = \sqrt{2}w_0$. Comparison of Eqs. (46) and (30) confirms the Hu and Whinnery derivation.

Equation (41) was used by Twarowski and Kliger (1977a) and by Swofford and Morrell (1978) to calculate the time-dependent thermal lens signal for comparison with experiment and for exploring the dependence of the signal on the various experimental parameters. Figure 5 shows the time-dependent signal calculated for the example of benzene above. In Fig. 5a is plotted $P_{\text{det}}(t)/P_{\text{det}}(0)$, the detected power normalized to the power detected in the absence of the thermal lens. This is the signal as would be seen in an actual measurement. Figure 5b replots this calculated signal according to Eq. (46).

D. Limitations of the Hu and Whinnery Technique

The method of Hu and Whinnery is being used extensively for the measurement of weak absorptivity in liquids. As a result, Eq. (32) has become generally accepted as a reliable prescription for calculating weak absorptivities. It is important, therefore, for the user to recognize the limitations of Eq. (32) which arise from the simplifying assumptions made in its derivation. The by-now familiar restriction on sample cell length applies to the present experiment, with the limitation that the sample cell length be small compared with the confocal length of the *focused* laser beam. The method is valid only for the measurement of small relative intensity changes by virtue of the simplifying assumption made on going

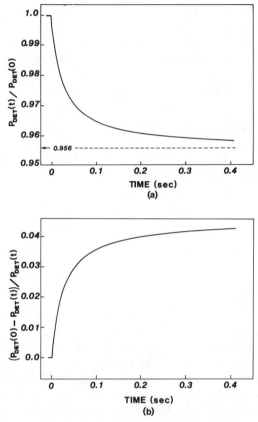

Fig. 5. Example of the time-dependent thermal lens signal calculated for the example of benzene (see text). (a) Detected power normalized to the power detected in the absence of the thermal lens, $P_{det}(t)/P_{det}(0)$. (b) Calculated signal in (a) replotted according to Eq. (46).

from Eq. (29) to (30) in Hu and Whinnery's derivation or from Eq. (44) to (45) in the *ABCD*-law treatment. Thus while a relative intensity change of 10% would lead to an error of 0.25% in the calculated α, a relative intensity change of 25% would lead to an error of 1.5% in α. And finally, the necessity of measuring $P_{det}(t = \infty)$ requires an intrinsic delay in the experiment, since the time scale for the approach to equilibrium for the thermal lens depends on the value of the critical time t_c. The delay can be minimized by using a relatively short-focal-length lens, being mindful of the cell length limitation imposed by the confocal length of the focused laser beam. In a typical situation, a 60-cm-focal-length lens might produce a beam spot size w_0 of approximately 0.01 cm, which gives a confocal distance of $b = 5.2$ cm. For benzene, this results in a critical time in the sample of $t_c = w_i^2/4D = 2w_0^2/4D = 5.3 \times 10^{-2}$ sec. It is seen from Eq. (46) that 80% of the signal develops in a time $t = 2t_c$, while a time $t = 9t_c$ is required to observe 95% of the total signal. As will be seen in a later section, the limitation on sample cell length can be removed by accounting directly for the changing laser beam size in the sample, thus providing the possibility of considerable signal enhancement at the expense of more complicated calculational techniques.

E. Wavelength Dependence of the Signal Strength

In anticipation of the spectroscopic application of the Hu and Whinnery technique, it should be noted that the minimum spot size w_0 produced by the input focusing lens in Fig. 3 depends on the wavelength λ of the laser beam owing to the variation in the beam parameters of a laser resonator with wavelength. The beam parameters of the laser resonator are discussed in detail by Yariv (1975), and the interested reader is referred to that treatment. Note that the discussion by Yariv and the summary below are based on a laser cavity without a gain medium which, if present, would complicate the treatment of the problem beyond the scope of the present discussion. Briefly, the fundamental parameters of a laser cavity are the confocal length b_c, which is determined by the fixed geometrical characteristics of the cavity, and the output beam size w_c, which varies according to the relationship $b_c = \pi w_c^2/\lambda$. Thus it is seen that w_c scales as $\lambda^{1/2}$. Furthermore, it can be shown that the focal spot size w_0 of a lens of focal length f placed a distance z from a laser of wavelength λ and cavity beam size w_c is given by

$$w_0^2 = w_c^2(1 - z/f)^{-2} + f^2(\lambda/\pi w_c)^2 = (\lambda f^2/\pi b_c)[1 + b_c^2/(z - f)^2].$$

It is seen that the size of the focal spot w_0 scales as $\lambda^{1/2}$ but that the confocal length of the focused beam remains constant with changing laser wavelength. It can also be shown that the position of the beam focus z_0 is given by $z_0 = f + (z - f)f^2/[(z - f)^2 + b^2]$, so that the position of the focused spot does not change with λ. We therefore conclude upon inspection of the first expression of Eq. (46) that, while the confocal distance is constant for varying λ, there is a dependence of the thermal lens signal since w_i^2 scales as λ. Thus absorption spectra measured by the Hu and Whinnery technique [Eq. (32)] are correctly scaled by λ as shown, since at constant $\alpha(\lambda)$, the magnitude of the observed signal increases with decreasing λ. In addition, the time dependence of Eq. (46), which is determined by $t_c = w_i^2/4D$, also includes a wavelength dependence since w_i^2 scales as λ. Thus careful comparison of time-resolved experimental measurements with Eq. (46) must account for the "hidden" λ dependence of t_c.

In contrast to the cw experiments described here, for the case of pulsed laser absorption, Twarowski and Kliger (1977b) showed how the focusing parameters of the laser beam introduce additional wavelength dependence which must be accounted for. We will return to this point in the later discussion of the pulsed thermal lens.

F. Refinements to the Experimental Technique

The most extensive application and refinement of the thermal lens technique of Hu and Whinnery has been the very careful work of Dovichi and Harris (1979, 1980, 1981a,b; Harris and Dovichi, 1980). The reports by those authors contain valuable insights into the careful procedures required for reliable, reproducible measurement of thermal lens signals.

The goal of their work is the application of the thermal lens technique for sensitive trace analysis by use of readily available discrete-wavelength cw lasers (e.g., He–Ne, Ar ion, He–Cd) and the appropriate chromogenic reagents selective for the trace species of interest. The authors recognized that the fractional change in signal intensity as described by Eq. (31) depends on the absolute magnitude of the ratio $(dn/dT)/k$ of the solvent. Thus the use of a solvent such as carbon tetrachloride, benzene, or acetone is preferred. In contrast, the magnitude of the thermal lens signal in water is a factor of 45 smaller than in carbon tetrachloride under otherwise identical conditions.

Dovichi and Harris (1979) described in detail their experimental setup, which uses a 4-mW He–Ne laser, a 13.6-cm-focal-length biconvex lens, and a 1-cm-pathlength sample cell. (Note that such a strong input lens produces a small focused spot size. For example, using the beam parameters of a commercial laser, we calculate a beam size w_0 of 3.5×10^{-3} cm and a confocal length b of 0.63 cm.) The 9-m optical path between the cell and detector should produce an initial beam size at the detector of approximately 5 cm. A 0.01-cm aperture is used to sample the intensity in the center of the beam, and the power passing through the aperture is detected by a vacuum photodiode (RCA 929). Since considerable uncertainty exists in the location of the laser beam focus created by the input lens, some means of careful translation of the sample cell along the beam direction is required in order to locate the position of maximum thermal lens signal. A reference solution of methylene blue in 2:1 methanol:water is used as an aid in locating this optimum cell position.

Additional precautions are taken to avoid the difficulties often inherent in the detection of coherent optical beams. All plane parallel faces in the laser beam path, particularly the cell windows, are tilted slightly to avoid the creation of interference fringes in the beam incident on the detector aperture. Otherwise, small mechanical vibrations cause large fluctuations in the detector signal. Similar difficulties might arise from the unavoidable surface scatter from optical components, which results in random interference spots or "speckle" in the beam profile at the detector. It is advisable to use a diffuser behind the aperture so that the detector views an evenly illuminated image of the aperture, free of speckle and free of interference rings which may be created by the edges of the aperture itself. Loss of detector signal is of little concern, since the measurement is definitely source-noise limited. In fact, conventional phototubes are found to be quite satisfactory for this application if care is taken to provide even illumination of the photocathode.

Beam perturbation may also be caused by concentration inhomogeneities ("stir lines") or by suspended dust or air bubbles being moved about in the sample liquid. Samples should be filtered (we use 0.2-μ polytetrafluoroethylene membrane filters from Millipore Corporation) and allowed to equilibrate in the sample cell for a few minutes before the experiment begins. Thermally induced air currents in the beam path should also be minimized. Of particular concern is the beam-blocking shutter, whose blades are often black. We use small electromechanical shutters ("Uniblitz" by Vincent Associates, Rochester, NY), available with either black or reflective blades. The shutters can be actuated to unblock a 0.1-cm laser beam in approximately 200 μsec. The input lens, exposed continuously to the cw laser, should not produce significant thermal perturbations. Cell window absorption, however, may be a troublesome source of heating, since the generated heat can be conducted into the liquid, where it creates a thermal lens. One way to minimize this difficulty is to avoid thin sample cells, in which the heat from the windows would diffuse a significant fraction of the way into the sample liquid. Optical glass quality varies dramatically among vendors, but in general, we find fused-silica cells to be less absorptive than borosilicate glass. Some lasers may emit additional weak lines which can be troublesome. This would be particularly important for the 632.8-nm He–Ne laser, since any 3.39-μ emission present would be strongly absorbed in water and perhaps in some hydrocarbons.

G. Data Acquisition in Thermal Lens Experiments

Dovichi and Harris (1979) used sample-and-hold amplifiers and a digitizer to record the initial intensity of the laser, $P_{det}(t = 0)$, and after an appropriate delay the steady-state intensity, $P_{det}(t = \infty)$. The source intensity was also recorded in order to correct for any long-term drifts in laser performance. In later work, Dovichi and Harris (1981a) recognized the increase in speed of measurement which can be a~' 'eved by recording the time-dependent buildup of the signal rather than just the initial and final amplitudes. Recording the time dependence of the signal avoids the need to wait until the equilibrium thermal lens is established. An additional benefit to be gained by repeated measurements is the averaging out of short-term source fluctuations. And finally, the short-term kinetic response of even a relatively strongly absorbing solution is well behaved, allowing measurement of the thermal lens formation in situations where the equilibrium lens is badly distorted or aberrated (see Fig. 2). An absorbance detection limit for a 1-cm pathlength in CCl_4 using a 0.16-W Ar^+ laser is reported to be $A = 7 \times 10^{-8}$ or $\alpha = 1.6 \times 10^{-7}$ cm^{-1} (Dovichi and Harris, 1981a).

Another clever innovation in the thermal lens technique was introduced by Dovichi and Harris (1980). Those authors recognized the significance of the antisymmetric dependence of the thermal lens effect on the position of the thermal lens relative to the laser beam waist, a feature reported without comment by Hu and Whinnery (1973). The antisymmetric behavior of the lens is easy to understand. If the thermal lens is placed *before* the laser beam waist, the diverging thermal lens acts to move the position of the beam waist further along the beam path. The laser has less distance in which to diverge, and the beam intensity at the detector actually *increases* as the thermal lens forms. The effect of the thermal lens changes from a focusing to a defocusing one as the sample is moved through and beyond the focus of the input lens. What Dovichi and Harris were the first to recognize is that the antisymmetric behavior of the thermal lens could provide a *differential* response for canceling the background absorbance of the sample matrix or the solvent. One cell, filled with solvent, is placed one confocal length before the beam focus, while the sample cell is placed one confocal length after the beam focus. A cancellation of nearly 99% of the "common-mode" thermal lens signal is demonstrated when the two cells are filled with identical weak absorbers ($A = 1.0 \times 10^{-4}$). The minimum detectable absorbance in CCl_4 using a 0.2-W Ar^+ laser is reported as $A = 6.3 \times 10^{-7}$ or $\alpha = 1.45 \times 10^{-6}$ cm^{-1}. It is to be emphasized that this is a *differential* detection limit, which may be achieved even in the presence of "common-mode" absorption several orders of magnitude stronger. But as Dovichi and Harris (1980) pointed out, the technique is restricted by the requirement that the thermal lenses be sufficiently weak for the divergence of the laser beam to respond linearly to the absorption in each cell.

H. Quantitative Standards for Thermal Lens Measurements

Harris and Dovichi (1980) raised an important issue in their discussion of the quantitative behavior of the thermal lens effect. As those authors demonstrated, the general features of the thermal lens behavior can be characterized by simple equations [e.g., Eqs. (24), (45), and (46)]; and those equations can be extremely valuable in the design of instrumentation and the prediction of feasibility. But one should not presume that experimental and molecular parameters are known with sufficient accuracy to allow *quantitative* prediction of instrument performance. The particular measurement setup to be used *must* be standardized by use of samples of known absorption which can be conveniently prepared by careful dilution of a more concentrated solution whose absorption has been determined by conventional spectrophotometric techniques. The thermal lens technique thus becomes a *calibrated* absorption technique, but it is not an *absolute* technique, as has been incorrectly inferred by other workers.

Several appropriate absorption standards are available for use in calibrating the thermal lens technique. The primary requirements are that the standard be photochemically stable over the power range and spectral region of the lasers used; that its absorption spectrum in solution be broad and featureless, particularly if continuously tunable lasers are used; and that its quantum efficiency for radiationless dissipation of optical excitation be nearly 1.0 or at least be a known value. In a similar fashion, the solvent must be available in very pure form and be relatively free of absorption features. This latter requirement almost completely rules out solvents which contain hydrogen, since harmonics of hydrogen-stretching vibrations which occur in the visible spectrum can have absorptivities as high as 10^{-3} cm^{-1} (Fang and Swofford, 1980). One of the solvents best suited for use in thermal lens measurements is CCl_4. A number of studies report the use of I_2 as a standard solute, since I_2 has spectral features throughout the blue–green region which overlap with the powerful emission lines of the argon laser. Fresh solutions must be prepared for calibration measurements, since the solution is not photochemically stable. In the red spectral region, several metal phthalocyanine dyes are available which can serve as suitable standards for thermal lens measurements.

IV. THE DUAL-BEAM THERMAL LENS TECHNIQUE

The experiments described so far have used a single laser source to provide both the sample excitation and the means for probing the heat produced by the absorption process. In many cases, however, the use of separate laser sources for the "pump" and "probe" beams can provide significant benefits in performance. The small intensity modulation that can be repetitively imposed on a well-behaved continuous laser beam by the creation of a thermal lens can be sensitively monitored by modern signal averaging devices (i.e., lock-in amplifiers, boxcar averagers, and transient recorders). In addition, detection optics and detectors can be optimized for a single, convenient probe laser wavelength rather than require detection in what might be a more difficult spectral region.

The first reports of dual-beam thermal lens measurements were the work of Grabiner et al. (1972) and Flynn (1974), who used a He–Ne probe laser to measure the time-resolved formation of the thermal lens as a monitor of energy transfer in gas-phase molecules pumped by a pulsed CO_2 laser. The first application of the dual-beam thermal lens technique to spectroscopic absorption measurements was the work of Long et al. (1976), who used a repetitively chopped tunable cw dye laser to provide the pump beam and a He–Ne laser to provide the probe.

The optical arrangement used by Long et al. is shown in Fig. 6. The vertically polarized heating beam is chopped at 13 Hz and is combined collinearly at the surface of an uncoated beam splitter with the cw probe laser. The two beams are focused by a lens of unspecified focal length (laboratory records indicate a 60-cm focal length lens is used), pass through the 1-cm sample cell, and are incident on the detector aperture.

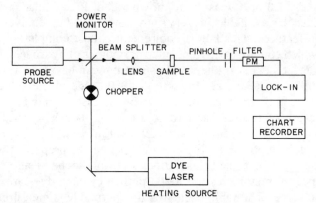

Fig. 6. Dual-beam thermal lensing spectrometer. The heating beam from the dye laser is chopped and is incident on the beam splitter, where a fraction of the power is directed toward the sample, with the remainder going to the power monitor. The beam is focused by a lens before entering the sample in which a "thermal lens" is formed. The probe laser beam passes collinearly with the heating beam through the sample. The dye laser is blocked from the detector by a filter which passes the probe beam. The probe laser senses the fluctuating thermal lens, and the intensity at the center of the probe beam is detected by the pinhole, photomultiplier (PM), and lock-in amplifier. [From Swofford and Morrell (1978).]

Suitable filters block the heating beam so that only the probe laser is incident on the photomultiplier. The detected probe power, modulated at the chopping frequency by the fluctuating strength of the thermal lens, is monitored by a frequency-tracking lock-in amplifier synchronized to the chopper frequency. The magnitude of the lock-in output is recorded on a strip-chart recorder, along with the output of a detector monitoring the fraction of the dye-laser power which is transmitted by the beam splitter, as the dye laser is scanned in wavelength by a clock motor attached to the laser tuning mechansim. Point-by-point division of the lock-in signal by the average dye-laser power gives a "corrected" absorption spectrum of the sample. That the spectrum of liquid benzene so obtained agrees quite well with a spectrum obtained point by point using the single-beam method of Hu and Whinnery is taken as validation of the simple, first-order power correction. No attempt is made to explore the parametric dependence of the signal strength nor to optimize the optical arrangement.

Several features of the work of Long *et al.* are worthy of note. The most important point concerns the choice of the 13-Hz modulation frequency, which is selected on the basis of the practical low-frequency limit of both the lock-in amplifier and the rotating-blade chopper. The critical time t_c for thermal lens formation under the experimental conditions is estimated to be 50 msec, although insufficient information is given in the report to allow the calculation. The chopper is open for less than $1t_c$, during which time the growing thermal lens causes a rapid decrease in the detected probe laser power to approximately 60% of the equilibrium change in P_{det}. When the chopper closes, the lens "relaxes," and the signal magnitude approaches the initial value. Deviations in the chopper speed will lead to significant fluctuations in the signal amplitude, since if the chopper were open for $2t_c$, the change in P_{det} would be 80% of its equilibrium change. Therefore, careful attention must be paid to stable chopper performance. An alternative solution would be to use a tighter beam focus in order to reduce t_c while maintaining the confocal length longer than the sample cell. On the other hand, while slower chopping rates may lead to larger signals, signal-to-noise performance may be seriously affected by the probe-source "$1/f$" or "flicker" noise. Although this problem can be overcome by narrow-bandwidth lock-in operation, the consequently long amplifier settling time constant increases the time required for the measurement. Thus careful consideration must be given to the timing trade-offs affecting performance.

The description of the optical arrangement for the experiment of Long *et al.* is not given in sufficient detail to allow quantitative evaluation of the behavior of the thermal lens; however, it is possible that additional optimization of the setup can be achieved. Of particular interest is the sensitivity of the signal magnitude to the location of the probe beam waist with respect to the pump beam waist. Additional considerations include the relative sizes of the two beams and the choice of pump and probe laser wavelengths. One final comment pertains to the use of a lock-in amplifier. Since the device measures the rms fundamental of the Fourier-analyzed signal, any details of dynamical signal behavior are lost, and comparison of experiment with theory becomes more difficult. We will return to this point during the discussion of the mathematical formulation of the dual-beam technique in a later section.

A.   Detection Alternatives for the Dual-Beam Technique

The convenience of lock-in amplifier detection of the dual-beam thermal lens signal has led to its widespread acceptance. Nevertheless, the search for alternative detection techniques has continued. The most direct technique is the digital recording of the buildup of the thermal lens signal. The experiment is repeated and the individual recordings averaged to improve the signal-to-noise characteristics, in a fashion similar to that of the previously described work of Dovichi and Harris (1981a). Digital processor oscilloscopes are available which make this an extremely convenient task. In addition, since the incident probe laser beam intensity is quite stable, a differential input amplifier plug-in with a dc voltage comparator as one input can be used to provide the full dynamic range of the digitizer across the transient change in probe beam intensity which occurs with thermal lens formation. The full temporal behavior of the thermal lens can be fit to the time-dependent equation, and the sample absorptivity obtained from the fitting parameters.

Another approach to data recording was described by Miyaishi *et al.* (1981). Those authors recognized the difficulty inherent in the analog lock-in technique of Long *et al.*, where the analog time constant circuitry performs an exponentially weighted average of the fluctuating probe-beam intensity. Thus noise fluctuations which have occurred in the recent past are weighted more heavily than fluctuations which occurred much earlier. Miyaishi *et al.* demonstrated a "digital lock-in" technique which uses a voltage-to-frequency (V/F) converter at the detector. The V/F converter output is switched between two counters by a circuit monitoring the chopper cycle. An appropriate delay is provided in the switching circuit so that one counter is active primarily during the time period when the pump beam is blocked, and hence that counter measures the P_{det} ($t = 0$) signal. The other counter is active primarily during the period when the pump beam is unblocked, and hence that counter measures the

P_{det} ($t = \infty$) signal. The delay is adjusted for maximum difference between the counters. Such a system can thus provide "lock-in" operation with arbitrarily long time constants and no performance degradation.

B. Optical Considerations for the Dual-Beam Experiment

In addition to the general optical improvements which have already been discussed for the thermal lens technique, there are several considerations introduced by the optical arrangement for the dual-beam method. One of the most important requirements is the stability of the probe source. Fortunately, many of the relatively inexpensive low-power cw lasers have low rms noise specifications. Very elegant lasers are available with light-regulated feedback control loops, but these are probably not necessary in the thermal lens application. Equally important with intensity stability is the pointing stability requirement of the probe laser and the requirement that relative motion of the two beams be minimized.

In such sensitive measurements other, often overlooked, factors can cause difficulties. Absorptive filters for blocking the pump laser light at the detector (see Fig. 6) can generate large spurious signals as they are heated by the beam. It is known that polished glass filters can act as low-finesse etalons, creating moving fringe patterns when heated by a powerful laser. The problem is much less severe when the filters are placed *behind* the aperture, as in Fig. 6. There are thin-film interference filters (often called "notch filters") available for the individual lines of cw probe lasers, and their use is highly recommended.

The optical arrangement in Fig. 6, which uses a single input lens for both the pump and the probe laser beams, may introduce optical distortion in the probe beam, particularly if a relatively high-power pump laser is used. In such cases, separate lenses for the two beams can be used without too much difficulty. Multicomponent lenses which have elements cemented together must be tested carefully for optical quality under experimental conditions.

Other optical arrangements for the dual-beam thermal lens technique have also been suggested. For example, Moacanin *et al.* (1979) showed the use of pump and probe beams counterpropagating through the sample. Such an arrangement requires extreme caution to avoid having the pump laser beam enter the probe laser cavity, where serious mode-coupling might influence the probe-beam stability. The situation can be avoided by use of a dispersive element in front of the probe laser. In particularly troublesome cases, a dispersive element can also be introduced between the sample and the detector aperture to aid in attenuation of the pump beam at the detector.

Finally, the control of the spatial mode quality of the pump laser is of utmost importance. It has been found in this laboratory that with the use of a cw dye laser for thermal lens measurements, an intracavity aperture near the output beam waist significantly improves the mode quality of the Gaussian beam. Without such an aperture, higher-oder modes appear as the laser is tuned across the gain curve. It is important to note that while these higher-order cavity modes may increase the total output power of the laser, they do not enhance thermal lens formation since many of the modes have no significant power in the center of the Guassian beam profile. Thus a correction of the observed thermal lens signal for total pump beam power may actually *degrade* the signal-to-noise ratio of the measurement. In severe cases, a beam power monitor with a spatial filter may be required for proper power correction of the observed thermal lens signal. Because the experiment measures the thermal lens formed by the *center* of the pump laser, spatial filtering of the pump laser beam might not be required.

C. Multielement Detection of the Thermal Lens

Many of the optical interference and distortion effects which cause problems in the point intensity measurement technique of Hu and Whinnery (1973) might be less troublesome in schemes which monitor the intensity profile of the entire probe beam. Early reports of the single-beam thermal lens technique (Leite *et al.,* 1964; Carman and Kelley, 1968) used just such a scheme, the former using a slowly scanning pinhole and detector to map out the equilibrium beam intensity profile, and the latter using a movie camera to record the time development of the beam image, which is then measured frame by frame from the film. Whinnery (1974) noted that in the former work, a more rapid measurement might be achieved with a fixed aperture and detector with a rotating mirror to sweep the beam across the detector. At the present time, both one-dimensional array detectors (e.g., the Reticon® linear diode array) and two-dimensional array detectors (e.g., Princeton Applied Research OMA® vidicon) are available for measuring the intensity profile of the beam. Such data can be fit to previously measured beam intensity profiles or directly fit to the expected Gaussian profiles, thus eliminating interference or aberration effects which could seriously affect the single-point intensity measurement. In addition, such methods might be less sensitive to beam-pointing stability, since the intensity profile fitting algorithm could include beam center location. (Of course, the beam-pointing problem still exists in the perpendicular dimension for the linear-array case. The use of a cylindrical lens to image the beam profile onto the detector can alleviate that difficulty, but only at the expense of additional computational requirements.)

It should be recognized that even the relatively simple one-dimensional image acquisition and processing leads to a considerable increase in computing requirements for the experiment. To our knowledge, there have been no recent reports of successful application of these beam-scanning techniques to the thermal lens measurement. The additional complexity of the method does not appear to be warranted, particularly since most of the reported high-sensitivity measurements are limited by absorption background or matrix effects and not by beam-quality problems.

One final comment concerning beam-profile measurements is in order. As has been noted earlier, the development of the model of thermal lens behavior has considered only the effect of the beam intensity *curvature* at its center. Thus profile fitting across the entire Gaussian beam is bound to be in error. Any attempts to fit the laser beam profiles should be restricted to the center portion of the beam, perhaps to a small fraction of the beam's "$1/e^2$" intensity diameter.

D. Mathematics of the Repetitively Chopped Thermal Lens

The use of the repetitively chopped dual-beam thermal lens technique can provide a significant increase in data collection efficiency, particularly if the experiment avoids waiting for the approach to the equilibrium strength of the thermal lens (Dovichi and Harris, 1981a). In order to develop a better understanding of the repetitively chopped experiment, it is necessary to extend the model of Gordon *et al.* (1965) to account for the cumulative effect of the chopped heating beam on the sample (Swofford and Morrell, 1978).

With the continuously chopped heating source, the temperature rise in the sample is caused not only by the action of the heating beam during the current chopper cycle but also by a contribution from each of the previous chopper cycles. Figure 7a shows the timing sequence for calculating the temperature rise in the sample at a time of observation ($t' = 0$) during the *m*th chopper cycle. The time scale is seen to run "backward" in time, where $t' = t_m$ is the time in the past at which the *m*th chopper "on" cycle began. Equation (18) can be modified to describe the temperature rise at t_m into the *m*th chopper "on" cycle as

$$T(r, t_m; m) = \int_0^{t_m} dt' \int_0^\infty dr' \, \dot{G}(r, r', t')$$
$$+ \sum_{n=1}^{m-1} \int_{A_n}^{A_n+\tau} dt' \int_0^\infty dr' \, \dot{G}(r, r', t_n'), \qquad (47)$$

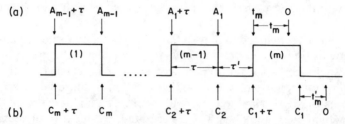

Fig. 7. Timing sequences for the calculation of the strength of the pulsating thermal lens in the two cases: (a) during and (b) following the mth chopper "on" pulse. In (a), the instant of observation occurs a length of time t_m after the beginning of the mth chopper "on" pulse, while in (b) the observation is a length of time t'_m after the end of the mth chopper "on" pulse. The times τ and τ' are the chopper "on" and "off" times, respectively. The time parameters in (a) and (b) are defined following Eqs. (47) and (48) as $A_n = t_m + \tau' + (n - 1)$ $(\tau + \tau')$ and $C_n = t'_m + (n - 1)(\tau + \tau')$. The time scales in (a) and (b) run from zero at the instant of observation back to the beginning of the first chopper "on" pulse. [From Swofford and Morrell (1978).]

where $A_n = (t_m + \tau') + (n - 1)(\tau + \tau')$, with τ and τ' the "on" and "off" times, respectively, of the chopper. The summation is over the $m - 1$ previous "on" pulses of the chopper, and the implicit dependence of T on m has been indicated. Note the value of t_m in Eq. (47) is limited to the range $0 \leqslant t_m \leqslant \tau$, where τ is the length of the "on" period of the chopper. In a similar manner, Eq. (18) can be modified to describe the temperature fall following the mth chopper "on" pulse (see Figure. 7b) as

$$T'(r, t'_m; m) = \sum_{n=1}^{m} \int_{C_n}^{C_n + \tau} dt'_n \int_0^\infty dr' \, \dot{G}(r, r', t'_n), \tag{48}$$

where $C_n = t'_m + (n - 1)(\tau + \tau')$. The limits of the time integration in Eq. (48) account for the fact that the chopper has been off for a time t'_m before the instant of observation. The value of t'_m in Eq. (48) is limited to the range $0 \leqslant t'_m \leqslant \tau'$, where τ' is the length of the "off" period of the chopper.

Following the solution developed for the model of Gordon *et al.* (1965) discussed previously [see Eq. (22)], we see that the desired quantity is $(\partial^2 T/\partial r^2)_{r=0}$. The integrals over r' in Eq. (47) are evaluated first, and the second derivative of the resulting expression with respect to r is evaluated at $r = 0$ to give

$$\frac{\partial^2}{\partial r^2} \left(\int_0^\infty dr' \, \dot{G}(r, r', t') \right)_{r=0} = \frac{\alpha P}{\pi J k} \frac{-8D}{(w_1^2 + 8Dt')^2}. \tag{49}$$

The integrals over t' in Eq. (47) are evaluated to give

$$\left(\frac{\partial^2 T}{\partial r^2}\right)_{r=0} = \frac{-\alpha P}{\pi Jkw_1^2}\left[\frac{1}{1 + t_c/2t}\right.$$

$$\left. + \sum_{n=1}^{m-1}\left(\frac{1}{1 + 2A_n/t_c} - \frac{1}{1 + 2(A_n + \tau)/t_c}\right)\right]. \qquad (50)$$

In similar fashion

$$\left(\frac{\partial^2 T'}{\partial r^2}\right)_{r=0} = \frac{-\alpha P}{\pi Jkw_1^2}\sum_{n=1}^{m}\left(\frac{1}{1 + 2C_n/t_c} - \frac{1}{1 + 2(C_n + \tau)/t_c}\right). \qquad (51)$$

The above equations provide a means of calculating the time behavior of the curvature of the temperature profile at the center of the laser beam during the chopped illumination of the sample by the heating laser beam. By use of the expression for the focal length of the thermal lens [Eq. (16)] and the *ABCD* law for Gaussian beams [Eq. (41)], one can calculate the time-dependent thermal lens signal and hence extract values of the absorptivity α.

An important feature of the *ABCD*-law technique is the ability to account separately for the propagation of the heating laser and the probe laser from their beam waists formed by the input lens, through the sample cell, and finally to the pinhole and detector. One cannot (indeed, *must not!*) assume in the experimental setup of Fig. 6 that the locations of the two lasers' beam waists are identical. The beam parameters must therefore be carefully measured in order to obtain an accurate estimate of the performance of a particular experimental setup. In this regard, it may be advisable to investigate the use of one or two beam-expanding telescopes to ensure proper parameter matching of the two laser beams.

Because the repetitively chopped thermal lens experiment does not reach equilibrium, it is convenient to redefine the expression for the thermal lens signal from Eq. (43). The signal is expressed as the fractional change in power relative to the *initially* detected power as

$$\frac{P_{\text{det}}(0; 0) - P_{\text{det}}(t; m)}{P_{\text{det}}(0; 0)} = 1 - \frac{w_2^2(0; 0)}{w_2^2(t; m)}, \qquad (52)$$

where $w_2^2(t; m)$ is given by Eq. (41) with $F(t; m)$ from Eq. (16) and either Eq. (50) or (51) as appropriate. The value of the signal in Eq. (52) is always positive, in contrast to the expression used in the earlier work of Swofford and Morrell (1978).

As shown by Swofford and Morrell (1978), the form of Eq. (52) does not result in as convenient an expression for the thermal lens signal as does Eq. (43). This is of little concern, since the primary use of Eq. (52) has been for numerical computer investigation of parameter sensitivity of the experimental design.

As seen in the earlier discussion of the Hu and Whinnery technique, the beam size dependence of Eqs. (50) and (51) enters as w_1^2. Thus at the confocal distance past the lens focus, the conditions leading to Eq. (45) obtain and the strength of the thermal lens scales as λ. It should be noted, however, that the critical time t_c which enters Eq. (52) scales as w_1^2 and hence as λ. For chopping frequencies where the chopper "on" time τ is approximately equal to t_c, there can be an "artificial" wavelength dependence introduced into the experiment. In conclusion, for appropriately chosen experimental conditions, the observed thermal lens signal for constant $\alpha(\lambda)$ scales as $1/\lambda$ and can be properly corrected by a factor of λ as in Eq. (32).

The calculated time dependence of the detected laser beam power is shown as $P_{det}(t; m)/P_{det}(0; 0)$ in Fig. 8a and as $[P_{det}(0; 0) - P_{det}(t; m)]/P_{det}(0; 0)$ in Fig. 8b. In the earlier example of the single-beam measurement of benzene (Fig. 5a), the maximum change in the detected power relative to the equilibrium value is 0.045, so that the minimum expected value of $P_{det}(t; m)/P_{det}(0; 0)$ in the present experiment is 0.957.

It should be noted that the calculation leading to Fig. 8, as well as the work of Swofford and Morrell (1978), assumes that the beam waists w_0 and the propagation characteristics of the pump and probe beam are identical, so that both beam waists are located at the same position. In that case the pump z' and probe z' are interchangeable in Eq. (41), and the simplifying assumption $z' = b$ in Eq. (46) is valid. A more complete treatement of the probe-beam behavior would require separate accounting of pump and probe beam z' and b values in Eq. (41), where it must be remembered that the magnitude of $F(t)$ from Eqs. (16), (50), and (51) is in terms of the beam size w_1 of the *pump* laser.

Based on the above considerations, it can be immediately recognized that, while maximum detection *sensitivity* requires that the thermal lens be located at one confocal distance beyond the beam waist of the *probe* beam, the maximum *strength* of the thermal lens [Eq. (24)] is achieved by focusing the pump beam *inside* the sample cell. This point has not been previously recognized by users of the dual-beam technique. Such an arrangement of beam focusing is difficult to accomplish with a single lens, but it may be achieved with minor additional difficulty by placing separate lenses in the pump and probe beams before the beam splitter (see Fig. 6). By locating the chopper after the pump beam lens, one may avoid creating in the probe-beam path fluctuating optical inhomogeneities which might be created in the single-input-lens configuration by the heating action of the chopped pump laser. Since the pump laser beam size w_1 at one confo-

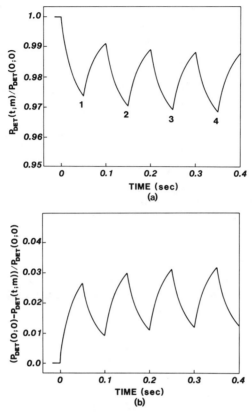

Fig. 8. Example of the calculated time dependence of the detected probe-beam power in the repetitively chopped dual-beam thermal lens experiment: (a) calculated signal plotted as $P_{det}(t; m)/P_{det}(0; 0)$; (b) calculated signal in (a) replotted according to Eq. (52).

cal length beyond its minimum size w_0 is given by $w_1 = \sqrt{2}w_0$, it is seen from Eqs. (50) and (51) that the strength of the lens formed in a sample located at the pump beam waist is a factor of two stronger than one generated with the sample at P_1 (see Fig. 4). In some cases, the added sensitivity available may warrant the added experimental complication.

   In the experiment, attention must be given to the proper measurement of $P_{det}(0; 0)$. The measurement can be made either by direct observation of the dc signal voltage presented to the lock-in amplifier or by inserting the chopper into the probe beam and measuring with the lock-in the square-wave signal magnitude of the detected probe beam in the absence of the heating beam. One must be aware in the latter case, however, that the lock-in amplifier is calibrated to indicate the magnitude of the rms fundamental signal and not the full peak-to-peak signal magnitude of the squarewave. The appropriate correction of $\sqrt{2}$ should be used to convert the lock-in reading to the true $P_{det}(0; 0)$.

For the repetitively chopped thermal lens, the measurement of interest is the magnitude of the ac signal component, i.e., the difference in detected probe beam power at the beginning and end of a chopper "on" cycle. The magnitude of the ac signal during the mth chopper pulse relative to $P_{det}(0; 0)$ is

$$\frac{\Delta P_{det}(0, \tau; m)}{P_{det}(0; 0)} = \frac{P_{det}(\tau; m) - P_{det}(0; 0)}{P_{det}(0; 0)} - \frac{P_{det}(0; m) - P_{det}(0; 0)}{P_{det}(0; 0)}. \quad (53)$$

Note that since the experimental measurements described by Eqs. (52) and (53) are source-noise limited, it is the *relative change* in the detected power and not the absolute magnitude of that change that should be optimized in the experimental design. It should also be recalled that the lock-in amplifier measures the rms fundamental of the Fourier-analyzed signal represented by Eq. (52). Thus only the sine-wave component of the signal in Fig. 8a is detected by the lock-in. A more complete quantitative comparison of the signal calculated from Eq. (52) with the signal detected by the lock-in would require a Fourier analysis of Eq. (52). This comparison has not yet been carried out in detail.

E. Dependence of the Thermal Lens Signal on Experimental Parameters

Swofford and Morrell (1978) investigated the dependence of the magnitude of the thermal lens signal on the parameters of the experimental design shown in Fig. 6. In their experiment, with $w_0 = 5.7 \times 10^{-3}$ cm and $\lambda = 600$ nm, one calculates $b = 1.7$ cm and $t_c = 1.69 \times 10^{-2}$ sec. Thus the use of a 1-cm sample cell in the experiment is questionable. In addition, the (asymmetric) chopper cycle gives $\tau = 4.5 \times 10^{-2}$ sec and $\tau' = 1.2 \times 10^{-2}$ sec, so that approximately 85% of the maximum available signal is developed during the first chopper "on" cycle. But since the chopper "off" time τ' is less than t_c, the thermal lens does not completely "relax" before the next illumination. Thus it is estimated that approximately half the maximum available signal is produced in the experiment. As expected from the discussions of the previous section, the observed signal is maximized for $z' < b$, in this case $z' = 0.4b$.

The work of Swofford and Morrell (1978) demonstrates the value of a careful analysis of the experimental parameters in the thermal lens measurement. The principal conclusion of their analysis is that the thermal lens signal defined by Eq. (53) should be less than 0.1 in order to assure linearity of the signal in the absorbed power. One constraint which remains is the limitation on sample cell length necessary to ensure the validity of the expression for the thermal lens focal length in Eq. (16). In the next section, it will be shown how this major limitation can be removed.

F. Removing Restrictions on Sample Cell Length

One of the primary motivations of the dual-beam lock-in technique has been the development of a high-sensitivity absorption spectrometer. As has been demonstrated, one of the fundamental limitations on available signal is the restriction of sample cell length so that the model of a "thin" thermal lens applies. Recently, it has been shown by Fang and Swofford (1979) that such a restriction can be removed by an approach which avoids altogether the "thin-lens" approximation. Those authors recognize that a thin slice of optical material with an index of refraction which is a quadratic function of radius [Eq. (16)] can act as a lens. Following Yariv (1975), those authors write the $ABCD$ matrix for the propagation of a Gaussian beam through a thickness dz of lenslike material as

$$A = \cosh(X^{1/2}\,dz), \quad B = X^{-1/2}\sinh(X^{1/2}\,dz),$$
$$C = X^{1/2}\sinh(X^{1/2}\,dz), \quad D = \cosh(X^{1/2}\,dz) \tag{54}$$

where

$$X = \frac{1}{n(0;\,T)}\frac{dn}{dT}\left(\frac{\partial^2 T}{\partial r^2}\right)_{r=0} \tag{55}$$

and $(\partial^2 T/\partial r^2)_{r=0}$ or $(\partial^2 T'/\partial r^2)_{r=0}$ is given directly by Eq. (50) or (51).

It is seen that the treatment of the sample as slices of lenslike material (Weiss, 1980) allows the calculation to account for the changing pump-laser beam size and for the resultant variation in the strength of the thermal lens which occurs with the use of longer sample cells. The calculation may be conveniently performed by use of an iterative technique. First the pump beam size incident on the cell is calculated. Next the magnitude of the quadratic index gradient created in the sample by the pump beam is determined. Then both laser beams are propagated through a thin slice of the medium, where the beam sizes increase not only because of the divergences of the Gaussian beams but also because of the action of the thermal lens. Finally the beam sizes at the exit of the first slice of sample are calculated and used as inputs to the second slice, and the iteration is continued through the sample cell. The number of iterations required for convergence of the calculation depends on the ratio of the sample cell length to the confocal distance of the pump laser. When that ratio is unity, then ten iterations produce a satisfactory convergence in the case of a relatively weak absorber ($\alpha < 10^{-3}$ cm^{-1}). Of course, for stronger absorbers, more iterations may be required.

It was shown by Fang and Swofford (1979) that an enhancement of the observed thermal lens signal is obtained by use of a sample cell whose length is several times the beams' confocal length (see Fig. 9). As expected, however, for sample cells longer than approximately $2b$, no significant enhancement of the signal is observed, since the strength of the thermal lens becomes weaker as the pump beam expands and the effect of the thermal lens on the probe beam diminishes beyond one confocal length from the probe-beam focus. Most importantly, Fang and Swofford showed that the signal remains linear in absorbed power for a number of input lens and cell length combinations. It is still seen, however, that for signal levels above approximately 0.2 as defined in Eq. (53), nonlinearity can become significant, as shown in Fig. 10.

G. The Thermal Lens Effect with an Elliptic Gaussian Beam

Up to this point, the model has assumed that both the pump and probe lasers have circularly symmetric beam intensity profiles. The assumption is valid for the linear two-mirror probe-laser resonator, but it is not valid for the folded three-mirror dye-laser resonator. Although a major portion of the dye-laser cavity astigmatism is compensated by the Brewster-angle dye jet (Kogelnik *et al.*, 1972), experimental measurements in this laboratory indicate that the dye-laser beam spot size in the far field has an ellipticity of approximately 1.4, with the major axis in the vertical (y) direction. The measurements were performed with several laser dyes

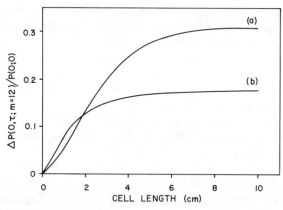

Fig. 9. Example of the calculated peak-to-peak ac signal $\Delta P_{\text{det}}(0, \tau; m)$ relative to the initial dc value, $P_{\text{det}}(0; 0)$, calculated from Eq. (53) for input lenses of (a) 40 and (b) 20 cm focal length, as a function of sample cell length. Under the conditions of the example here, $P = 10^{-2}$ W, $\tau = \tau' = 5 \div 10^{-2}$ sec, $m = 12$, $\alpha = 2.3 \times 10^{-3}$ cm^{-1}, and in (a) $w_0 = 8.6 \times 10^{-3}$ and $b = 3.9$ cm, while in (b) $w_0 = 3.8 \times 10^{-3}$ cm and $b = 0.74$ cm. [From Fang and Swofford (1979).]

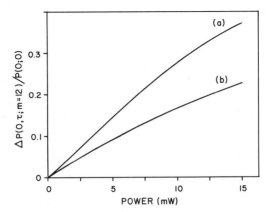

Fig. 10. Example of the calculated peak-to-peak ac signal $\Delta P_{\text{det}}(0, \tau; m)$ relative to the initial dc value $P_{\text{det}}(0; 0)$, calculated from Eq. (53) for input lenses of (a) 40 and (b) 20 cm focal length, as a function of input pump laser power. The curves were calculated under conditions identical with those of Fig. 9, and for a sample cell length of 5 cm. [From Fang and Swofford (1979).]

under a variety of pump-power and laser alignment conditions. Although the problem has not been carefully analyzed, it appears that the off-axis laser pumping configuration is a major source of the astigmatism since the profile of the optically pumped gain medium is elliptical and its astigmatism is uncompensated.

The elliptic Gaussian beam profile of the pump laser can be easily included in the present model of the thermal lens. Yariv (1975) showed how the fundamental Gaussian beam solution conveniently separates into orthogonal problems for the vertical (y) and horizontal (x) dimensions of the elliptic Gaussian beam. Thus we can calculate separately the thermal lensing in the two dimensions by applying the $ABCD$ law as outlined above. All that remains is to combine the two probe-beam dimensions at the target pinhole and calculate the detected power signal. The detected probe laser power is given by

$$P_{\text{det}} = 4 \int_0^a dx \int_0^a dy\, I(x, y) \tag{56}$$

where a is the radius of the target pinhole and

$$I(x, y) = \frac{2P_{\text{pr}}}{\pi w_{2x\,\text{pr}} w_{2y\,\text{pr}}} \exp\left(-\frac{2x^2}{w_{2x\,\text{pr}}^2} - \frac{2y^2}{w_{2y\,\text{pr}}^2}\right) \tag{57}$$

is the intensity of the now-elliptical Gaussian probe beam. Inserting Eq. (57) into (56), one obtains

$$P_{\text{det}} = P_{\text{pr}}\, \text{erf}\left(\frac{\sqrt{2}a}{w_{2x\,\text{pr}}}\right) \text{erf}\left(\frac{\sqrt{2}a}{w_{2y\,\text{pr}}}\right) \tag{58}$$

where erf is the error function. Since by design $a \ll w_{2pr}$, the series expansion

$$\text{erf}(u) = \frac{2}{\pi} \sum_{n=0}^{\infty} \frac{(-1)^n u^{2n+1}}{n!(2n+1)} \tag{59}$$

reduces to the leading term. Thus in place of Eq. (52), one obtains for the fractional change in the detected probe-beam power relative to its initial value

$$\frac{P_{\text{det}}(0; 0) - P_{\text{det}}(t; m)}{P_{\text{det}}(0; 0)} = 1 - \frac{w_{2x\,pr}(0; 0)w_{2y\,pr}(0; 0)}{w_{2x\,pr}(t; m)w_{2y\,pr}(t; m)}. \tag{60}$$

H. Calculated Fit to Thermal Lens Measurement

Fang and Swofford (1979) demonstrated very good agreement between the measured and calculated signal in the repetitively chopped thermal lens experiment, as shown in Fig. 11. It should be noted that the example in Fig. 11 uses a sample cell nearly two orders of magnitude longer than one for which the previous model was shown to be valid. This is a significant improvement since it allows more sensitive measurement of lower sample absorptivities than is possible with short sample pathlengths while still ensuring that the observed signals scale linearly in absorbed power.

There are several important features in Fig. 11 worthy of note. Under the experimental conditions, one calculates a confocal distance of 0.7 cm

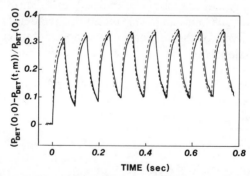

Fig. 11. Observed thermal lensing signal (—) expressed as the fractional change in the detected probe laser power $P_{\text{det}}(t; m)$ relative to its initial value $P_{\text{det}}(0; 0)$. The signal is captured and averaged 200 times by a digital processor oscilloscope. In this example, the signal from a sample of pure benzene in a 10-cm cell was measured under the following conditions: $\tau = \tau' = 50$ msec, $w_{0x\,pu} = 3.76 \times 10^{-3}$ cm, $w_{0y\,pu} = 2.56 \times 10^{-3}$, $w_{0\,pr} = 3.38 \times 10^{-3}$ cm, $P_{pu} = 19$ mW, $\lambda_{pu} = 607$ nm, and $\lambda_{pr} = 632.8$ nm. The dashed line (---) is computed with Eq. (52) for a value of $\alpha = 2.3 \times 10^{-3}$ cm^{-1}. [From Fang and Swofford (1979).]

in the horizontal (xz) plane. Thus at the entrance to the sample cell, the critical time is calculated to be approximately 7.5×10^{-3} sec. Since the beam size varies as $w = w_0[1 + (z/b)^2]^{1/2}$, the critical time in the sample increases as $1 + (z/b)^2$. Thus at the exit window of the sample, the beam size is approximately fourteen times larger and the critical time is approximately 0.1 sec. It is the ability to account for these large changes in experimental conditions that is the power of the calculational technique of Fang and Swofford.

A second feature of Fig. 11 is the fact that the magnitude of the fluctuating signal as defined by Eq. (53) is approximately half that available in the cw experiment of Hu and Whinnery. Thus one trades off a fraction of the signal against the improved signal-to-noise ratio available in the repetitively chopped thermal lens experiment.

V. CREATION OF A THERMAL LENS BY A PULSED LASER

Although the experiments described in the previous sections have used continuous or chopped cw laser heating sources, it is also possible to carry out thermal lens measurements with pulsed pump lasers (see, e.g., Longaker and Litvak, 1969). The experiments of necessity employ the dual-beam technique, with a continuous or quasicontinuous probe laser used to monitor the creation and relaxation of the thermal lens which is formed virtually instantaneously by the pulsed laser.

It is instructive to calculate the pulse energy that can create a thermal lens of equivalent strength to one created by a chopped cw laser of the kind used in the previous examples. A reasonable estimate of the energy required for the creation of a measurable thermal lens is the integrated energy that is absorbed by the sample under continuous illumination for the time equal to the "half-life" for the thermal lens, i.e., one-half the critical time t_c for the particular setup. If we assume as in the previous example $t_c = 5.3 \times 10^{-2}$ sec for $w_1 = 0.014$ cm and a 1-cm liquid sample with absorptivity $\alpha = 10^{-3}$ cm^{-1} illuminated by 10^{-2} W of laser power, then 2.65×10^{-7} J are absorbed in the time $\frac{1}{2}t_c$. For a pulsed laser to generate an equivalent thermal lens, the incident pulse must contain 2.65×10^{-4} J of energy. If the pulsed laser beam is derived from a Q-switched device with an output pulse duration of 10^{-8} sec, then the pulse power must be 2.65×10^4 W. If a flash-lamp-pumped dye laser with a pulse duration of 10^{-6} sec is used, then the required pulse power is only 265 W. It is seen that in order to create a thermal lens of equal strength to one formed by a cw laser of power P, a pulsed laser of power $\frac{1}{2}Pt_c/t_p$ is required, where t_p is the pulse duration. It should also be noted that, because of the relatively long time constant of the thermal lens experiment, the process can effectively integrate the individual heat pulses generated by a repetitively pulsed laser when the time between pulses is considerably shorter than $\frac{1}{2}t_c$. In the limit of very high repetition rates, of course, the laser behaves as a cw source in the creation of a thermal lens and the measurement may be performed conveniently by use of a chopper.

Given the problem of pulse-to-pulse instability of many pulsed lasers, such a laser is not very competitive with cw lasers as a pump source for thermal lens measurements except in cases in which the desired spectral tuning region or energy delivery capability is not available with cw sources. As will be seen, the major benefit of pulsed lasers occurs in experiments where high peak power is important, i.e., experiments concerned with the nonlinear absorption of light. An additional difficulty in the use of a pulsed laser for the creation of a thermal lens arises from the generally poorer quality of the spatial intensity profile of a pulsed beam compared to a cw laser beam. The pulse-to-pulse changes in the laser beam intensity profile can be particularly troublesome, as can the systematic changes in beam characteristics that occur when a tunable pulsed laser is scanned across its spectral gain curve. Pulsed laser configurations which have an "unstable resonator" design are more difficult to use in thermal lens measurements than are lasers with stable resonator designs. The output beam of a tunable laser oscillator–amplifier system pumped by a laser with the "unstable resonator" configuration may have a very complicated spatial intensity profile, particularly if the system has a final amplifier stage which is pumped longitudinally rather than transversely. In summary, it is very important that a pulsed laser which is to be used in a thermal lens experiment be fully characterized in terms of its beam propagation parameters and its transverse spatial intensity profile.

A. Mathematics of the Pulsed Thermal Lens

The relatively slow thermal response of the sample to heating by the pulsed laser simplifies the mathematical description of the pulsed thermal lens (Twarowski and Kliger, 1977a; Bailey *et al.*, 1978, 1980). The propagation function describing the temperature rise in the sample has already been given as Eq. (17a). In a manner similar to Eq. (19), one can write the instantaneous heat pulse generated per unit length of the sample by the absorption of the pulsed Gaussian laser beam as

$$Q(r')dr' = \frac{2\alpha}{\pi J k w_1^2} \exp\left(-\frac{2r'^2}{w_1^2}\right)dr' \int_0^{t_p} P(t')dt', \qquad (61)$$

with all symbols as in Eq. (19). The total effective heating energy H of the pulsed laser is

$$H = \int_0^{t_p} P(t')dt' \tag{62}$$

where H is in joules and t_p is the laser pulse duration. Note that the magnitude of the heat pulse is independent of the functional form of $P(t)$ so long as $t_p \ll \frac{1}{2}t_c$. In Eq. (61), the parameter t_p has been introduced to avoid confusion with Eq. (17a), where t specifies the time during which the temperature profile evolves *after* an instantaneous heat pulse. The form of Eq. (61) allows one to write immediately, following Eq. (22), the expression for the time evolution of the curvature of the temperature distribution in the sample as

$$\left(\frac{\partial^2 T'}{\partial r^2}\right)_{r=0} = \frac{-8\alpha HD}{\pi Jk(w_1^2 + 8DT)^2} = \frac{-8\alpha HD}{\pi Jkw_1^4(1 + 2t/t_c)^2}$$

$$= \frac{-\alpha H/\frac{1}{2}t_c}{\pi Jkw_1^2(1 + 2t/t_c)^2}, \tag{63}$$

where T' indicates, as in Eq. (48), the temperature *fall* after the heating beam is off. Note that the time integration over the duration of the heating beam in the present case is included in Eq. (61). This results in different functional forms in Eqs. (23) and (63).

The form of Eq. (63) has been chosen to emphasize the similarities with Eq. (23). The "effective heating power" of the pulsed laser beam is the equivalent pulse energy being delivered in a time $\frac{1}{2}t_c$. It is seen that the strength of the pulsed thermal lens, Eq. (63), at $t = 0$, is the same as the strength of the equilibrium thermal lens ($t \to \infty$) generated by a cw source of power $P = H/\frac{1}{2}t_c$ [compare Eqs. (23) and (63)].

As in the earlier case, one can write the time-dependent focal length of the thermal lens from Eqs. (16) and (63) as

$$F(t) = \frac{\pi Jkw_1^4(1 + 2t/t_c)^2}{-8\alpha HDl(dn/dT)}$$

$$= \frac{\pi Jkw_1^2(1 + 2t/t_c)^2}{\alpha(H/\frac{1}{2}t_c)l(dn/dT)}. \tag{64a}$$

Following the earlier treatment of the cw thermal lens [see Eqs. (43–45)], one can express the time dependence of the pulsed thermal lens signal as the fractional change in the detected probe-beam power relative to the power detected in the presence of the pulsed thermal lens,

$$\frac{P_{det}(t = \infty) - P_{det}(t)}{P_{det}(t)} = \frac{w_2^2(t)}{w_2^2(t = \infty)} - 1 \approx \frac{-2z'}{F(t)}. \tag{64b}$$

The expression for the thermal lens signal in Eq. (64b) is a convenient one for investigating analytically the behavior of the pulsed thermal lens signal (Twarowski and Kliger, 1977a). Alternatively, one can express the signal in the experimentally more convenient form [see Eq. (52)] as the fractional change in the detected probe-beam power relative to the power detected in the absence of the thermal lens. In most cases, where $P_{det}(t = 0)$ is approximately equal to $P_{det}(t = \infty)$, the use of Eq. (64b) for studying the signal dependence on the experimentally adjustable parameters is quite satisfactory.

It is interesting to note that in the present case, the wavelength dependence of Eq. (64a) results from the factor of w_1^4 and hence differs from the earlier case of the Hu and Whinnery experiment [see Eq. (46)]. Thus both the signal magnitude, determined by $H/\frac{1}{2}t_c$ and w_1^2, and the time behavior of the thermal lens relaxation must be corrected for the wavelength dependence of w_1, which scales as $\lambda^{1/2}$. From Eqs. (64a, b) it is seen that for constant $\alpha(\lambda)$, the magnitude of the signal as defined by Eq. (64b), with $t = 0$, scales inversely as λ^2 [compare Eq. (46)]. This result is in agreement with the earlier work of Twarowski and Kliger (1977a), where, however, an additional factor of λ appears owing to those authors' definition of H [Eq. (62)] in photons per pulse.

As an alternative to Eq. (64a), one can treat the sample as an extended optical medium with a transverse quadratic index profile and use the *ABCD* matrix elements given in Eqs. (54) and (55) to compute how the probe-beam propagation is affected by the pulsed thermal lens.

Figure 12a shows the pulsed thermal lens signal plotted as $P_{det}(t)/P_{det}(t = \infty)$ for the example of benzene with a pulsed laser energy of 2.65×10^{-4} J. As expected, the signal strength at $t = 0$ is equal to the equilibrium signal strength of the cw thermal lens. In the present case, note that the relaxation of the pulsed thermal lens is more rapid than the buildup of the cw thermal lens (see Fig. 5a) because of the different time dependence of Eqs. (24) and (64a). Figure 12b plots the signal according to Eq. (64b).

Since commonly available pulsed lasers operate at repetition rates of 10–20 Hz, it can be recognized upon inspection of Fig. 12 that the signal observed in such an experiment will be a series of narrow pulses with a relatively small baseline deviation. Although the lock-in amplifier has been the most commonly used detection device in these experiments, it is clear that the signal in Fig. 12 has more high-frequency components than the cw thermal lens signal in Fig. 8. Signals such as the one in Fig. 12 may be more appropriately measured by use of a gated integrator, often called a boxcar averager. Modern boxcar averagers have useful features such as baseline correction, dual-channel operation, and an analog output proportional to the ratio of the two channels. Thus one can monitor the pulsed

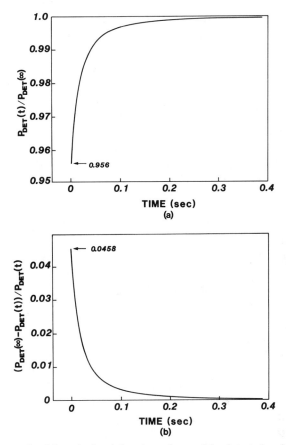

Fig. 12. Example of the calculated time dependence of the detected probe-beam power in the pulsed dual-beam thermal lens experiment: (a) calculated signal plotted as $P_{det}(t)/P_{det}(t = \infty)$; (b) calculated signal in (a) replotted according to Eq. (64b).

laser energy on one channel and the thermal lens signal on the other and record the ratio signal representing the strength of the thermal lens normalized to the average pulse energy of the laser. This normalization may significantly improve the signal-to-noise ratio of the measurement.

B. Thermal Lens Formation by Pulsed
Two-Photon Absorption

The availability of pulsed lasers has made possible the study of a wide variety of nonliner optical effects, i.e., molecular behavior which no longer scales linearly with the incident laser intensity (Swofford and

Albrecht, 1978). One of the simplest of these phenomena is two-photon absorption, by which a molecule is promoted to an excited state at twice the frequency of the incident light as a result of the simultaneous absorption of two photons. Since the rate of this very weak process depends quadratically on the laser intensity, the use of a pulsed laser combined with a sensitive detection method is required. (See Chapter 2 by Birge). The thermal lens technique was recognized as a possible detection scheme for two-photon absorption by Long *et al.* (1976) and was first demonstrated by Twarowski and Kliger (1977b).

It is instructive to calculate the amount of energy that can be deposited in a sample by two-photon absorption. We begin by writing Beer's law for the power of a laser beam transmitted through a sample with two-photon absorptivity $\alpha^{(2)}$ (cm^{-1}),

$$P_{\text{trans}} = P \exp(-\alpha^{(2)}l) = P \exp(-\delta CIl), \qquad (65)$$

where δ (cm^4 sec photon^{-1} molecule^{-1}) is usually called the two-photon cross section, C (molecules cm^{-3}) is the sample concentration, and I (photons cm^{-2} sec^{-1}) is the intensity of the incident laser. The definition in Eq. (65) is the conventional one, but it differs by a factor of two from the one used by Twarowski and Kliger (1977a). Note that the effective absorptivity of the sample, $\alpha^{(2)} = \delta CI$, depends on the incident intensity. Thus focusing the laser beam to a small spot size w_0 can produce higher sample absorptivity; however, the size of a tightly focused beam will vary considerably within the sample, further complicating the analysis. It should also be noted that the time dependence of the laser pulse power is important in determining the total magnitude of two-photon absorption (see Chapter 2 by Birge). For simplicity, the present treatment assumes a square pulse of constant power P and time duration t_p. A more appropriate pulse shape is used by Birge, by Twarowski and Kliger (1977a), and by Swofford and McClain (1975), who write $P(t) = (2P/\pi t_p)\exp(-t^2/t_p^2)$.

In the limit of low two-photon absorptivity, which is usually the case, the fraction of laser power absorbed is

$$(P - P_{\text{trans}})/P = \Delta P/P = \delta CIl \qquad (66)$$

where l (cm) is the sample pathlength. Let us consider the laser pulse of the previous example, with a pulse duration of 10^{-8} sec, a pulse power of 2.65×10^4 W for a total pulse energy of 2.65×10^{-4} J, and a wavelength of 600 nm focused to a minimum radius of $w_0 = 10^{-2}$ cm in the center of a 1-cm sample cell containing a neat liquid with $C = 6.7 \times 10^{21}$ molecules cm^{-3} and $\delta = 0.1 \times 10^{-50}$ cm^4 sec photon^{-1} molecule^{-1}. Since the confocal distance for the beam $b = 5.2$ cm, we are assured of a reasonably constant beam size in the sample. A 600-nm photon has energy $E = hc/\lambda$, with $h = 6.62 \times 10^{-34}$ J sec, $c = 3 \times 10^{10}$ cm sec^{-1}, and $E = 3.31 \times 10^{-19}$ J photon^{-1}, so the 2.65×10^4 W laser produces $\dot{N} = P/E = 8.0 \times 10^{22}$ photons sec^{-1}, or $N_0 = 8.0 \times 10^{14}$ photons per pulse, and a focused intensity of 2.55×10^{26} photons cm^{-2} sec^{-1}. From Eq. (56), the fraction of incident photons absorbed from the pulse is $\delta C I l = 1.7 \times 10^{-3}$, so that the total number of photons absorbed from the pulse is $N_{abs} = N_0 \delta C I l = 1.36 \times 10^{12}$ photons or 4.5×10^{-7} J. Thus it is seen that the energy absorbed by a two-photon process compares favorably with the energy absorbed by a typical one-photon process, calculated earlier as 2.65×10^{-7} J.

If the laser beam is more tightly focused, then it becomes necessary to account for the varying beam intensity in the sample and the resultant variation in the two-photon absorptivity. Swofford and McClain (1975) showed how to calculate the number of photons absorbed from a focused Gaussian laser beam, which has the form

$$I(r, z) = \frac{2P/E}{\pi w^2(z)} \exp\left[- \frac{2r^2}{w^2(z)}\right] \tag{67}$$

where P (W) is the beam power, $E = hc/\lambda$, $w(z) = w_0 [1 + (z/b)^2]^{1/2}$ (cm) is the beam radius, and $b = \pi w_0^2/\lambda$ (cm) is the confocal parameter of the beam. The minimum beam radius w_0, located at $z = 0$ in the focused beam, is in the center of a sample cell of length l. The number of photons absorbed per second, $\dot{N}_{abs}$, is given by the integral form of the two-photon Beer's law,

$$\dot{N}_{abs} = \delta C \int_{-l/2}^{l/2} \int_0^\infty I^2(r,z) 2\pi r \, dr \, dz$$
$$= (C\delta P^2/2\lambda E^2)[\tan^{-1}(l/2b) - \tan^{-1}(-l/2b)] \tag{68}$$

In a tightly focused beam, $l \gg b$ and $\tan^{-1}(l/2b) = \frac{1}{2}\pi$; therefore

$$\dot{N}_{abs} = \pi \delta C P^2/2\lambda E^2 \tag{69}$$

is the maximum number of photons per second which can be absorbed by the sample. In the example used previously, in which $P/E = 8.0 \times 10^{22}$ photons sec^{-1}, $\dot{N}_{abs} = 1.12 \times 10^{21}$ photons sec^{-1}. Thus during the 10^{-8}-sec laser pulse, a total of $N_{abs} = 1.12 \times 10^{13}$ photons or 3.71×10^{-6} J are absorbed. It should be noted from Eq. (69) that fully half of the total absorption, or 5.6×10^{12} photons, occurs within a sample length $l = 2b$, since $\tan^{-1}(1) = \frac{1}{4}\pi$. One can calculate the beam size implied in the present example using $l = 1$ cm $= 2b$, or $w_0 = 3.1 \times 10^{-3}$ cm and compare with the previous example, where $w_0 = 1 \times 10^{-2}$ cm. It is seen that the three times smaller beam size results in approximately eight times larger energy absorption.

C. Mathematics of the Two-Photon Thermal Lens

The instantaneous heat pulse generated per unit length of the sample by two-photon absorption of the pulsed Gaussian laser beam can be written by modifying Eq. (61) with $\alpha^{(2)} = \delta CI$ and Eq. (67) to give

$$Q(r')dr' = \left(\frac{2}{\pi w_1^2}\right)^2 \frac{\delta C}{J} \exp\left(\frac{-4r'^2}{w_1^2}\right) dr' \frac{1}{E} \int_0^{t_p} P^2(t')dt', \qquad (70)$$

where all symbols are as before. It is interesting to note that the heat distribution in Eq. (70) is still Gaussian, but with an effective radius of $w_1/\sqrt{2}$, i.e., more sharply peaked than the heat distribution of Eq. (61). The time integral over the pulse power in Eq. (70) is

$$H' = \int_0^{t_p} P^2(t')dt' \qquad (71)$$

where H' is in Watt–Joules or Joules squared per second. Again following Eqs. (22) and (63), one can write the following expression for the time evolution of the curvature of the temperature distribution in the sample as

$$\left(\frac{\partial^2 T'}{\partial r^2}\right)_{r=0} = \frac{-32\delta CH'D}{\pi Jkw_1^6 E(1 + 4t/t_c)^2}$$

$$= \frac{-\delta CH'/\pi w_1^2 E \frac{1}{2} t_c}{\pi Jkw_1'^2 (1 + 2t/t_c')^2} \qquad (72)$$

where all symbols are as before, with $w_1' = w_1/\sqrt{2}$ and $t_c' = w_1'^2/4D$. By comparing Eqs. (63) and (72), it can be recognized that the two-photon thermal lens has a time dependence which is twice as fast as that of the one-photon lens. This time dependence is a result of the more sharply peaked heat distribution for the two-photon case. We have purposely written Eq. (72) to highlight that behavior. The numerator contains the "effective heating power" times the two-photon absorptivity and the fact that it is the total energy deposited in the sample in the time $\frac{1}{2}t_c$ that creates the thermal lens. The denominator contains the terms that determine the physical extent of the heated material, its curvature, and the time response of the thermal lens relaxation. As mentioned earlier, the effective radial size of the heated liquid is $w_1' = w_1/\sqrt{2}$, i.e., a more sharply peaked Gaussian distribution than the one expected for one-photon absorption. The effective critical time $t_c' = w_1'^2/4D$, indicates that this distribution relaxes more quickly than the one expected for one-photon absorption. It should be noted that the different spatial and temporal behavior of the two-photon thermal lens may be of benefit in distinguishing it from the one-photon thermal lens. It is immediately recognized from Eq. (72) that for the same total energy deposited, the strength of a two-photon thermal lens would be twice that of a one-photon thermal lens.

As in the previous cases, one can write the time-dependent focal length of the thermal lens from Eqs. (16) and (72) as

$$F(t) = \frac{\pi J k w_1^6 E(1 + 4t/t_c)^2}{-32\delta C H'D \ l \ (dn/dT)}$$

$$= \frac{\pi J k w_1'^2 (1 + 2t/t_c')^2}{-\delta C(H'/\pi w_1^2 E^{\frac{1}{2}} t_c) \ l \ (dn/dT)}. \tag{73}$$

It is immediately recognized that the strength of the two-photon thermal lens is considerably enhanced by arranging for the pump-beam focus to be in the center of the sample cell. An additional benefit of such an arrangement is that it avoids high laser intensities near the cell windows, where serious sample damage can occur. Twarowski and Kliger (1977b) reported just such difficulties. Equation (73) must be used with caution in such a tightly focused configuration, since the effective sample cell length is reduced to approximately twice the confocal length of the pump beam. Tight focusing may also introduce additional heat-producing nonlinear phenomena such as a multiphoton ionization or plasma formation in the sample. These strong effects may completely mask the much weaker two-photon absorption thermal lens. It is also apparent from the earlier discussion of the cw dual-beam experiment that the influence of the thermal lens on the probe beam is strongest when the sample is centered one confocal length past the probe beam waist. Finally, it should be noted that the wavelength dependence of Eq. (72) appears in the factor $w_1^6 E$, which scales as λ^2. Thus for constant $\delta(\lambda)$, the thermal lens signal as defined by Eq. (64b) with $t = 0$ scales inversely as λ^2. This result is in agreement with Twarowski and Kliger (1977a), where, however, an additional factor of λ^2 appears due to those authors' definition of H' [Eq. (71)] in photons squared per pulse. In contrast, the present definition is in terms of the experimentally measurable pulse power and pulse energy. Of course, as shown by Twarowski and Kliger, the observed signal must also be corrected for the peak power and the pulse energy. Several approaches to this correction have been discussed in detail by Swofford and McClain (1975).

D. Two-Photon Absorption with a High-Repetiton-Rate Pulsed Laser

A more recent innovation in tunable pulsed lasers is the synchronously mode-locked dye laser, capable of producing a continuous train of pulses at approximately 80 MHz, each 10^{-11} sec long and an average power of 0.2 W. Since the duty factor of the laser (i.e., the ratio of on time to off time) is 8×10^{-4}, the individual pulses have a peak power of 250 W. For the example conditions used earlier [following Eq. (56)], where $w_0 = 10^{-2}$ cm, we calculate that a single laser pulse produces $N = 7.55 \times 10^{20}$ photons sec^{-1} or $N_0 = 7.55 \times 10^9$ photons per pulse and a focused intensity of 2.4×10^{24} photons cm^{-2} sec^{-1}. The fraction of photons absorbed from the pulse is $\delta CIl = 1.6 \times 10^{-4}$, so that the total number of photons absorbed is $N_{abs} = N_0 \delta CIl = 1.2 \times 10^6$ photons per pulse or 3.97×10^{-13} J per pulse. This is a hopelessly small heat pulse to detect by the thermal lens technique. On the other hand, since the critical time in the present example $t_c = 2.65 \times 10^{-2}$ sec, then during a time $\frac{1}{2}t_c$, a total energy of 4.21×10^{-7} J is absorbed in the sample from the repetitively pulsed laser. In contrast, a sample with one-photon absorptivity $\alpha^{(1)} = 10^{-13}$ cm^{-1} would absorb 2.65×10^6 J from the average 0.2-W beam power in $\frac{1}{2}t_c$. Thus it is seen that the creation of a thermal lens by two-photon absorption of a synchronously mode-locked dye laser may compete favorably with the creation of a weak one-photon absorption ($\alpha^{(1)} < 10^{-4}$ cm^{-1}) thermal lens in the same sample.

There are several additional points worthy of note. First is the fact that the relatively slow response of the thermal lens in effect provides an integration of the heat produced by the individual pulses of picosecond two-photon absorption. This allows the detection of a relatively strong thermal lens by use of the dual-beam technique, where the picosecond laser is chopped as in the conventional experiment.

The second point to note is that for the example given the two-photon absorptivity $\alpha^{(2)} = \delta CI = 1.6 \times 10^{-4}$ cm^{-1} leads immediately to the result that a fraction $\alpha^{(2)}l = 1.6 \times 10^{-4}$ of the average heating laser power P_{avg} is absorbed by the sample. Thus it is seen that while the peak intensity is important in determining $\alpha^{(2)}$, it is the product $\alpha^{(2)}lP_{avg}$ which determines the total amount of energy absorbed. If the laser has pulse width t_p, focused beam radius w_0, and pulse repetition frequency ν_R, the energy absorbed can be written as

$$\alpha^{(2)}lP_{avg} = \delta Cl \frac{2P/E}{\pi w_0^2} \exp\left(-\frac{2r^2}{w_0^2}\right) P_{avg}$$

$$= \frac{2\delta Cl}{\pi w_0^2 E} \exp\left(-\frac{2r^2}{w_0^2}\right) \frac{P_{avg}}{t_p \nu_r} P_{avg}. \tag{74}$$

Equation (74) shows the interesting result that the energy absorbed by the two-photon process can be enhanced at constant average power by reducing ν_R. A cavity dumper can perform just such a function in a synchronously mode-locked laser. In the laser of the present example, a cavity dumper can be used to reduce v_R by a factor of twenty while reducing P_{avg} by a factor of two, leading to a factor of five increase in energy absorbed by the two-photon process. It should also be noted that the reduction in P_{avg} also reduces the heat absorption due to the one-photon process, leading to a further enhancement of the two-photon absorption process over the background one-photon absorption. Thus in order to maximize the ratio of two-photon to one-photon absorption, one must maximize the ratio P_{avg}/v_R.

One final point involves the competition between one- and two-photon absorption in the sample. The total energy absorbed in the sample is

$$\alpha^{(1)}lP_{avg} + \alpha^{(2)}lP_{avg} = \alpha^{(1)}lP_{avg} + [\delta Cl/(\pi w^2 t_p v_R)] P_{avg}^2. \tag{75}$$

If the deposited energy is monitored as a thermal lens signal of magnitude S for varying values of P_{avg}, then a plot of S/P_{avg} gives an intercept of $\alpha^{(1)}l$ and a slope of $2\delta Cl/\pi w^2 t_p v_R$. If $\alpha^{(1)}$ is known from independent measurements, then δ may be calculated in absolute units. This technique may thus provide a valuable quantitative method for determining δ. The value of P_{avg} may be varied conveniently with the use of a cavity dumper by changing the rf drive level applied to the acousto-optic Bragg cell. Such an adjustment does not significantly affect the laser alignment. The relative contribution of the two-photon thermal lens to the total signal S is twice that implied by Eq. (75), owing to the double strength of the two-photon thermal lens as indicated in Eq. (73).

Preliminary measurements performed in this laboratory (Fang et al., 1982) indicated the very high sensitivity of the two-photon thermal lens technique that uses the synchronously mode-locked dye laser.

VI. CONCLUSION

We have attempted to present a systematic description of the thermal lens technique to include both the essential experimental details and the theoretical development of the effect. The thermal lens is being applied to the study of a number of weak absorption effects, as reviewed recently by Kliger (1980). As can be seen from the examples of the present chapter, even very weak absorption can be conveniently measured with good signal-to-noise. The sensitivity of the technique compares quite favorably with that of other ultrasensitive techniques discussed in this volume, particularly photoacoustic (see, e.g., Brueck et al., 1980; Patel and Tam, 1981). Certain features of the thermal lens make it quite attractive for absorption and energy relaxation measurements. It is an extrinsic property of the sample, unlike photoacoustic spectroscopy, and hence longer sample paths can enhance signal levels. It employs optical detection in a probe-source-noise-limited regime, and hence further optical source improvements may allow even higher-sensitivity measurements. It is a slowly responding phenomenon, and hence can effectively integrate excitation pulses such as those produced by a synchronously pumped mode-locked dye laser. It works best for liquids owing to their high dn/dT values, although the technique has been used successfully to measure the one-photon (Grabiner $et\ al.$, 1972; Flynn, 1974) and two-photon (Nieman and Colson, 1978; Vaida $et\ al.$, 1978) absorption in gas-phase samples. And finally, it requires only apparatus commonly available in modern laser research labs. No unusual detection apparatus is required.

In recent years, several comments have appeared regarding the relative sensitivity of thermal lensing and photoacoustic measurements (Brueck $et\ al.$, 1980; Patel and Tam, 1981). Patel and Tam summarized the advantages of the photoacoustic technique and pointed out that under optimum conditions the technique can detect 10^{-10} J of energy absorbed. This is better than the sensitivity of current thermal lens measurements; however, it is rarely absolute sensitivity but rather background absorption that limits the measurement. Thus such claims may be a bit misleading. On the face of it, methods of optical detection, which offer interferometric sensitivity, should be equal to or better than mechanical techniques like photoacoustic or extensiometric measurement. In this regard, the recent work of Miles $et\ al.$ (1980) is significant. Those authors reported the use of a thin fiber element for interferometric detection of the photoacoustic signal in a gas. Thus it is seen that the combination of mechanical and optical techniques may provide a new round of increased sensitivity measurements. On the other hand, the heterodyne interferometric measurements of Davis (1979; Davis and Petuchowski, 1981) promise significant improvement of the photothermal technique.

In conclusion, we feel that the technique provides an important new ultrasensitive spectroscopic method, and we look forward to its expanding use.

REFERENCES

Abbate, G., Bernini, U., Brescia, G., and Ragozzino, E. (1981). $Opt.\ Laser\ Technol.$ **13**, 97–98.
Abramowitz, M., and Stegun, I. A. (1965). "Handbook of Mathematical Functions," p. 486, Integral 11.4.29. Dover, New York.
Akhmanov, S. A., Krindach, D. P., Sukhorukov, A. P., and Khokhlov, R. V. (1967). $Zh.$ $Eksp.\ Teor.\ Fiz.,\ Pis'ma\ Red.$ **6**, 509–513, $JEIP\ Lett.$ ($Engl.\ Transl.$) **6**, 38–42 (1967).

Akhmanov, S. A., Khokhlov, R. V., and Sukhorukov, A. P. (1972). *In* "Laser Handbook" (F. T. Arecchi and E. O. Schulz-Dubois, eds.), Vol. II, pp. 1151–1228. North-Holland Publ., Amsterdam.

Aleshkevich, V. A., Migulin, A. V., Sukhorukov, A. P., and Shumilov, E. N. (1972). *Zh. Eksp. Teor. Fiz.* **62,** 551–561; *Sov. Phys.—JETP (Engl. Transl.)* **35,**292–297 (1972)

Aung, H., and Katayama, M. (1975a). *Jpn. J. Appl. Phys.* **14,** 82–86.

Aung, H., and Katayama, M. (1975b). *Chem. Phys. Lett.* **33,** 502–505.

Bailey, R. T., Cruickshank, F. R., Pugh, D., and Johnstone, W. (1978). *Chem. Phys. Lett.* **59,** 324–329.

Bailey, R. T., Cruickshank, F. R., Pugh, D., and Johnstone, W. (1980). *J. Chem. Soc., Faraday Trans. 2* **76,** 633–647.

Banilis, R., Burgos, A., Mancini, H., and Quel, E. (1973). *Jpn. J. Appl. Phys.* **13,** 486–487.

Bell, A. G. (1880). *Proc. Am. Assoc. Sci.* **29,** 115–136.

Beysens, D., and Calmettes, P. (1977). *J. Chem. Phys.* **66,** 766–771.

Boccara, A. C., Fournier, D., and Badoz, J. (1980). *Appl. Phys. Lett.* **36,** 130–132.

Born, M., and Wolf, E. (1970). "Principles of Optics," 4th ed., p. 124. Pergamon, Oxford.

Boyd, R. D., and Vest, C. M. (1975). *Appl. Phys. Lett.* **26,** 287–288.

Brueck, S. R. J., Kildal, H., and Bélanger, L. J. (1980). *Opt. Commun.* **34,** 199–204.

Burgos, A., Mancini, H., Quel, E., and Westfreid, J. (1977). *Opt. Commun.* **20,** 434–437.

Callen, W. R., Huth, B. G., and Pantell, R. H. (1967). *Appl. Phys. Lett.* **11,** 103–105.

Calmettes, P., and Laj, C. (1972). *J. Phys. (Paris)* **33,** C1-125–C1-129.

Carman, R. L., and Kelley, P. L. (1968). *Appl. Phys. Lett.* **12,** 241–243.

Carslaw, H. S., and Jaeger, J. C. (1963). *"Operational Methods in Applied Mathematics."* Dover, New York.

Craddock, H. C., and Jackson, D. A. (1968). *J. Phys. D* **1,** 1575–1577.

Dabby, F. W., Gustafson, T. K., Whinnery, J. R., and Kohanzadeh, Y. (1970). *Appl. Phys. Lett.* **16,** 362–365.

Da Costa, G., and Calatroni, J. (1978). *Appl. Opt.* **17,** 2381–2385.

Davis, C. C. (1979). *IEEE J. Quantum Electron.* **QE-15,** 26–29.

Davis, C. C., and Petuchowski, S. J. (1981). *Appl. Opt.* **20,** 2539–2554.

Dovichi, N. J., and Harris, J. M. (1979). *Anal. Chem.* **51,** 728–731.

Dovichi, N. J., and Harris, J. M. (1980). *Anal. Chem.* **52,** 2338–2342.

Dovichi, N. J., and Harris, J. M. (1981a). *Anal. Chem.* **53,** 106–109.

Dovichi, N. J., and Harris, J. M. (1981b). *Anal. Chem.* **53,** 689–692.

Fang, H. L., and Swofford, R. L. (1979). *J. Appl. Phys.* **50,** 6609–6615.

Fang, H. L., and Swofford, R. L. (1980). *J. Chem. Phys.* **73,** 2607–2617.

Fang, H. L., Gustafson, T. L., and Swofford, R. L. (1982). Submitted for publication.

Flynn, G. W. (1974). *In* "Chemical and Biochemical Applications of Lasers" (C. B. Moore, ed.), Vol. I, pp. 163–201. Academic Press, New York.

Fujishima, A., Masuda, H., Honda, K., and Bard, A. J. (1980). *Anal. Chem.* **52,** 682–685.

Gordon, J. P., Leite, R. C. C., Moore, R. S., Porto, S. P. S., and Whinnery, J. R. (1964). *Bull. Am. Phys. Soc. [2]* **9,** 501.

Gordon, J. P., Leite, R. C. C., Moore, R. S., Porto, S. P. S., and Whinnery, J. R. (1965). *J. Appl. Phys.* **36,** 3–8.

Grabiner, F. R., Siebert, D. R., and Flynn, G. W. (1972). *Chem. Phys. Lett.* **17,** 189–192.

Gupta, M. C., Hong, S.-D., Gupta, A., and Moacanin, J. (1980). *Appl. Phys. Lett.* **37,** 505–507.

Harris, J. M., and Dovichi, N. J. (1980). *Anal. Chem.* **52,** 695A–706A.

Helander, P., Lundström, I., and McQueen, D. (1981). *J. Appl. Phys.* **52,** 1146–1151.

Hu, C., and Whinnery, J. R. (1973). *Appl. Opt.* **12,** 72–79.

Jackson, W. B., Amer, N. M., Boccara, A. C., and Fournier, D. (1981). *Appl. Opt.* **20**, 1333–1344.

Jamieson, D. T., Irving, J. B., and Tudhope, J. S. (1975). "Liquid Thermal Conductivity. A Data Survey to 1973." H. M. Stationery Office, Edinburgh.

Kern, D. Q. (1950). "Process Heat Transfer." McGraw-Hill, New York.

Kliger, D. S. (1980). *Acc. Chem. Res.* **13**, 129–134.

Kogelnik, H. W., and Li, T. (1966). *Proc. IEEE* **54**, 1312–1329.

Kogelnik, H. W., Ippen, E. P., Dienes, A., and Shank, C. V. (1972). *IEEE J. Quantum Electron.* **QE-8**, 373–375.

Kohanzadeh, Y., and Auston, D. H. (1970). *J. Quantum Electron*, **6**, 475–477.

Leite, R. C. C., Moore, R. S., and Whinnery, J. R. (1964). *Appl. Phys. Lett.* **5**, 141–143.

Long, M. E., Swofford, R. L., and Albrecht, A. C. (1976). *Science* **191**, 183–185.

Longaker, P. R., and Litvak, M. M. (1969). *J. Appl. Phys.* **40**, 4033–4041.

Miles, R. B., Gelfand, J., and Wilczek, E. (1980). *J. Appl. Phys.* **51**, 4543–4545.

Miller, A., Hussmann, E. K., and Mclaughlin, W. L. (1975). *Rev. Sci. Instrum.* **46**, 1635–1638.

Miyaishi, K., Imasaka, T., and Ishibashi, N. (1981). *Anal. Chim. Acta* **124**, 381–389.

Moacanin, J., Gupta, A., and Hong, S.-D. (1979). NTIS N-79-17683.

Murphy, J. C., and Aamodt, L. C. (1980). *J. Appl. Phys.* **51**, 4580–4588.

Nieman, G. C., and Colson, S. D. (1978). *J. Chem. Phys.* **68**, 2994–2996.

Patel, C. K. N., and Tam, A. C. (1981). *Rev. Mod. Phys.* **53**, 517–550.

Pelletier, M. J., Thorsheim, H. R., and Harris, J. M. (1982). *Anal. Chem.* **54**, 239–242.

Rieckhoff, K. E. (1966). *Appl. Phys. Lett.* **9**, 87–88.

Severin, P. J., and van Esveld, H. (1981). *Appl. Opt.* **20**, 1833–1839.

Sinclair, D. C. (1970). *Appl. Opt.* **9**, 797–801.

Solimini, D. (1966a). *J. Appl. Phys.* **37**, 3314–3315.

Solimini, D. (1966b). *Appl. Opt.* **5**, 1931–1939.

Stone, J. (1972). *J. Opt. Soc. Am.* **62**, 327–333.

Stone, J. (1973). *Appl. Opt.* **12**, 1828–1830.

Swofford, R. L., and Albrecht, A. C. (1978). *Annu. Rev. Phys. Chem.* **29**, 421–440.

Swofford, R. L., and McClain, W. M. (1975). *Chem. Phys. Lett.* **34**, 455–460.

Swofford, R. L., and Morrell, J. A. (1978). *J. Appl. Phys.* **49**, 3667–3674.

Tada, Y., Harada, M., Tanigaki, M., and Eguchi, W. (1978). *Rev. Sci. Instrum.* **49**, 1305–1314.

Twarowski, A. J., and Kliger, D. S. (1977a). *Chem. Phys.* **20**, 253–258.

Twarowski, A. J., and Kliger, D. S. (1977b). *Chem. Phys.* **20**, 259–264.

Vaida, V., Turner, R. E., Casey, J. L., and Colson, S. D., (1978). *Chem. Phys. Lett.* **54**, 25–29.

Vest, C. M. (1974). *Phys. Fluids* **17**, 1945–1950.

Weiss, J. D. (1980). *J. Opt. Soc. Am.* **70**, 375–380.

Wesfreid, J., Burgos, A., Mancini, H., and Quel, E. (1977). *Opt. Commun.* **21**, 413–418.

Whinnery, J. R. (1974). *Acc. Chem. Res.* **7**, 225–231.

White, K. I. (1976). *Opt. Quantum Electron.* **8**, 73–75.

Yariv, A. (1975). "Quantum Electronics," 2nd ed., pp. 110–114. Wiley, New York.

Zitter, R. N., Koster, D. F., Cantoni, A., and Ringwelski, A. (1980). *High Temp. Sci.* **12**, 209–214.

4 LASER IONIZATION SPECTROSCOPY AND MASS SPECTROMETRY

David H. Parker

Department of Chemistry
University of California
Santa Cruz, California

233

I. INTRODUCTION

Charged particles are very desirable objects for detection by any ultra-sensitive technique owing to their high collection efficiency and their direct conversion to laboratory signal. Since 1895 it has been known that photons (x rays in the early studies) can create charged particles in gases. Here the photon energy obviously exceeds the atomic or molecular ionization potential (IP). Following the work of Bohr in 1913 it was clear that photons that could excite an atom to a high-lying electronic state could then bring that excited state past the IP in a sequential absorption process. This can only be accomplished, however, at high enough photon flux that excited-state photoionization occurs at a rate competitive with other processes which remove excited-state population. Predissociation is a common population loss mechanism in polyatomic molecules, with rates as high as 10^{14} sec^{-1}. Because of this flux requirement, resonant two-photon ionization (R2PI) was not actively investigated until the advent of high-intensity pulsed lasers in the middle 1960s.

Intense lasers can also cause nonlinear or multiphoton events, which take place simultaneously ($\sim 10^{-15}$ sec) and follow an I^n dependence (where I is the photon flux and n the number of photons absorbed). Multiphoton absorption may terminate at energies above the IP, leading to multiphoton ionization (MPI). Polyatomic molecules have dense excited-state manifolds; thus resonance is likely, for example, at the second or third photon in a four-photon ionization process. This is called resonance-enhanced MPI (REMPI) because it is more probable than the corresponding MPI event. The purpose of this chapter is to summarize what information can be gained from the study of resonant multiphoton ionization—R2PI and REMPI—in polyatomic molecules, and the methods by which this information is obtained.

The discussion is best begun by distinguishing the various ionization processes which may occur at different laser wavelengths from the vacuum UV to the visible. Figure 1 shows the energy levels of benzene, which often serves as a prototypic polyatomic molecule. Since the first IP lies 9.2 eV above the ground state X, direct one-photon ionization (a)

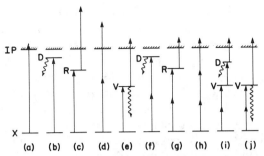

Fig. 1. Multiphoton ionization processes: (a) one-photon ionization, (b) below-threshold excitation to a dissociative state D, (c) R2PI through a Rydberg state R, (d) two-photon ionization, (e) R2PI through a valence state V, (f) below-threshold two-photon excitation to a dissociative state D, (g) REMPI through a Rydberg state, (h) three-photon ionization, (i) REMPI through a valence state, and (j) two-color REMPI.

begins at 134 nm. This electric-dipole-allowed absorption can, of course, be observed with weaker, nonlaser, sources such as rare-gas continuum lamps or electron synchrotrons. Below threshold (b), resonant states often dissociate (jagged lines) long before another photon is absorbed. Rydberg states R, however, are less coupled to the molecular core and are often more stable with respect to dissociation. R2PI through a Rydberg state (c) may compete at high laser flux with other loss processes. At even higher flux nonresonant two-photon ionization (d) may occur. The lowest valence state V of benzene lies at greater than one-half the IP, and within the convenient tuning range of dye-laser systems. R2PI in this case (e) is usually efficient since the lower excited states are often (although not always) long lived. Fluorescence is often observable from such states. Processes (f–j) also lie within dye-laser wavelengths but require three (or more) photons to reach the ionization threshold. Since two-photon absorption takes place at higher flux, increased competition with dissociation (f) may be expected once the resonant level is reached. REMPI through Rydberg states (g) is more probable than three-photon ionization (h). REMPI through the lowest valence level (i), however, is a four-photon process where the third photon lies in regions in which competition from dissociation is expected. This problem may be overcome by two-color ionization (j) where the lowest excited state is populated by two-photon absorption of visible light and ionized by a UV photon. Other two-color schemes are possible, as are other REMPI processes such as three-photon resonant, four (or higher) photon ionization. The (c) and (e) R2PI processes; (g) and (i) REMPI processes; and two-color schemes such as (j) form the main body of laser ionization events discussed in this text.

The important observations to be made from the preceding discussion are that resonant states may enhance ionization (often by several orders of magnitude), and that ionization begins essentially from the resonant state rather than the ground state. At present, most activity in laser ionization research centers around three fields which stress different aspects of the above observations. These fields are (i) excited-state spectroscopy and dynamics, (ii) the study of photoionization and subsequent events originating from an excited state, and (iii) the coupling of resonant multiphoton ionization with mass spectroscopy for development of a molecule-specific, state-selective ultrasensitive detector.

There are several advantages to the study of excited electronic states by laser ionization. While it is difficult to measure direct absorption quantitatively at high resolution, ions arising via R2PI through the same state are easily detected and with resolution comparable to the best absorption spectrometers. Multiphoton absorption follows different symmetry selection rules than those of one-photon absorption; thus new states can be detected via REMPI, using visible rather than ultraviolet photons. In addition, excited-state lifetimes and relaxation mechanisms can be determined by time-delaying the ionizing photon in the two-color ionization scheme.

Most excited states involved in REMPI or R2PI lie within one visible or near-UV photon of the ionization potential. Ionization processes from these states may thus be studied with high-resolution lasers. Photoionization mass spectroscopy, photoelectron spectroscopy, and other conventional techniques now being applied to REMPI and R2PI systems provide information on the mechanism of ionization and fragmentation of neutral molecules and ions in intense laser fields.

Two-dimensional mass spectroscopy is another important asset of resonant multiphoton ionization. Ions are created efficiently only when the laser is tuned to specific excited states of the selected molecule. The ionization spectrum can be used to characterize the initial-state distribution (ground and/or excited) of the molecule, and mass analysis of the resulting ion(s) also allows direct identification. Charged particles can be detected with very high sensitivity, even with mass dispersion. The possibility arises that very small concentrations of molecules in any specified rovibronic state may be measured at a defined time in a defined volume in the presence of a large background. Efforts in this direction form the third branch of laser ionization research.

A number of promising applications of resonant multiphoton ionization have been outlined. While many have been demonstrated, obstacles to complete realization of these potentials abound. There are problems related to the laser and the molecule, and to the interpretation of the ionization mechanism. Implicit in this introduction is the most simple mechanism—where each absorption step, whether multiphoton or one photon, is separable. Whether this attractive view is correct depends greatly on the characteristics of the resonant state and on the various properties (intensity, pulse length, coherence, etc.) of the laser. Resonant multiphoton ionization of polyatomics is a relatively new field of research with uncertainties and complications which have yet to be resolved. A substantial portion of this text is devoted to discussion of these issues. Areas where caution is needed will be defined, along with the inherent limitations of the technique. A few illustrative examples of the spectroscopic, photoionization, and sensitive detection applications will be given.

A. Multiphoton Processes—Historical Background

Most of what is known about the interaction in light with molecules comes from the study of one-photon processes such as absorption and emission. The laser has now made possible the observation of many novel and exotic events involving the absorption of more than one photon. These are called multiphoton transitions, of which there are three types: (i) those in which the final state is identical to the initial state, for example nth-harmonic generation; (ii) those in which the final state is bound but different than the initial state, e.g., the Raman effect; and (iii) those in which the final state is a continuum, such as multiphoton ionization and dissociation. All three types have their own areas of practical and theoretical interest, and each should be considered in any MPI study. This chapter is concerned with multiphoton ionization and with two-photon absorption of the second type. Besides the specific topics of this work, MPI has a more general implication. Multiphoton ionization becomes the dominant primary process in all systems as the light intensity increases, regardless of wavelength (Bunkin and Prokhorov, 1964). Intensities have now reached the point where ionization should be considered in any experiment using present-day pulsed lasers.

This experimental ability to cause MPI has been only a recent development in multiphoton research. Nonlinear processes were first recognized over 50 years ago with the interpretation by Kramers and Heisenberg (1925) of two-photon events that depend linearly on the incident light intensity, e.g., resonant scattering (Rayleigh and Raman scattering). Only three years later, Raman and Krishnan (1928) observed what is now known as the Raman effect. Two-photon absorption, predicted by Göppert-Mayer as early as 1931, follows a quadratic intensity dependence. Brossel et al. (1953) made the first observation of two-photon absorption, in the radio frequency region of the spectrum. The detection of two-photon absorption in the visible region was delayed until the development of the laser when, in 1961, Franken *et al.* observed second-harmonic generation in quartz.

Multiphoton ionization was first detected in the form of a spark at the focal point of a focused ruby laser by Maker *et al.* (1964). MPI was predicted by Bunkin and Prokhorov in 1964, and was first rigorously treated by Keldish in the same year. The first direct observation of MPI was reported by Voronov and Delone (1965) in xenon atoms. The early investigations of the MPI process were mainly confined to ionization of noble gases with high-intensity ruby lasers. With the introduction of tunable but lower-power dye lasers, MPI research expanded to the easily ionized alkali-metal atoms, metal dimers, and to polyatomic molecules.

Early work on molecular MPI was reported by Collins *et al.* (1973) on Cs_2, and by Lineberger and Patterson (1972) on two-photon photodetachment of negative ions. The demonstration of multiphoton ionization as a new spectroscopic tool is credited to Johnson and co-workers (1975) for work on NO, and to Dalby and co-workers (Petty *et al.*, 1975) for work on I_2, both in 1975. One of the earlier molecular R2PI studies was reported by Mainfray and co-workers (Held *et al.*, 1972) on the detection by time-of-flight mass analysis of Cs_2^+ in an atomic beam of Cs. Work on similar easily ionized systems followed (Granneman *et al.*, 1977). A great step forward in two-laser R2PI was made by Letokhov and co-workers (Andreyev *et al.*, 1977) on two-step photoionization of formaldehyde, quickly followed in the next year by mass-analyzed R2PI of various molecules in molecular beams. In 1978, Schlag and co-workers (Boesl *et al.*, 1978) reported R2PI of a molecular beam of benzene with mass analysis, and Bernstein and co-workers (Zandee *et al.*, 1978) reported REMPI with mass analysis of I_2 and also benzene (Zandee and Bernstein, 1979a,b).

B. Perturbation Theory

More than 30 years passed between the prediction and observation of multiphoton absorption in molecules. The reasons for this gap can be appreciated by considering the results of time-dependent perturbation theory for nonresonant multiphoton absorption. For an *n*-photon transition the lowest nonvanishing term in the perturbation expansion is of the *n*th order. This term predicts a transition rate (per second) of

$$W_n = \sigma_n I^n \tag{1}$$

where I is the laser flux (photons/cm² sec) and σ_n is the *n*th-order cross section (units $cm^{2n} sec^{n-1}$). A simple I^n dependence for an *n*-photon tran-

sition holds only when no intermediate resonances exist. With the introduction of resonances the intensity dependence goes as I^{n_0}, where n_0 is
usually a nonintegral number less than n.

The cross section σ_3 for three-photon ionization is given by perturbation theory as

$$\sigma_3 \propto \sum_i \sum_j \frac{\langle f|r \cdot \varepsilon|j\rangle\langle j|r \cdot \varepsilon|i\rangle\langle i|r \cdot \varepsilon|g\rangle}{(E_j - 2h\nu_L - E_g)(E_i - h\nu_L - E_g)} \tag{2}$$

where g and f are the ground and final levels and i and j are the two
intermediate levels. These "virtual" intermediate levels are formed from
energy-weighted summations $\sum_i\sum_j$ over all real bound and continuum
states of the system. The electric dipole term $r \cdot \varepsilon$ represents the molecule radiation interaction and $h\nu_L$ is the photon energy. The two terms in
the denominator are weighting factors determined by the energy separation of the real and virtual states. Provided the three electric dipole transition moments are known or can be estimated in the summation over the
important excited states, (1) and (2) can be used to predict the ionization
rate.

Equations (1) and (2) also account for the long delay in multiphoton
research. Consider Fig. 2 where the ionization rate w_n of Eq. (1) is plotted
versus flux (photons/cm^2 sec) for a typical range of one-, two-, and three-
photon cross sections. Conventional one-photon absorption cross sections vary from 10^{-16} to 10^{-22} cm^2 and conventional light sources produce
as many as 10^{15} photons/cm^2 sec within a useful spectroscopic width.
Even weak one-photon absorption can be detected under these conditions. At the same laser flux, however, a two-photon transition is 10^{20} less
probable. Two-photon cross sections are known experimentally and by
calculations similar in form to Eq. (2) to range from 10^{-48} to 10^{-57} cm^4 sec.
A three-photon transition is less probable by 10^{30} at this flux, which explains why multiphoton transitions could not be observed before the laser.

The type of lasers used in the molecular MPI apparatus can easily
produce a flux of 10^{28} photons/cm^2 sec during a 5-nsec pulse, as is discussed in Section II. Figure 2 shows that at this level two- and three- (and
higher-order) photon processes become not only observable but in many
cases most probable or even saturated. Above 10^{35} photons/cm^2 sec perturbation theory is no longer valid since the electric field $E = (4\pi I h\nu/c)^{1/2}$
exceeds the binding energy of an electron to the nucleus. In the other
extreme, Lambropoulos (1976) has estimated the minimum photon flux
necessary to observe multiphoton effects as $\sim 10^{20}$ photons/cm^2 sec. With
typical laser intensities of 10^9 W/cm^2 (which is related to the flux I and the
wavelength λ in nanometers by W/cm$^2 = 2 \times 10^{-16} I/\lambda$), perturbation
theory [Eqs. (1) and (2)] should be valid.

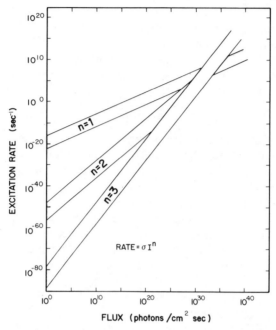

Fig. 2. Plot of the excitation rate (sec^{-1}) vs the instantaneous laser flux (photons/cm^2 sec) for one-, two-, and three-photon absorption processes.

C. Survey of Atomic MPI Research

The study of MPI in atoms is now a large and mature research field as reviewed most recently by Lambropoulos (1980) and by Mainfray and Manus (1980). Active topics include general MPI theory, the effects of photon statistics, resonances, polarization, angular momentum, and the influence of the above on the resulting electrons.

Photon statistics or field correlations are important in multiphoton transitions owing to the necessity of absorbing another photon during the exceedingly short ($\sim 10^{-15}$ sec) lifetime of a virtual state. A transition is much more likely when the photons come in "bunches" than when they arrive equally spaced, as in a single-mode laser. For an N-photon transition a chaotic light source is known (see chapter 2 by Birge for a discussion) to give an $N!$ enhancement over a pure coherent source. This effect has been dramatically exhibited (Lecompte *et al.*, 1975) in the 11-photon ionization of Xe where an essentially chaotic (~ 1000 modes) Nd laser gave a $10^{6.9 \pm 0.3}$ ($\approx 11!$) stronger signal than a single-mode Nd laser of the same average intensity.

Resonances considerably complicate the description of the MPI process. Major efforts have gone toward quantitative calculations of resonance-enhanced cross sections and intensity dependences, and conversely, the laser- and ionization-induced shifting and broadening of real intermediate levels. The proportionality of n-photon ionization to the nth power of the photon flux is correct only in the absence of resonant intermediate states. Note that Eq. (2) approaches infinity when some integer number of photons corresponds with a transition energy. To account for resonance behavior, second-order perturbation theory may be used to introduce intensity-dependent corrections for the atomic level energy $\Delta\omega_j$ and width $\Delta\gamma_j$ as

$$\Delta\omega_j(I) = a_j I, \qquad \Delta\gamma_j(I) = b_j I \qquad (3)$$

where a_j and b_j are proportionality constants. If, for example, level R is two-photon resonant, then on inserting these linewidths and shifts into Eq. (2) the transition rate can be found proportional to

$$W_n \propto I^n[(E_R - E_g - 2h\nu_L + C'_R I)^2 + (\gamma_R + d'_R I)^2]^{-1} \qquad (4)$$

where $C'_R = C_R - C_g$ and $d'_R - d_g$, and the remaining parameters are field-free values. The main feature of Eq. (4) is that the simple I^n dependence no longer holds, and that the order of nonlinearity will vary as the laser is tuned through resonance. If level broadening is greater than the shift, the measured nonlinearity will be less than n. If level shifting is greater than level broadening, however, on one side of the peak the laser field will pull the transition into resonance—giving $n_{measured} > n$—and on the other side the field will shift the level further away, giving $n_{measured} < n$. This rather dramatic behavior has been observed in MPI of atoms (Bakos et al., 1972) and molecules (Johnson, 1980a). The former case has also been observed in atoms (Delone et al., 1972) and undoubtedly occurs in many molecules.

The enhancement of the MPI signal by a resonance has a definite time dependence as illustrated (Lompre et al., 1976) in the multiphoton ionization of xenon, krypton, and argon, where resonance conditions are known to be satisfied for the Nd–glass laser tuning range. While a large enhancement was observed with a 10-nsec laser pulse, no resonance enhancement was detected with 30-psec laser pulses. In a later study by the same group (Lompre et al., 1978) on the four-photon ionization of Cs resonance effects were clearly observed even with 1.5-psec pulses. These differences were explained in terms of a characteristic ionization time required to reach a steady-state population of the resonant state.

Other topics of interest in atomic MPI include autoionizing states and the effects of different MPI selection rules on their behavior, interference phenomena in two-photon ionization, and multiphoton events in laser-induced inelastic collisions. In summary, the above subjects and especially the active theoretical efforts in atomic MPI represent a healthy research field which is very relevant to REMPI and R2PI processes in polyatomic molecules.

D. REMPI, R2PI of Polyatomic Molecules—Reviews

Figure 1 has introduced various schemes for laser ionization of polyatomic molecules. The similarity of two different resonance processes, R2PI and REMPI, has been emphasized. A distinction should be made between REMPI and R2PI which is not always observed in the literature. Since no two-photon absorption through virtual states takes place in R2PI processes, they are more properly called "multiple-photon" rather than "multiphoton" events. Processes consisting of one-photon steps (multiple-photon events) predominate at lower flux. R2PI and REMPI spectroscopy developed almost at the same pace, however, mainly because doubling of tunable lasers used for REMPI to frequencies useful for R2PI is only a few percent efficient. Signal levels are thus similar for both processes. Advancements in laser technology are now being made toward increased intensity and tuning range in the ultraviolet, favoring R2PI studies as a whole and REMPI of small molecules.

REMPI has proven extremely useful as a spectroscopic probe of the excited electronic structure of many polyatomic molecules. Reviews of REMPI spectroscopy have been presented by Johnson and Otis (1981), Johnson (1980a,b), and Parker et al. (1978). Just as photoionization follows resonant excitation, further absorption may follow ionization. This can lead to extensive fragmentation of the parent ion, especially under high-flux conditions. The processes which take place after resonant-state excitation have been directly investigated only lately, via mass spectroscopy, photoelectron, and photoion energy spectroscopy. Robin (1980) and Bernstein (1982) have reviewed the most recent progress in understanding the systematics of laser ionization-fragmentation of polyatomics.

Antonov and Letokhov (1981) have reviewed R2PI and REMPI spectroscopy and mass spectroscopy, in addition to discussing the development of laser ionization mass spectroscopy as an analytic tool. Their review most closely covers the material presented in this text. Lichtin et al. (1981a) have discussed the potential analytical aspects of REMPI mass spectroscopy. A very relevant and rigorous treatment of analytical resonant ionization spectroscopy of atoms has been presented by Hurst et al. (1979). The reader is also referred to the excellent monograph of Berkowitz (1979) on photoionization and photoelectron spectroscopy.

II. EXPERIMENTAL CONSIDERATIONS

A critic who might describe laser ionization research as the easiest experiment one could do with a state-of-the-art laser system is not far off the mark. Resonant multiphoton ionization, particularly R2PI and REMPI spectroscopy, is a simple technique, mainly because of the fact that ions are detected. Several of the important and unique types of information available via laser ionization have been mentioned in Section I and are further discussed in the following sections. As an example of the simplicity and content of REMPI spectroscopy consider Fig. 3, a schematic of two-photon resonant three-photon ionization. Several torr of an organic molecule are placed in a parallel-plate ionization cell, biased by 100 V/cm. A pulsed tunable dye-laser beam is focused between the plates and tuned through a portion of the visible spectrum. When the laser is resonant with a two-photon allowed state $v^\dagger$, at the energy sum of two dye-laser photons an increase in the ion current appears. On resonance, currents as high as 10^{-9} A are generated, and measured with a simple electrometer. Two-photon states, which are often impossible to detect by any other technique, are simply traced out as resonances in the ion current spectrum. Other laser ionization experiments can, of course, become much more complex than this, particularly those where mass analysis is required.

In this section the general laser ionization experiment will be discussed in terms of three components: (i) the laser, (ii) the sample, and (iii) the ion detection apparatus. Optimization of the experiment will be emphasized, and the inherent limitations will be pointed out.

A. The Laser

In a REMPI process the initial multiphoton absorption is usually rate limiting. By considering a two-photon cross section of 10^{-50} cm^4 sec it is

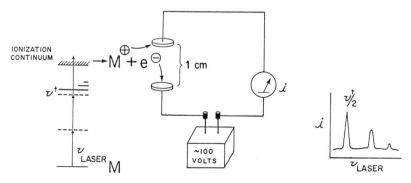

Fig. 3. Schematic of two-photon resonant three-photon ionization spectroscopy.

reasonable that the laser should produce an intensity on the order of 10^{-25} cm^4 sec. Most lasers can be focused down to a beam waist of ~20 μm or ~3 × 10^{-6} cm^2; thus a 1-W cw laser at 500 nm (2.5 × 10^{18} photons/sec) should cause two-photon absorption. In practice, however, working so close to the limits is inconvenient, and cw lasers are not used.

Pulsed lasers can easily produce such intensities. For example, a typical N_2-pumped dye-laser output of 1 mJ/pulse, 5-nsec pulse length at 500 nm can be focused tightly to generate 10^{29} photons/cm^2 sec, or an intensity of 5 × 10^{10} W/cm^2. It is usually advantageous to work with less tightly focused beams than this, in order to increase the number of molecules in the focal volume.

Tunable dye lasers, pumped by N_2, eximer, or Nd–YAG lasers are used almost exclusively in molecular REMPI experiments. R2PI experiments often utilize the frequency-doubled output of these lasers, although some studies are done without wavelength optimization using Nd–YAG laser harmonics (355, 266 nm) or the direct eximer outputs (XeCl 306, KrF 250, ArF 193 nm). Recently, a picosecond laser R2PI fragmentation study has been reported (Hering *et al.*, 1981) using single-shot ~50-μJ, 20-psec (~4 × 10^{11} W/cm^2) pulses.

One of the least desirable aspects of short-pulse lasers is their coherence qualities. As discussed previously, chaotic sources are $n!$ more efficient than coherent (single-mode) sources in n-photon processes. For both types of sources, average intensity—which can be directly measured—is a good indication of the instantaneous intensity, which determines the multiphoton absorption rate. Unfortunately, present-day pulsed dye lasers fall in between the two limits, operating simultaneously on several transverse and axial modes with random coupling between each mode. This results in substantial spatial and temporal fluctuations, making accurate intensity calculations questionable. Short coherence times do allow use of less complicated rate equations, however, as discussed in Section III.

Another coherence-caused problem arises due to scanning the dye laser, as was first described by Johnson (1976). Both the number of modes and the pulse length increase as the laser is tuned from the edge to the center of the dye profile. Pulses generated at the edge of the dye profile are thus more effective in causing multiphoton transitions than pulses generated at the highest dye gain (attenuated to give the same integrated intensity). It is best to work at the same energy over threshold across the entire dye-gain profile, which is accomplished by feeding the dye-laser intensity measurement back into a servo-controlled attenuator of the pump beam to maintain constant dye-laser output. Intensity normalization through qualitative equations such as (1) and (4) is thus avoided.

Intensity dependences at constant wavelength are found by attenuating the dye-laser output directly.

Spectral bandwidths of ~ 1 cm^{-1} are common for tunable pulsed dye lasers, and since the dye-laser tuning regions of interest for many REMPI studies can exceed 20,000 cm^{-1}, calibration is an important consideration. By deflecting a portion of the laser beam through a hollow-cathode lamp and onto an etalon with ~ 15 GHz free spectral range, the wavelength can be measured very accurately [using the optogalvanic effect (King *et al.*, 1977)] and almost continuously (by monitoring etalon fringes with a photodiode) simultaneous with the ionization spectrum. Less elegant and more tedious calibration can, of course, be done with a high-resolution monochromator.

Proper focusing is very important in multiphoton experiments, not only in optimizing the intensity to focal volume ratio, but also in avoiding geometric saturation effects. At the lowest intensities most REMPI processes follow the overall order of the process, i.e., I^4 for four-photon ionization. Kinetic saturation can take place quickly, however, yielding less than integral intensity dependences. It is often very easy to ionize every molecule within the confocal region of a tightly focused beam. As the laser intensity increases, the completely ionized region grows out into the less focused beam (dog-bone geometry) and then follows an $I^{3/2}$ power law regardless of the number of photons absorbed (Speiser and Jortner, 1976). Often, the collection geometry is exceeded (e.g., with crossed laser–molecular beams, ions are formed only up to the width of the molecular beam), at which point a log I dependence is found which reflects the growing laser beam waist. These processes occur gradually and are often difficult to detect and avoid. The best approach is to use the longest-focal-length lens which still allows acceptable signal-to-noise ratios.

B. The Sample

Laser ionization experiments have been carried out with wide ranges of sample conditions, from skimmed, collision-free supersonic molecular beams to pressures as high as 50 atm. Liquid-phase REMPI and R2PI studies have also been reported, as have studies of unstable free radicals, nascent photofragments, and reactive scattering products, as described in the following sections. Even media where ions are already present, such as flames, are approachable by using time-gated detection.

Simple laser ionization chambers operate from several Torr down to the sample vapor pressure at the cell temperature. Although laser ionization in proportional counters can be sensitive to extremely small amounts of molecules, most experiments are not attempted with less than a few millitorr of sample. High-pressure cells have been equipped with parallel-plate electrodes to monitor pressure-dependent broadening and shifting of resonant Rydberg states in the REMPI process (Robin, 1980).

High electrode bias (several thousand volts per centimeter) is needed to collect a substantial amount of photoions in liquid-phase laser ionization. Exceptional care must thus be taken in purifying solvents of charge carriers to avoid large dc signals. Since ionization potentials are generally lowered in condensed media, photoionization of impurities becomes much more important than in the gas phase, because R2PI of impurities can often mask REMPI of the desired sample. Spectral selectivity is of course quite compromised in condensed phases because of broadening.

Even in the gas phase, spectral selectivity of large polyatomic samples mixed with impurities of similar size is usually lost because of broad resonance levels present in R2PI and to a lesser extent in REMPI. Such broadening is often caused by spectral congestion, i.e., overlapping transitions from the extremely large number of rovibrational levels populated at room temperature. Supersonic jets, formed by expanding a carrier gas (He, Ar, . . .) mixed with a small percentage of sample through a small orifice into vacuum, offer very efficient solutions to this problem. Collisional cooling of translational, rotational, and to a lesser degree vibrational temperatures takes place in these expansions and results in tremendous spectral simplification. Rotational temperatures less than 20 K are routine; thus continuous 100-cm^{-1}- wide rovibronic bands reduce to a few narrow lines. By concentrating the ground-state population into a few rotational levels, a significantly larger number of molecules can absorb within the laser bandwidth. By careful selection of carrier gas, expansion nozzle temperature, and backing pressure, clustering in the jet beam arising from van der Waals complexes can usually be avoided (Amirav et al., 1980).

Since pulsed lasers operate at low repetition rates (less than 100 Hz) it is best to work with pulsed nozzles. Common pulsed valves (see, for example, Otis and Johnson, 1980) operate at 10–20 Hz, producing 20–200-μsec-long pulses and may be heated to ~100°C. Several advantages come with the use of pulsed valves: total throughput into the vacuum is greatly reduced, allowing smaller pumping systems; with short pulses (~20 μsec) wall reflections are avoided, thus smaller vacuum chambers are feasible; and in addition, sample is conserved.

Supersonic beams do not solve all problems in understanding resonant spectra, or achieving state selectivity. Rotationally cold spectra of complex molecules are not trivial to interpret, especially when torsional modes become more visible. In determining excited-state structure, more than the first few J levels available in cooled spectra often need to be measured. Cooling does not guarantee that the transitions lie further apart and are easily resolved; thus narrow-bandwidth lasers can become necessary, requiring slower scans and loss of intensity. Supersonic beams also complicate the otherwise simple apparatus, to a degree dependent on the extent of cooling desired. If a clean sample can be introduced into the ion cell at a reasonable vapor pressure ($\sim$1 Torr) then rotational cooling may not be necessary for spectroscopic identification. Almost all other experiments can benefit from supersonic expansions.

C. Detection Apparatus

A property of the laser is usually measured versus a property of the ions in most laser ionization experiments. REMPI spectroscopy, for example, monitors total ion current versus laser wavelength (at constant laser intensity). Other commonly varied properties of the laser are intensity, polarization, and, for two-laser experiments, the time between the first and second laser pulses. Electrons, cations, and anions may be produced by laser excitation and three different characteristics of each are measured: (i) total amount, (ii) the mass-to-charge ratio M/e, and (iii) the initial kinetic energy distribution. Other possibilities include the internal energy distribution of molecular ions, and the laboratory angular distribution of the various charged species, along with other types of concurrent measurements including excited-state and photofragment fluorescence and neutral-fragment identification by electron impact mass spectroscopy, or by other laser techniques such as stimulated Raman scattering. Many such combinations of measurements have been reported and are described in the following sections, especially those which help elucidate REMPI and R2PI mechanisms.

1. Total Ion Current

Ionization chambers, which are some of the simplest, oldest, and most sensitive devices available, are used for monitoring free electrons, negative ions, and positive ions. Since these ions are created during the short laser pulse length, time-gated electronics can be set to detect only when the ion pulse drifts to the collection electrodes, thus eliminating background ionization by cosmic rays or other processes. Electrons drift faster than heavy cations and anions, and can usually be distinguished. Two standard ion cell configurations—parallel plate and cylindrical—are shown in Fig. 4. The proper operating conditions for each type are discussed later. Ion cells are constructed of almost any material, although glass or metal cells are easiest to handle. Windows should be replaceable

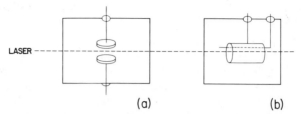

Fig. 4. Ionization cells: (a) parallel plate, (b) cylindrical.

since laser-caused damage from focusing, or from pyrolysis of sample on the inside surface, is sometimes difficult to avoid. Only moderate evacuation between samples ($\sim 10^{-4}$ Torr) is usually needed. Flowing the sample through the cell can avoid buildup of photochemical product. However, in some molecules this can initiate a glow discharge, even at low field bias (Lahmani and Srinivasan, 1976). Ion cells require a reasonable knowledge of gaseous electronics for most effective use (see, for example, Huxley and Crompton, 1974), especially in choosing the proper voltage bias.

Parallel-plate cells (Fig. 4a) have the least constricted geometry, allowing access for crossed laser beams, laser–molecular beams, or for monitoring fluorescence simultaneously. These cells should be operated in the plateau region, where all of the initially formed ions are collected, but without gain. With gated detection and careful shielding these cells are sensitive down to a few hundred ion pairs. In practice, $\sim 10^5$ ion pairs (10^{-14} A) are more typical limits for parallel-plate cells.

Higher collection voltage causes amplification of the initial free electrons by collisional cascade. In this mode the cell is a proportional counter, as invented in 1908 by Rutherford and Geiger. The best field design uses a cylinder surrounding a thin collection wire, mounted along the cylinder axis either on or off center, with the laser also focused along the axis as shown in Fig. 4b. Normally, a gain medium—100 Torr of P-10 counting gas (90% Ar + 10% CH_4)—is used with several torr of sample, which with +1000 V on a 0.05-mm collection wire gives $\sim 10^4$ gain. The ion pulse height is proportional to the number of initially formed electrons in the ideal case; however, in practice, spectral distortion is more likely than with parallel-plate cells. Proportional counters can easily record single thermal electrons.

A typical apparatus used for laser ionization spectroscopy is shown in Fig. 5, where a N_2-pumped dye laser is polarized circularly or linearly (by a Fresnel rhomb) and then focused by a simple 5–20 cm focal-length lens into a parallel-plate ion cell biased by 100 V. The signal from a simple photodiode, which monitors the laser through the cell, is processed by a

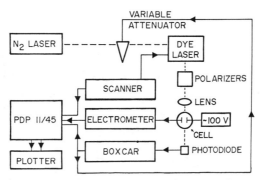

Fig. 5. REMPI spectroscopy apparatus using parallel-plate cell, Fresnel rhomb polarizers, and servo-controlled attenuator for constant dye-laser output during wavelength scans.

boxcar averager and sent back to a servo-controlled attenuator, to maintain constant dye-laser output (intensity, pulse-length, and pulse shape) across the dye profile, by blocking to variable degrees the N_2 laser beam. Adjustment is needed if the monitor is energy sensitive rather than properly flux sensitive, otherwise the red regions of the spectra will be undercorrected. Also shown in Fig. 5 is a sensitive electrometer and a computer for calibration and plotting of the spectra. With lower ionization currents a fast amplifier–boxcar averager can replace the electrometer.

With pressures below $\sim 10^{-5}$ Torr electron multipliers and similar devices can be used, with gains as high as 10^8. Because of the extra vacuum requirements, and usually less stable gain, electron multipliers are employed only when higher vacuum is already present, e.g., with molecular beams or mass spectrometers. Channeltrons are preferable to Cu–Be electron multipliers because of their more stable gain and lower sensitivity to pressure cycling. Multichannel plates are optimal for TOF mass spectrometry owing to their very short (<1 nsec) rise times.

Figure 6 shows a pulsed-valve seeded supersonic beam apparatus used for REMPI spectroscopy of benzene (Aron *et al.,* 1980). Several Torr of benzene mixed into ~ 5 atm of He were expanded in ~ 200-μsec pulses at 10 Hz into an ejector-pumped chamber. The central portion of the jet was skimmed 3 cm downstream into a diffusion-pumped chamber, where the laser beam (2–3 GHz bandwidth, 400 μJ/pulse) intersection takes place. A −600-V plate repels the cations toward a channel electron multiplier located in a separate differentially pumped chamber (3×10^{-6} Torr). Flight paths and ion kinetic energies were set to collect total ions without mass resolution. Spectroscopic results of this study are described in Section IV.

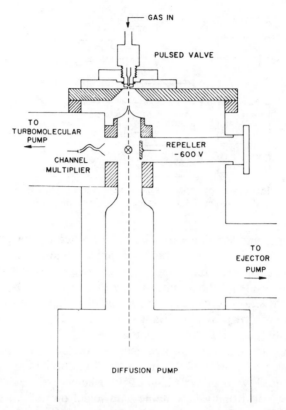

GAS IN

PULSED VALVE

TO
TURBOMOLECULAR
PUMP

CHANNEL
MULTIPLIER

REPELLER
-600 V

TO
EJECTOR
PUMP

DIFFUSION PUMP

Fig. 6. Pulsed-valve seeded supersonic-beam REMPI apparatus (Aron *et al.*, 1980).

2. Mass Analysis of Laser-Produced Ions

Charge-to-mass ratios M/e were the second basic characteristic of laser-produced ions to be measured. With nanosecond pulsed lasers, time of flight (TOF) is the natural technique for mass analysis for two reasons. First, ions are created in a reasonably small volume in a very short time—affording a built-in $t = 0$ for the time-of-flight equation. Second, with low-duty-factor pulsed lasers, obtaining a complete mass spectrum for each pulse is both possible and the only efficient means of running the experiment. The major disadvantage of time of flight is in expensive signal processing electronics—for good mass resolution a ~5-nsec per channel transient recorder and a spectrum storage/processing device such as a lab computer or signal averager is required. Another approach for single-shot mass spectra uses magnetic deflection and spatial (instead of temporal) resolution by a linear array of detectors. This technique is yet to be used in laser MPI ionization studies, and is not clearly more cost effective than TOF. (Of course, a photograph of an oscilloscope trace retains the essential features of the fragmentation pattern.)

Mass filters, such as quadrupoles or fixed-slit magnetic deflectors, are effective for measuring intensity or frequency dependences at single mass. These have the advantage of electronic control of resolution, but require time-gated detection (boxcar averager). Quadrupoles found extensive use in earlier R2PI studies, owing mainly to their availability. Another interesting technique affording mass resolution combines laser ionization with gaseous electrophoresis (known commercially as plasma chromatography) (Lubman and Kronick, 1982). Here, laser-produced ions are formed in a drift tube of N_2 carrier gas, repelled by an electric field, and mass separated according to their mobility. Separation and detection takes place in the millisecond regime in a very simple apparatus, albeit with poor mass resolution.

a. Laser Ionization–Time-of-Flight Mass Spectrometry

Time-of-flight mass analysis is, as stated, one of the most efficient techniques for nanosecond-laser-produced ions and has accordingly received the most attention and development. In this section the apparatus and different factors affecting resolution by TOF will be briefly discussed.

Two approaches to the design of laser–TOF mass spectrometers have been taken. For best mass resolution, the apparatus closely resembles (or often is) a commercial TOF mass spectrometer utilizing electron impact ionization. The plate voltages, timing, and physical dimensions of these spectrometers have long been optimized (Wiley and McLaren, 1955), and with an electron impact source available, calibration and comparison is quite convenient. Fitting an additional laser beam and especially a molecular beam into these commercial machines, however, is quite difficult. The other approach involves redesign of the ionization–acceleration region to allow room for additional beams and differential pumping, at the expense of mass resolution. Such "home-built" spectrometers are, of course, less complex, cheaper, and can be designed to allow slower processing electronics. One advantage is that information on initial kinetic-energy distributions (Section III.C.3) can sometimes be extracted from the individual peak widths.

Figure 7 shows a laser ionization–time-of-flight mass spectrometer in operation at Columbia University. The laser (Quanta Ray DCRIA Nd:YAG laser pumped, PDL dye laser) generates 7-mJ pulses at 10 Hz in the 370-nm region by mixing the dye-laser output with the YAG-laser fundamental. A 50-cm lens (not shown) focuses the laser into the ioniza-

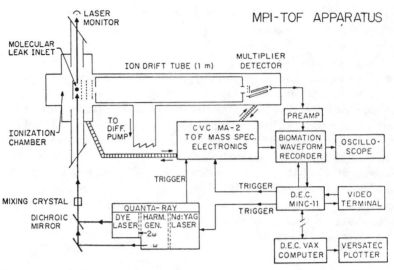

Fig. 7. Laser ionization TOF mass spectroscopy apparatus in use at Columbia University.

tion region of a commercial TOF mass spectrometer with a local static sample pressure of $\sim 10^{-4}$ Torr. Ions formed are accelerated by ~ 3000 V and drift down the 1-m tube pumped by a Hg diffusion pump to an electron multiplier. Flight times for a mass-150 ion are ~ 20 μsec, requiring a fast waveform recorder (Biomation 6500, 2 nsec/channel) for single-mass resolution. The entire waveform (mass spectrum), or portions thereof, may be displayed on an oscilloscope, and stored for addition and processing in a laboratory computer. Alternatively, a boxcar averager can replace the waveform recorder for single-mass laser scans. A very reasonable spectrum is obtained by summing spectra of ~ 50 laser pulses—a process requiring less than one minute, after which the laser wavelength or intensity is changed. Timing of the apparatus is controlled by the computer, which can accept 1024 channel scans at rates of up to 40 Hz. Laser–electron beam sequences are also possible, allowing detection of laser–produced neutral fragments, or laser photodissociation of electron-impact-created ions (Newton and Bernstein, 1982).

Supersonic beams can help recover mass resolution lost owing to the additional spacing in the ionization region needed to accommodate them. This is because the resolution of the TOF spectrometer is determined mainly by the size of the ionizing region and the velocity distribution of the molecules prior to ionization (Wiley and McLaren, 1955). Tighter focusing minimizes the focal volume (i.e., the laser beam waist) and the extremely low translational temperature of supersonic beams eliminates the initial spread of velocities. Figure 8 shows a laser ionization–time-of-flight apparatus which utilizes this concept (Dietz *et al.*, 1980). A, B, and C are the repeller plate, draw out, and flight tube grids, respectively (3100, 2660, and 0 V); E is the flight tube (1 m long) and I the ion multiplier. The A–B distance is 25.4 mm, and a heavy black arrow shows the molecular beam path. In order for ions created in the beam to reach the detector, 120 V must be applied to the deflector plates D, which counteract the translation energy of the beam. Very few room-temperature molecules move at the same speed and in the same direction as the beam; thus the configuration eliminates "hot" background. The authors found resolution to be limited by the time dispersion in the detector and signal processing electronics, not by a time spread in the ion packet.

Time spread can also arise from excess energy in the photodissociation event. Portions of any energy above the dissociation threshold can translate into kinetic energy along the TOF axis, resulting in asymmetric peak spreading. Commercial spectrometers are designed to minimize such spreading, through field design and time-delayed pulses of the draw-out, acceleration bias. Laser ionization mass spectrometers which do not utilize such designs can give information on ion-fragment kinetic-energy release by analyzing peak width (Carney and Baer, 1982). More direct techniques are discussed in the next section.

The time-of-flight equation can be useful in pointing out which factors most affect resolution in laser ionization time-of-flight mass spectrometers (Carney, 1981). If an ion of mass m formed at $t = 0$ is accelerated through a distance X_a by an applied electric field ϵ and then drifts down

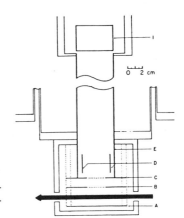

Fig. 8. Vertical cross section of a laser ionization TOF mass spectrometer with supersonic molecular beam (Dietz *et al.*, 1980).

the flight tube of length X_d, the total TOF τ is given by

$$\tau = \frac{1}{\sqrt{a}} \left(\sqrt{2X_a} + \frac{X_d}{\sqrt{2X_a}} \right) \tag{5}$$

where a is the acceleration, $a = q\epsilon/m$ where q is the ion charge, and the initial velocity is assumed to equal zero. Equation (5) shows that the TOF is proportional to $\sqrt{m/q}$, and also to $\sqrt{1\epsilon}$. The mass resolution of the system is controlled by the time resolution of the experiment, which is effected by (i) timing jitter of the electronics—this becomes important at high draw-out fields. (ii) Pressure and ion space-charge effects—the flight tube should be $<10^{-6}$ Torr, and the sample region $\gtrsim 10^{-4}$ Torr to avoid collisional effects such as ion–molecule reactions. Ion densities should be $\gtrsim 10^8$ ions/cm³ to avoid space-charge effects. (iii) Thermal or kinetic effects: these can be avoided with supersonic molecular beams as discussed previously. With static pressures at room temperature, molecules have a thermal kinetic energy E_{th} and corresponding velocity components positive and negative along the TOF axis. An ion initially traveling away from the detector must stop and turn around, in a time which is roughly estimated as

$$(\Delta\tau)_{th} = \frac{2m}{q\epsilon} \sqrt{\frac{2E_{th}}{M}} \tag{6}$$

where m is the mass of any fragment ion and M is the parent-ion mass, assuming no kinetic energy is released in fragmentation (thus fragment ions have the same velocity distribution as parent ions). Since $\Delta\tau$ varies as $1/\epsilon$ thermal effects are smaller at high draw-out fields. (iv) Spatial effects: since the focal point is of finite width ions will be created in different regions and thus experience different draw-out voltages. Different flight times result with a time spread given as

$$(\Delta\tau)_{sp} = \frac{d\tau}{dX_a} \Delta X_a = \sqrt{\frac{1}{2aX_a}} \left(1 - \frac{X_d}{2X_a} \right) \Delta X_a. \tag{7}$$

In this case the time spread follows an $\epsilon^{-1/2}$ dependence. At lower draw-out fields or larger beam waists spatial effects can be quite substantial. This analysis is somewhat oversimplified and may not apply for all TOF mass spectrometers. The essential features have at least been mentioned, and most effects are shown to be minimized by higher draw-out fields. This results in faster flight times, however, and so better (more expensive) time-resolving waveform recorders are needed. Since in most laser ionization experiments mass resolution is not usually a limiting factor, compromise in flight time is usually made.

3. Kinetic-Energy Distribution

The kinetic energy initially imparted to the photoions is the third type of characteristic to consider. While less applicable in an analytical sense, such measurements, particularly for photoelectrons, give detailed dynamical insight, since peaks seen in the photoion and photoelectron energy distribution can often be ascribed to the different ionization fragmentation mechanisms.

In the studies reported thus far, simple TOF drift measurements are made for cations and anions, and more involved devices such as electrostatic energy analyzers are needed for photoelectrons. An apparatus designed for both TOF mass analysis plus photoelectron and photoion kinetic-energy measurement, shown in Fig. 9, is in use at Oak Ridge National Labs (Miller and Compton, 1981a), where a focused N_2-pumped dye-laser beam crosses an effusive molecular beam in the region shown. A voltage V draws positive ions into the drift tube for TOF mass detection by a channeltron (top) and at the same time repels photoelectrons into the entrance of a 160° spherical-sector kinetic-energy analyzer. The energy spectrum is scanned by floating the analyzer with a ramp voltage to accelerate or retard the incoming electrons to the analyzer transmission energy. The electron energy resolution was ~150 meV. By simply reversing the voltage bias, energy spectra of positive ions can also be measured in a similar fashion. By turning off the bias voltage, energy distributions can be taken using the TOF drift tube. DeVries *et al.* (1981) have described such a technique and applied it in a very complete study of fragmentation–ionization of molecular iodine. When TOF is done without applied

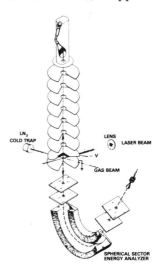

Fig. 9. Multiphoton ionization mass spectrometer and photoelectron energy analyzer (Miller and Compton, 1981a).

fields, fragment angular dependences (laser polarization versus detection angle) can also be measured. In these studies the deconvolution of initial thermal velocity distributions and detection steric factors can become very involved. An important observation of this study was that ion-fragment acceleration caused by space charge occurs at ion densities as low as 10^6 ions/cm^3. Careful magnetic shielding and avoidance of stray charges are, as expected, mandatory in MPI photoelectron spectroscopy.

4. Two-Laser Experiments

Two-color (usually meaning two wavelengths produced by one pump laser), and two-laser (two separate pump lasers) techniques offer the greatest potential for laser ionization experiments, not only in making ionization possible by avoiding loss mechanisms from high-lying resonance levels, but also in the greater information content separate tunable sources with variable time delays can provide. The two-laser apparatus, although often almost twice as expensive, is not much more complicated than the normal (single-laser) ionization apparatus.

Figure 10 shows a two-laser apparatus used for REMPI spectroscopy of the caged amine triethylenediamine (Parker and El-Sayed, 1979). The resonant state of this molecule is long lived (~ 1 μsec); thus two separate lasers with electronic time delay are needed to measure the excited-state lifetime. Since the resonant state fluoresces efficiently, emission can be monitored simultaneously with ionization for comparison. At fixed time

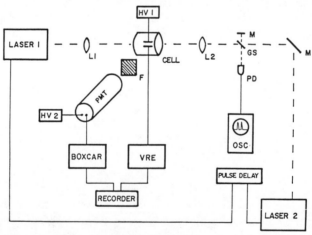

Fig. 10. Two-laser REMPI apparatus: (L1, L2) lenses, (M) mirror, (GS) glass slide, (PD) photodiode, (F) filter, (VRE) vibrating reed electrometer, (PMT) photomultiplier tube (Parker and El-Sayed, 1979, courtesy of North-Holland Publishing Company).

delay the exciting wavelength can be scanned to trace out the excited-state ionization potential. The two beams can be crossed at right angles, mixed collinearly with dichroic mirrors, or counterpropagate as shown in Fig. 10. The latter arrangement allows different (and nonachromatic) focusing lenses for the two beams. Alignment and optimization of the signal with the proper lenses can be tedious and often best solved by trial and error. Zakheim and Johnson (1980) have described the experimental optimization of two-photon versus multiphoton signals using rate-equation modeling.

The resonant-state lifetime determines whether two-color or two-laser methods should be used. The timing jitter of most pump lasers exceeds a few nanoseconds; thus lifetimes on this time scale are difficult to measure by two-laser techniques. Optical delays can be used in the two-color method as long as the ionizing laser is not focused. Another factor is noise. Ionization in multiphoton events with uncorrelated noise sources (due to pulse output fluctuations of the pump laser) is often less noisy than with single pump lasers (i.e., two-color experiments). However, noise is introduced by jitter in the time-delay electronics in two-laser experiments.

III. DYNAMICS OF RESONANT MULTIPHOTON IONIZATION

REMPI was recognized and utilized as a spectroscopic tool well before the underlying dynamics were even qualitatively understood. Although complete comprehension of resonant MPI dynamics in molecules has not been attained, significant progress has been made over the past few years, guided by similar efforts in atomic MPI research. The goal is to find the most tractable formalism valid for the actual experimental conditions, or, more directly, to determine when simple population rate equations (PRE) properly describe the ionization process.

A. Perturbation Theory and Density Matrices

Perturbation theory is the natural starting point for a nonresonant multiphoton process. Resonant intermediate states complicate the perturbation-theory approach. For example, the introduction of a two-photon state at $2h\nu_L$ in time-dependent perturbation theory causes the energy denominator $E_j - E_g - 2h\nu_L$ in Eq. (2) to approach zero; thus σ_3 diverges. As discussed for the atomic case, resonance effects may be included by introducing intensity-dependent energies and level widths in the cross section equation. The simple I^n intensity dependence is no longer preserved [Eq. (4)], however, and the inclusion of predissociation, intramolecular relaxation, and other processes specific to molecules is difficult. Perturbation theory has been used to predict ion currents for REMPI in atoms and simple molecules (Cremaschi *et al.*, 1978) with moderate success.

Besides the exceptional amount of data required in calculating σ_n, and especially in estimating linewidths, the use of perturbation theory is questionable on experimental grounds. The single-rate approximation of perturbation treatments (Fermi's Golden Rule) does not account for the effect of the long laser pulse length (5 nsec, compared to absorption, a $\sim 10^{-15}$-sec event). This objection can only be removed by returning to the general density-matrix picture from which perturbation theory is derived.

Two-photon resonant three-photon ionization is represented in the density-matrix concept as a three-level system composed of the ground state, the two-photon state, and the ion continuum. The populations of each state, represented by the on-diagonal elements of the matrix, are coupled at the Rabi frequency with the off-diagonal elements and driven coherently in time. The general expression for the Rabi frequency Ω is

$$\Omega = [(2dE/\hbar)^2 + (\omega_L - \omega_A)^2]^{1/2} \qquad (8)$$

where d is the transition dipole matrix element, E the laser field amplitude, ω_L the laser frequency, and ω_A the transition frequency. On resonance ($\omega_L = \omega_A$) the Rabi frequency is *linear* in the laser field amplitude, and so is quite different from the stimulated absorption rate $W = \sigma I$, which is *quadratic,* via the photon flux, in the field amplitude. By solving the Bloch equations for this system the overall ionization yield can be found. This requires an intimate knowledge of the dye-laser coherence, however, which is at present lacking. Even with a well-characterized source, the complexity of the Bloch equations severely limits their practical information content and attractiveness for molecular systems.

B. Rate Equations

Population rate equations correspond to the use of the Einstein A and B coefficients in describing the interaction of light with atoms or molecules. This ignores the phase-sensitive properties of the system, or, in effect, neglects the off-diagonal elements of the density matrix. Rate equations predict smooth monotonic buildup and depletion of level populations, while coherently driven systems exhibit rapid oscillations of the same populations. If the instantaneous ion current were measured, the actual case might be determined directly. With nanosecond-pulsed lasers, however, the observed quantity is the total ion count per pulse. A few practical arguments can be made as to which conditions support the rate-equation approximation.

First, the coherence or fluctuation time of the laser should be considered. A pulsed multimode dye laser with 1 cm^{-1} bandwidth $\Delta\nu_L$ has a coherence time $\tau_{coh} \sim 1/\Delta\nu_L = 33$ psec, which is short compared to the pulse length. If resonant ionization occurs on a time scale longer than this coherence time the overall process is no longer coherent. In a REMPI process the initial multiphoton absorption must still be coherently driven because of the short virtual-state lifetime, but the ionization step will be statistically decoupled from the (bound–bound) excitation step. With increasing intensity, however, the Rabi frequency of the excitation step and the ionization rate can become comparable to the bandwidth; thus some correlation between excitation and ionization may be expected. Rate equations are valid, then, in the limit of large bandwidths and low intensities (Zoller and Lambropoulos, 1980).

Another approach has been presented by Ackerhalt and Eberly (1976) which is independent of the laser coherence quality. Their general conclusion was that for the following conditions, kinetic equations can successfully describe the population changes. First, the successive transitions from the ground state to the ion should have successively increased rates. This avoids population bottlenecks. Second, the final step must irreversibly remove the molecule from the system. Thus, no matter how coherently driven the system is, the final step allows no cycling or pulsation of population.

The latter condition is satisfied due to the ionization step in the MPI experiment. The first condition should certainly be satisfied for two-photon resonant absorption followed by a third one-photon ionization step, since two-photon transitions are less probable than one-photon transitions at the dye-laser flux. Thus, when the rate of the ionization process is determined by the rate of the two-photon absorption, pulsation effects can be neglected, and rate equations can be safely used to describe the time dependence of each level's population

Excited-state ionization cross sections are not expected to exceed bound–bound one-photon cross sections. Population rate equations (PREs) are thus not automatically justified by the second argument for R2PI or for higher-order REMPI, e.g., two-photon resonant four-photon ionization. Ackerhalt and Shore (1977) and Eberly and O'Neil (1979) have reported further calculations that show PREs can very closely model the total ion count for a wide range of conditions, even when the ionization step does not have the greatest rate.

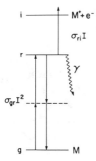

Fig. 11. Simple kinetic scheme for two-photon resonant three-photon ionization. γ represents any population loss mechanism.

C. Simple Example of Rate Equations—Two-Photon Resonance MPI

As an example of PREs consider two-photon resonant three-photon ionization, as shown in Fig. 11, where the resonant level r is populated from the ground state g at a rate $\sigma_{gr}I^2$ and depleted by ionization at a rate $\sigma_{ri}I$. Also included is a nonradiative relaxation rate γ, from the resonant state, which effectively removes population from the system. This may represent predissociation or curve crossing to a lower electronic state which does not ionize. Fluorescence from the resonance level is not included on the time scale (10^{-8} sec) of the laser pulse. The time dependence of the system is found from

$$dN_g(t)/dt = \sigma_{gr}I^2[N_r(t) - N_g(t)], \tag{9}$$

$$dN_r(t)/dt = \sigma_{gr}I^2[N_g(t) - N_r(t) - \sigma_{ri}IN_r(t) - \gamma N_r(t)], \tag{10}$$

$$dN_i(t)/dt = \sigma_{ri}IN_r(t). \tag{11}$$

For mathematical simplicity the laser pulse is approximated by a step function with height I between 0 and 5 nsec, and zero otherwise, which is reasonable for the above conditions.

After substituting the exponential decay of the ground state into dN_2/dt, one can find the population of the two-photon state as a function of time using the additional assumption of steady-state population in the resonant level, i.e., $dN_r/dt = 0$, as

$$N_r(t) = \frac{\sigma_{gr}I^2}{\sigma_{gr}I^2 + \sigma_{ri}I + \gamma} N_g, \tag{12}$$

and therefore the ion current is given as

$$\frac{dN_i}{dt} = \frac{\sigma_{gr}\sigma_{ri}I^3 N_g}{\sigma_{gr}I^2 + \sigma_{ri}I + \gamma}. \tag{13}$$

When the two-photon step is rate limiting, the steady-state assumption is valid, in which case

$$\sigma_{gr} I^2 \ll \sigma_{ri} I,$$

and the expression reduces to

$$\frac{dN_i}{dt} = \frac{\sigma_{gr}\sigma_{ri} I^3 N_g}{\sigma_{ri} I + \gamma} = \sigma_{gr} I^2 N_g \frac{\sigma_{ri} I}{\sigma_{ri} I + \gamma}, \tag{14}$$

i.e., the ion current is determined by the two-photon absorption rate times the fraction of molecules in the two-photon state which ionize. Note that if the ionization rate $\sigma_{ri} I$ dominates the radiationless transition rate γ, then the overall intensity dependence will be quadratic. If γ is competitive with $\sigma_{ri} I$, then the intensity dependence will be closer to cubic. This is a general result: competition between loss processes out of the resonant state and absorption of additional photons from this state tends to increase the intensity dependence of the total ion signal.

Several comments should be made concerning the steady-state approximation. In this very simplified model (Parker *et al.*, 1978) the steady state is reached with a laser flux of 10^{28} photons/cm^2 sec, as long as $\sigma_{ri} \geq 10^{-18}$ cm^{-2} or $\gamma \geq 10^{10}$ sec^{-1}, and $\sigma_{gr} \leq 10^{-49}$ cm^4 sec. Within these limits the population N_r of the resonant level remains a small fraction of the ground-state population N_g ($N_r < 10^{-5} N_g$ excited); this repopulation of N_g by stimulated two-photon emission (rate $\sigma_{rg} I^2 = \sigma_{gr} I^2$) or by fluorescence need not be included. Larger two-photon cross sections cause ground-state depletion (saturation), which should be avoided for useful spectroscopic information. Small ionization cross sections or slow relaxation rates result in a longer period before reaching a steady state. With the relaxation rates and cross sections expected and measured thus far, the steady-state approximation has not proven unreasonable. The distinct advantage is that a direct expression for the ion current is obtained which yields qualitative insight to the REMPI mechanism.

More sophisticated rate-equation treatments are easily managed. An extended two-level system has been presented by Zakheim and Johnson (1980), for n-photon resonant m-photon ionization with n, m any integer. Three-level (doubly resonant) equations were also described. Instead of the square-pulse approximation, smoothly varying functions which simulate the temporal and spatial qualities of the laser, including focusing, can also be included for more accuracy (Cervenan and Isenor, 1975). Rothberg *et al.* (1981) have recently presented an even more relevant extended two-level rate-equation treatment. A manifold of ground and excited states is incorporated in this model, with the laser resonant in a submanifold of each. Direct MPI is allowed and the effects of collisional thermalization were investigated. Results of these more sophisticated treatments are described in the next section.

D. Analysis of Rate Equations

Ideally, by inserting the experimental parameters (laser intensity, focusing conditions, sample concentration, etc.) and molecular parameters (cross sections, relaxation rates) into the rate-equation program an accurate ion current is calculated. Several obstacles have prevented this thus far: (i) ionization cross sections (e.g., σ_{ri}) are generally unavailable or quite uncertain (orders of magnitude); (ii) two-photon absorption cross-section measurements are also rare, three-photon absorption cross sections may only be available indirectly through REMPI; (iii) relaxation rates γ are also difficult to estimate—as discussed next; (iv) competing mechanisms (e.g., harmonic generation, fragment ionization, etc.) are sometimes very likely; (v) linewidth complications can arise—for example Stark shifting and lifetime broadening; and, most importantly, (vi) the instantaneous laser intensity is nearly impossible to know accurately—owing to random axial and transverse modes, imperfect optics, etc., as discussed in Section II. For these reasons quantitative rate-equation predictions for more than the simplest molecular system are unlikely in the near future, but qualitative progress has been achieved.

1. The Loss Rate γ and Molecular Rydberg States

Equation (14) predicts that the most stable excited states (those where γ is small) will be detected most efficiently. This is undoubtedly the reason for the relatively high sensitivity of MPI spectroscopy to molecular Rydberg states, so named because these states form a convergent series, similar to atomic series, the frequencies of which may be fit by the formula IP-R/$(n - \delta)^2$, where IP is the ionization potential to which the series converges, R is the Rydberg constant, δ is the quantum defect, and n is an integer ranging from some initial value to infinity. Valence-excited states at similar energies often have strong dissociative channels, resulting in a small quantum yield of ionization. In contrast to valence states, Rydberg states are nonbonding, and to a certain degree may be treated as a single electron orbiting a positively charged molecular core. With increasing n, this approximation becomes increasingly valid, for the size of the Rydberg orbital becomes very large compared to the size of the molecule. Because they are nonbonding and are largely uncoupled from the molecular core, Rydberg states have longer lifetimes with respect to predissociation than valence states. Other factors favoring detection of Rydberg states are the sharpness of Rydberg absorption bands and the fact that Rydberg–ion transitions are less Franck–Condon forbidden than valence–ion transitions owing to the similarity of Rydberg and ionic geometries. Ionization cross sections σ_{ri} for Rydberg states are expected to be smaller than those of valence states.

Other events besides predissociation are represented by γ. Simple radiationless transitions from the resonant state to vibrational levels of a lower electronic state can remove the system as effectively from ionization, since Franck–Condon factors for excitation to a vibrationally cold ion from a vibrationally hot valence state should be very unfavorable. Valence states are observed by REMPI, however, with the $^1B_{2u}$ state of benzene as an example. The absorption spectra of many molecules (e.g., ammonia, tertiary amines) are completely Rydberg in character, making the lack of valence sensitivity irrelevant. Section IV gives several examples of the power of REMPI as a spectroscopic tool.

Information on dissociative valence states may also be available via resonant multiphoton ionization spectroscopy in a process labeled depletion of nonresonant ionization (DNRI) (Rothberg *et al.*, 1981). Essentially, a dip may be observed in the broad nonresonant (MPI) background when resonant states that do not ionize are populated. This lack of ionization may be due to a low resonant-state ionization cross section, or to rapid predissociation from the resonant state before another photon is absorbed. Structured valleys in the ion current background are, of course, difficult to observe with narrow dye profiles and diffuse resonant states. Pressure may increase or decrease both REMPI and DNRI as discussed in the following subsection.

2. Intensity Dependences

Since intensity is the main variable once the laser is tuned to resonance, the intensity of the ion current is of major importance. Often, an unknown molecular constant (cross section or relaxation rate) is obtained by fitting the PRE ion current [e.g., Eq. (14)] to the observed i vs I plot. This process is most successful in R2PI since more parameters are likely to be known and unfocused laser beams are employed. Letokhov *et al.* (1977) have derived R2PI rate equations for optically thin ($\sigma_1 N_0 l \ll 1$, $\sigma_2 N_0 l \ll 1$) and optically thick ($\sigma_1 N_0 l \sim 1$, $\sigma_2 N_0 l \sim 1$) samples—where σ_1 is the excitation cross section, σ_2 the ionization cross section, N_0 the sample concentration, and l the optical length. Three limiting cases for optically thin samples are found from the PREs: (i) low-intensity case $\sigma_1 I \ll 1$, $\sigma_2 I$ $\ll 1$: $N_i = (\bar{n}/n_0) N_0^{\frac{1}{2}} \sigma_1 \sigma_2 I^2 \tau_L^2$; (ii) saturation of the first step $\sigma_1 I \ll 1$, $\sigma_2 I \gg$

1: $N_i = (\bar{n}/n_0)N_0\sigma_2 I\tau_L$; and (iii) saturation of the second step $\sigma_1 I \ll 1$, $\sigma_2 I \gg$ 1: $N_i = (\bar{n}/n_0)N_0\sigma_1 I\tau_L$. Here $\bar{n}/n_0$ is the fraction of ground-state molecules which can absorb the laser light, τ_L the pulse length, and N_i the total ion count. Resonant-state lifetimes are assumed to be long in this example—short lifetimes are easily included. Saturation of the first step occurs at the intersection of cases (i) and (ii) with the intensity $I_{sat}^{(1)} = 2/\sigma_1\tau_L$, assuming no collisional or radiative depopulation during τ_L. A similar expression is found for saturation of the second step. In these cases, unknown cross sections can be obtained directly from the regions of discontinuity in the i vs I plot. Focusing smears out these transition regions (e.g., I^2 to I dependence), thus lowering the quantity of PRE intensity dependence analysis.

An I^n dependence should be observed at the lowest-intensity case, where n is the number of photons needed for ionization. Several processes can alter this order of nonlinearity, including ionization of fragments produced by photodissociation of the sample during the pulse length, collisional ionization and field ionization of excited states below the IP, and the Stark effect. Fragmentation can often be isolated by mass analysis. Collisional processes, including collisional redistribution during the pulse length, are discussed in the next section and field ionization is discussed in Section IV.C. The Stark effect in molecules is discussed in the following subsection.

3. Collisional Effects

Collisions make rate equations valid, even with narrow-bandwidth light, if they cause the molecule to lose memory of how it was excited before further absorption or stimulated emission can take place, provided that the Rabi frequency is less than the collision rate, and that the laser bandwidth exceeds the power broadening bandwidth (see Section IV.E). A typical hard-sphere collision time of 50 Torr nsec for a medium-size molecule gives a collision rate of 2×10^9 sec^{-1} at 100 Torr, or about 10 collisions per 5-nsec pulse. The duration of such collisions is typically $<10^{-12}$ sec (Hurst *et al.*, 1979) so photoionization is unlikely during a collision.

Collisions can have several effects on resonant MPI and DNRI (Rothberg *et al.*, 1981). First, pressure may allow more molecules to reach the excited state because collisional redistribution during the pulse can bring ground-state molecules into resonance, and shift excited-state molecules before stimulated emission takes place. In addition, collisional redistribution can shift excited-state molecules into regions of higher or lower ionization probability owing to geometric effects (Franck–Condon factors), or to differing loss rates within the excited-state manifold.

Collisions can also produce ions if resonant states are populated near the IP. These high-lying states are often Rydberg states which occupy large orbitals for higher principal quantum numbers (e.g., ~ 10 nm for $n = 14$). Reaction rates for collisional ionization, charge transfer, and chemionization are likely to be very fast for such large orbitals, which will blur the distinction between n and $n - 1$ ionization. Ion pair formation $(AB + nh\nu \rightarrow A^+ + B^-)$ and predissociation into ion pairs $(AB + nh\nu \rightarrow AB^* \rightarrow A^+ + B^-)$ also produce ions below the IP.

4. MPI Polarization Information

An especially important advantage of two-photon spectroscopy over conventional methods is the ability to distinguish states of different symmetry directly, through the dependence of the two-photon cross section on the polarization of each exciting photon. The quantity of interest in two-photon spectroscopy with a single laser is the ratio Ω of the signal from a circularly polarized laser to that from a linearly polarized laser, i.e., $\Omega = \sigma_{gr}^{circ}/\sigma_{gr}^{lin}$). As shown in Section IV, the size of Ω gives specific information on the symmetry of any rovibronic two-photon transition. The measured quantity in the MPI experiment is Ω of the ion current, i.e., $\Omega^{MPI} = i^{circ}/i^{lin}$. Equation (14) gives two limits as to how Ω^{MPI} compares to Ω^{2P}:

(i) If $\sigma_{ri} I > \gamma$, then using (14), Ω^{MPI} is found to have the form $\Omega^{MPI} = \sigma_{gr}^{circ}/\sigma_{gr}^{lin}$. In this case, i.e., kinetic saturation, the third photon will have no effect on the observed value of Ω. When every molecule that reaches the two-photon state is ionized, the polarization ratio is only determined by the two-photon excitation. In physical terms, the intensity is high enough that even photons without the most favorable polarization are still able to saturate the transition. If one-photon ionization is forbidden owing to the photon polarization, the same result will occur with saturated two-photon (excited state) ionization.

(ii) If nonradiative relaxation from the two-photon state is faster than the ionization rate $(\gamma > \sigma_{ri} I)$, then

$$\Omega^{MPI} = \frac{\sigma_{ri}^{circ} \sigma_{gr}^{circ}}{\sigma_{ri}^{lin} \sigma_{gr}^{lin}}$$

Thus if the orientation (rotation) relaxation rate is slower than the ionization rate, Ω will reflect the polarization character of the third photon (σ_{ri}).

The relaxation rates of disorientation and ionization thus determine the behavior of Ω^{MPI}. At high rotational relaxation rates, one-photon ionization takes place over a randomly oriented sample and shows no dependence on the ionizing photon polarization. At high ionization rates ionization can become so probable that every excited molecule (besides those lost to the system due to processes represented by γ) is ionized regardless of its initial orientation or the polarization of the laser. When ionization exceeds disorientation without saturating the transition, Ω^{MPI} can differ from Ω^{2P}.

To attempt to prove (or disprove) qualitative arguments based on Eq. (14) over extremes of experimental condition would of course be foolish. Robin and co-workers (Robin, 1980), however, have pointed out, using more comprehensive rate equations that condition (i) (kinetic saturation) is difficult to attain experimentally without complications and that condition (ii)—i.e., working at lower intensities (where an I^3 dependence is found), is generally preferable.

To date, a number of two-photon resonant MPI polarization experiments have been reported with good success (see Section IV). In addition, three-photon polarization results have been used with good effect in the reassignment of the electronic structure of ammonia using REMPI.

Intensity- or pressure-dependent polarization ratios indicate regions where the MPI ratio is unreliable. Robin and co-workers (Heath et al., 1980) have reported the only case so far—acetaldehyde—in which the polarization ratio was not obtainable—since the I^3-dependent region was not reached with sufficient signal. Acetaldehyde is a fascinating but atypical molecule with an ionization efficiency many orders of magnitude larger (owing to one-photon resonance) than any other molecule—so large it has even been observed by REMPI in newly constructed ion chambers, never purposely exposed to the sample. This study does, however, reinforce the cautions in using REMPI polarization ratios.

E. Laser-Produced Line Broadening

Besides lowering the quality of REMPI spectra, line-broadening effects greatly confuse intensity dependences, polarization ratios, state-selective detection, etc. These effects, which arise from the high intensities required for nonresonant absorption, are often observed by scanning through resonance at varying laser powers, and can always be minimized by working at the lowest intensity (with a loss of signal). Three different types of laser-produced broadening are known: (i) geometric saturation due to tight focusing—which will bring up the less or nonresonant background, (ii) lifetime, or power broadening, and (iii) the Stark effect. Focusing considerations were discussed in Section II.

Lifetime broadening results from the MPI process itself: in order to detect a state it must be ionized before the pulse ends. A minimum linewidth of 200 MHz results from a 5-nsec pulse. Ionization is often kinetically saturated, however, which can yield substantially larger power broadening.

The Stark effect can yield much more dramatic broadening and shifting of the resonance by the laser. Otis and Johnson (1981) have observed a strong linear (in the laser flux) Stark effect in REMPI of rotationally cooled beams of NO, which behaves in a remarkably similar fashion to Stark broadening in atomic MPI (Section I). Although this effect is most easily observed in narrow-line collisionless regimes, it is likely a very general process occurring to some extent in all REMPI systems. Rate equations must be used with great caution if Stark broadening is suspected.

IV. SPECTROSCOPIC APPLICATIONS

In this section spectroscopic applications of laser ionization in polyatomic molecules will be reviewed. The examples are chosen either because they represent a significant advancement, or because they nicely illustrate the application of a technique. Because of the abundance of excellent work in this area, neglect of other important examples is unavoidable. Total ion current is the measured quantity in this section; however, mass analysis is used in some of the examples discussed to discriminate against background. Similarly, two-color studies are included which give information on resonant states rather than fragmentation mechanisms.

A. REMPI Spectroscopy

As mentioned in the Introduction, REMPI finds great use as a two-photon or three-photon excitation technique, giving information on optically forbidden transitions (new states) which is often unavailable by another method. Static gas ionization cells are the most common medium, although supersonic beams are becoming increasingly popular. Polarization analysis is also included in this section. Since REMPI spectroscopy has been reviewed recently by Johnson and Otis (1981), only a few examples will be discussed.

1. Gas-Cell Spectra

Gas ionization cells were the first and simplest medium for REMPI spectroscopy in molecules, with benzene the first polyatomic molecule to be reported (Johnson, 1975). REMPI spectra of NO (Johnson *et al.,* 1975) and I_2 (Petty *et al.,* 1975) were published previously in the same year and, interestingly, these three molecules have been centers of activity since then.

One of the most spectacularly successful applications of REMPI has been on ammonia. Virtually the entire excited electronic structure of this textbook molecule has been explored and for several states reassigned via REMPI, in studies by Colson and co-workers (Nieman and Colson, 1979; Glownia *et al.*, 1980). All of the ammonia excited states are believed to be of Rydberg type arising from excitation of a nitrogen lone-pair orbital. The lowest excited state is the 3s Rydberg Ã state which is reached by two-photon resonance three-photon ionization. The most intense higher excited states, which are reached by three-photon resonance, were found to arise from nd orbitals, which were not even considered in prior theoretical analysis. Supersonic beam cooling greatly assisted in these later assignments.

The 3s Rydberg state appears prominently in the REMPI spectra of many polyatomic molecules constructed from first-row atoms. In general, the first Rydberg member of any molecule lies at roughly three-quarters of the IP; thus two-photon resonant three-photon ionization is energetically allowed. Since s orbitals are totally symmetric, two-photon transitions from totally symmetric ground states are also allowed by symmetry. In addition, the lowest Rydberg state is usually low enough in energy to stand out from nonresonant or diffuse background, making it easily detected in the early REMPI studies.

Another 3s Rydberg state appears in the different spectra of the caged amine ABCO (1-azabicylo[2,2,2]octane) shown in Fig. 12 (Halpern *et al.*, 1982). One-photon absorption (OPA) and one-photon fluorescence excitation (OPFE) are essentially identical since this is the lowest excited state. Resolution in these one-photon spectra is instrument limited (using standard commercial spectrometers). Much higher resolution is easily attained with the laser-produced two-photon-resonance-enhanced MPI spectra and two-photon fluorescence excitation (TPFE) spectra also shown in Fig. 12. Sharp sequence structure due to torsional modes of the rigid cage structure are resolved in these spectra, with sequence band origins marked by dashed lines. Note the great similarity of the one-photon and MPI spectra, which is expected since selection rules for this C_{3v} symmetry molecule allow both one- and two-photon transitions. The TPFE spectrum is greatly distorted, however, owing to competition between ionization and fluorescence, as shown by Halpern *et al.* (1982). Similar but less striking effects were seen in the complementary caged amine triethylenediamine—DABCO (Parker and Avouris, 1979). These effects point out the more general nature of MPI—as the laser flux increases ionization will compete effectively with any relaxation process.

ONE PHOTON WAVELENGTH (nm)

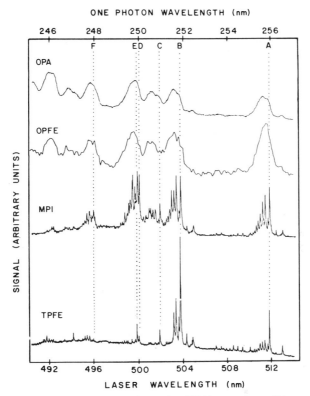

Fig. 12. Comparison of one-photon absorption (OPA), one-photon fluorescence excitation (OPFE), multiphoton ionization (MPI), and two-photon fluorescence excitation (TPFE) for the 3s Rydberg transition of quinuclidine (ABCO) (Halpern *et al.*, 1982).

Not only Rydberg states are detectable by REMPI. Goodman and co-workers (Rava *et al.*, 1981) have carefully studied the valence $^1B_{2u}-^1A_{1g}$ transition of benzene, its isomers, and substituted benzenes, using two-photon resonant REMPI. A number of significant findings were made on excited-state assignments and the vibronic mechanisms that produce this one- and two-photon-forbidden transition.

It now appears that the long-awaited era of small-molecule REMPI spectroscopy has arrived. The excited states of small molecules usually lie at more than twice the energy of the typical dye-laser fundamental tuning limit ($\sim$360 nm). Eximer-laser-pumped dye lasers and high-power frequency-doubled dye lasers have now extended this range well into the vacuum UV. One example is the molecule ethylene, which has recently been studied by two-photon resonant MPI (Gedanken *et al*, 1982a). The

3s Rydberg state of ethylene is symmetry allowed and the 3p Rydberg states symmetry forbidden in one-photon absorption. By frequency-doubling dye-laser outputs, two-photon resonant MPI was detected with tunable laser wavelengths from 280 to 380 nm, and the two out-of-plane 3p Rydberg states were located. Vibronically allowed two-photon transitions to the 3s state were also detected and used to clear up several uncertainties concerning excited electronic states of this important molecule.

Two-photon resonant three-photon ionization of H_2 has been reported by Marinero et al. (1982). Wavelengths in the 220–200 nm regions were generated by Raman-shifting frequency-doubled dye-laser outputs. Individual rotational levels can be selectively ionized and the ion signal mirrors the ground-state population. This technique has obvious applications in the study of scattering processes.

Laser ionization spectroscopy has also been carried out in flame environments, with work thus far centered on the NO molecule. Rockney et al. (1982) and Mallard et al. (1982) have reported the REMPI and R2PI flame spectra of NO respectively, with most attention given to analysis of temperature and collisional relaxation rates. Such work should be easily extended to the many unstable intermediates present in most flames.

2. REMPI Polarization Ratios

McClain (1971) has shown that with identical photons (single laser) the polarization ratio Ω must be 3/2 for nontotally symmetric states and less than 3/2 for the Q_Q branch of a totally symmetric transition. From this simple relation the symmetry of any rovibronic band can be assigned or confirmed. Since the symmetry of the highest filled orbital is usually known by calculation and photoelectron studies, the value of Ω will determine whether the excited orbital is different from ($\Omega = 3/2$) or identical to ($\Omega < 3/2$) the ground-state orbital symmetry. Conversely, if the ground-state assignment is unclear, a knowledge of Ω and the symmetry of the excited orbital will determine the initial orbital assignment.

Figure 13 (Parker et al., 1980) shows the vibrationless origin of the spin–orbit state in the $6s \leftarrow 5p\pi$ Rydberg transition of methyl iodide. Two-photon resonant three-photon ionization is used to detect this state—which is not observed in the one-photon spectra. The lower curve is the transition profile using circularly polarized light, and the middle curve uses linearly polarized light (laser wavelength ~371 nm). As seen, several bands within this transition are greatly attenuated under circular polarization, indicating Q_Q branches. This totally symmetric symmetry assignment ($\Omega < 3/2$) is supported by rotational line-shape fitting—shown as the upper curve in Fig. 13. The sharp branches are confirmed as Q_Q bands for

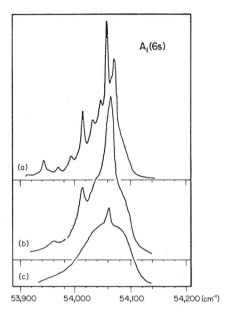

Fig. 13. Two-photon resonant MPI spectrum of methyl iodide: (a) rigid-rotor rotational line-shape simulation of the $^1A_1$ (6s) Rydberg origin, (b) REMPI spectrum—linear polarization, (c) REMPI spectrum—circular polarization. (Parker *et al.* 1980, courtesy of North Holland Publishing Company.)

different K levels and the fit roughly estimates the excited-state geometry. (The less than perfect quality is due to an overly simplified rigid-rotor approximation.) By locating and directly assigning this long-sought state, spin–orbit coupling mechanisms in methyl iodide were verified.

A direct comparison of REMPI polarization ratios and TPFE polarization ratios is possible in the caged amine DABCO (Parker and Avouris, 1979). The lowest excited state of this interesting molecule is two-photon allowed, one-photon forbidden, and totally symmetric—as proved by a polarization ratio from MPI of $\Omega_{MPI} < 1/40$. The TPFE ratio Ω_{TPFE} is identical. For the second-lowest excited state, which is both one- and two-photon allowed, Ω from both measurements was found to be 3/2, as expected since the state is nontotally symmetric.

Assignment of the 3s Rydberg states of several molecules has been confirmed by polarization analysis (Berg *et al.*, 1978). Ground-state orbital assignments in pyrazine and pyridine has also been possible using REMPI polarization, since the excited-state symmetry is known (Berg *et al.*, 1978). Three-photon resonance polarization analysis was successfully carried out by Colson and co-workers in their classic assignment of the ammonia spectrum (Glownia *et al.*, 1980). Several other studies have used REMPI polarization to good effect, including the first such measurement by Dalby *et al.* (1977) on REMPI of molecular iodine.

3. REMPI Spectroscopy in Supersonic Beams

Supersonic beams can be an excellent medium for REMPI spectroscopy, as first shown by Zakheim and Johnson (1978) for NO. The relief of spectral congestion and narrowing of the Doppler profile allows detailed insight into both the spectroscopy and photodynamics of molecules. Benzene and several halobenzenes expanded in a free jet were studied by Murakami *et al.* (1980), who found the technique sensitive to as few as ten ions per laser pulse. Effective ionization was seen above the energy range in which the fluorescence quantum yield drops dramatically—owing to a new dissipative process ("channel three")—in benzene. Using stronger expansion conditions in a skimmed supersonic-beam apparatus (Fig. 7) Aron *et al.* (1980) measured vibronic linewidths across this same region in benzene. Working at low laser flux to avoid Stark broadening, the natural linewidth showed channel-three relaxation to level off some 3000 cm^{-1} past the fluorescence cutoff, in support of an internal conversion mechanism. Many other studies of supersonic-beam REMPI spectroscopy have been reported, and because of the numerous advantages of the technique, more are expected in the future.

B. Liquid-Phase REMPI

Many interesting effects on electronic structure and spectra occur in transition from the gas to liquid phase. Rydberg orbitals are usually obliterated, or broadened extremely and shifted to higher energy; dissociative processes from valence states can be quenched by a surrounding solvent; and ionization potentials are often dramatically lower in condensed phases compared to the gas phase. These processes directly affect REMPI spectra in the liquid phase.

The first liquid-phase REMPI study was that of Vaida *et al.* (1978) which was motivated by a search for the elusive $^1E_{2g}$ state in benzene. Their spectra covered the region of the 3s Rydberg state in the neat liquid and showed only broad structureless ionization. Scott *et al.* (1979) made careful intensity-dependent measurements in the same spectral region and identified the ionization mechanism as two-photon resonant three-photon ionization. Polarization ratios were used to assign the transition symmetry at 355 nm, and time-delayed two-pulse (same-color) ionization indicated that the two-photon state relaxed to an intermediate state with a 10-

nsec or longer lifetime before one-photon ionization took place. The authors also found the technique to be a remarkably sensitive detector of impurity molecules with IP less than the host molecule. Interestingly, in a later study Scott and Albrecht (1981) did detect the 3s Rydberg state in the liquid-phase TPFE of dilute benzene in an alkane solution.

If the liquid-phase IP is sufficiently low, two-photon ionization can take place directly following two-photon absorption to a state above the IP. This has been observed by Braun *et al.* (1981) in the easily ionized diamine TMPD in 3-methylpentane solution. Biphotonic (R2PI) absorption is ruled out by polarization analysis since, in sufficiently fluid media, ion yields are insensitive to the laser polarization. Two-photon ionization, which competes with two-photon resonant three-photon ionization in the wavelength range studied, was found to originate from relaxation followed by autoionization in upper electronic states. Relaxation to the lowest excimer state following TPA in pure benzene precedes third photon absorption to superexcited levels (which then collapse to geminate charge pairs) according to a recent study by Scott *et al.* (1982).

C. R2PI Spectroscopy

R2PI has proven to be experimentally a more attractive approach to laser ionization than REMPI, in almost all aspects except for the detection of forbidden states. The reason is that since no multiphoton steps are involved, lower powers (10^6 W/cm^2 rather than the 10^9–10^{11} W/cm^2 common for REMPI) are used, allowing larger focal volumes or no focusing at all, and resulting in less fragmentation of the parent ion. When coupled with supersonic-beam expansions, state selectivity is also possible with R2PI. One disadvantage is that either a tunable UV source is needed, or fixed (nonselective) wavelengths such as the fourth-harmonic Nd$_7$YAG—266 nm, 4.7 eV, or KrF—248 nm, 4.8 eV, ArF—193 nm, 6.3 eV eximer lines must be resorted to. Advancement in tunable UV generation via frequency doubling and Raman shifting of high peak-power pulsed-dye lasers has almost alleviated this problem.

Another problem is simple energetics. For efficiency, the lowest excited singlet S_1 must lie at over one-half the energy of the IP; otherwise less desirable processes are forced, such as R3PI (which is less efficient owing to competition at the second photon level), or R2PI through higher excited states (which may also have fast dissipative processes). Stability of the lowest singlet is assumed, which is not always the case. Methyl iodide, for example, is an exception. Also, low S_1 states are perfect candidates for two-laser photoionization.

Even though the first step is common one-photon absorption, sensitivity makes R2PI an important spectroscopic tool, especially when coupled with supersonic expansions. Direct one-photon absorption measurements in supersonic beams, while possible (see Leopold *et al.*, 1981), are not optimal, owing to geometric constrictions. Excitation measurements are efficient, however, and photoionization is a general, sensitive excitation technique which is applicable in some regions where free-jet fluorescence excitation fails. In addition, compared to fluorescence excitation, maximum collection efficiencies are higher (100% compared to ~5%), and the entire ground-state population may be ionized, instead of just leveled. Finally, with nanosecond lasers ionization from the level of interest is not only possible, sometimes it is unavoidable.

Three conditions should be met for R2PI to most closely mimic the absorption spectrum: (i) saturation of the ground-state to excited-state transition should be avoided; (ii) the ionization cross section from the excited state should be wavelength independent over the region of interest; and (iii) the excited state should not decay faster than the up-pumping rate, or in an irregular manner. In practice, the latter two conditions are rarely rigorously met (or controllable). Other excitation techniques have similar conditions, however, and R2PI has been shown to follow the above for a number of molecules.

Figure 14 shows a direct comparison of R2PI and fluorescence excitation in the static gas phase $^1B_2$–$^1A_1$ aniline (Brophy and Rettner, 1979). Both spectra were obtained under the same conditions and clearly show the same wavelength dependence. An example of the spectral improvements possible with supersonic expansion is shown in Fig. 15 (Murakami *et al.*, 1981). The upper curve is an R2PI spectrum of toluene in a mildly expanded free jet, and the lower curve is the room-temperature absorption spectrum. Discrete bands appear in the supersonic-jet spectrum, while the room-temperature spectrum is essentially continuous. Isomeric and isotopic selectivity become possible under the discrete-band condition.

Dietz *et al.* (1980) have tested the applicability of R2PI excitation spectroscopy to molecules that do not fluoresce. No ion signal was observed when iodobenzene was excited into the first singlet state, while an excellent-quality excitation spectrum of the same transition was generated in bromobenzene. Mass analysis showed only parent-ion formation at 4×10^6 W/cm^2. The decay rate due to predissociation from the resonant state was estimated as greater than 10^{11} sec^{-1} for bromobenzene and greater than 4×10^{13} sec for iodobenzene. Since neither molecule fluoresces, the technique is remarkably successful for bromobenzene, and still informative for iodobenzene.

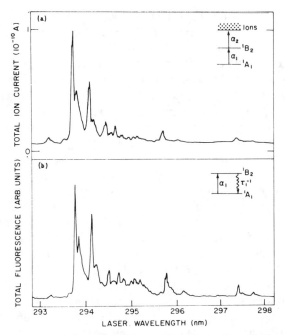

Fig. 14. Comparison of direct absorption and laser ionization of bulk gas-phase aniline; (a) R2PI spectrum of aniline, (b) laser-induced-fluorescence spectrum. (Brophy and Rettner, 1979, courtesy of North-Holland Publishing Company.)

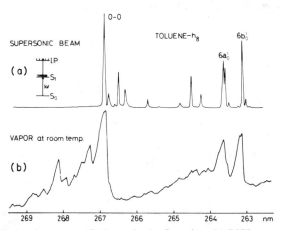

Fig. 15. Electronic spectra of toluene in the S_1 region: (a) R2PI spectrum in a supersonic jet, (b) absorption spectrum at room temperature. (Murakami *et al.*, 1981, courtesy of North-Holland Publishing Company.)

Low-lying triplet states have been detected by resonant three-photon ionization in pyrazine (Turner *et al.*, 1978) and by resonant four-photon ionization in CS_2 (Rianda *et al.*, 1980). In these examples the second (and third) photons may be only slightly resonant. Such states appear as extremely weak bands in the absorption spectra—indicating the high sensitivity of resonant ionization. Rothberg *et al.* (1981) introduced the CS_2 spectrum shown in Fig. 16 as possible evidence for DNRI. Structured valleys are seen in the 400-Torr gas-cell spectra which correlate with one-photon absorption to the lowest CS_2 triplet state. The low-pressure spectra of Rianda *et al.* for the same transition show only resonance enhancement. Collisional redistribution of excited-state population to levels where ionization is less probable was suggested as the mechanism and several rates and cross sections were extracted using a PRE analysis.

A number of other authors have used rate-equation modeling of REMPI (Cremaschi *et al.*, 1978; Fisanick *et al.*, 1980) and R2PI to extract molecular parameters. Reilly and Kompa (1980) found excellent experimental agreement with the PRE total ion count versus I curve in the R2PI of benzene by KrF (249 nm, 5.0 eV) radiation, where the lifetime, excitation and ionization cross sections, and laser intensity were known. The same model was used to predict the excited-state lifetime at ArF wavelengths (193 nm, 6.4 eV) as less than 5 psec.

Boesl *et al.* (1981) have used the R2PI limiting-case analysis presented in Section III.D.2 to obtain ionization cross sections from the lowest excited states of benzene, thiophene, naphthalene, toluene, and aniline. For the last two molecules, ionization cross sections were found to ex-

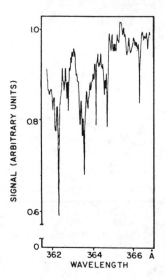

Fig. 16. REMPI spectrum of CS_2 at 400 Torr. The structured valleys correlate with one-photon DNRI transitions to the lowest triplet state (Rothberg *et al.*, 1981).

ceed the absorption cross section, resulting in high ionization efficiencies (25% for aniline at 10^7 W/cm^2). A frequency-doubled dye laser was tuned to resonance with these lowest states; thus long excited-state lifetimes (compared to the 3-nsec pulse length) and unfocused beams allowed use of the limiting-case treatment.

D. Dissociation—Detection of Nascent Photoproducts by Laser Ionization

In Section III resonance-enhanced ionization was shown to be very sensitive to loss processes from the resonant state taking place on the time scale of photoionization. Predissociation is an important loss process which can introduce new species to the system. The majority of observed REMPI spectra can be correlated entirely with the sample, but in some cases a few lines in the REMPI spectra belong to a dissociation product, and in other cases the entire spectra consists of dissociation-product resonances. Nascent photoproducts and unstable molecules generated by other means have also been detected by REMPI and R2PI.

1. Parent-Molecule Dissociation—Ionization

In this process, the sample dissociates following resonant-state excitation and the neutral fragments undergo ladder climbing through their own resonant states to ionize and fragment. This was first detected by Tai and Dalby (1978), who observed atomic iodine lines in the molecular-iodine REMPI spectra, apparently following dissociation from a dissociative resonant state. NH and NH_2 radicals were also observed in the REMPI static-cell spectra of NH_3 by Glownia *et al.* (1980). Figure 17 shows a comparison of NH_3 static-cell spectra and the same spectra obtained in a supersonic molecular jet. Beam spectra can often be used to discriminate against, or clarify, multiple-product REMPI. In the ammonia system, collisions taking place during the laser pulse length in the 5-Torr cell may relax NH*, formed by dissociation at the third photon level of NH_3, to ground-state NH, which then undergoes three-photon resonant four-photon ionization to produce the distinctive feature of the cell spectra. Collisional relaxation does not occur in the beam; thus the hydride spectrum does not appear. Clusters were blamed for the broad background (which follows the dye profile) seen in the beam spectra.

Even more complex behavior was observed by Morrison *et al.* (1981) in REMPI of NO_2; they also compared static-cell and beam spectra, with mass resolution for the pulsed-supersonic-beam condition. Above 500 nm the spectrum shows several lines which are assigned to three-photon

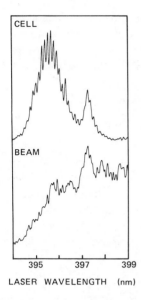

Fig. 17. Comparison of the static-cell and molecular beam REMPI spectra of NH_3 in the 394–399 nm region (Glownia *et al.*, 1980).

resonances of NO_2, plus a broad background due to $NO^* \rightarrow NO^+$ transitions. Below 500 nm the spectrum is pressure dependent, with the static-cell spectra showing only one-photon resonances of NO_2, and the low-pressure spectra showing only REMPI resonances of NO. Pressure-dependent dissociation branching ratios were proposed to explain the different spectra.

Gedanken *et al.* (1982b) have recently investigated REMPI of methyl iodide across the region of the well-known dissociative $\bar{A}$ state. As soon as the laser wavelength reaches the $\bar{A}$-state wavelengths, two-photon resonant MPI of CH_3I ceases abruptly, replaced by resonances of atomic iodine and, tentatively, the methyl radical. Resonance enhancement of three-photon levels with the second photon in the $\bar{A}$ band is possible, however, indicating that a longer-lived bound component of the A state is accessed.

The transition-metal carbonyls and organometallics seem to be a special class of molecules where almost total dissociation occurs before ionization. Dense lines seen in the REMPI spectra of these molecules belong to the bare metal atom, which at the start of the pulse had multiple ligands attached (Duncan *et al.*, 1979; Gerrity *et al.*, 1980; Fisanick *et al.*, 1981). The extreme width of the absorption bands reached by two-photon excitation indicates rapid dissociation ($>10^{12}$ sec^{-1}) with which ionization cannot compete. Stuke and Wittig (1981) found multiply charged molecular ions in the REMPI mass spectra of UF_6, with U^+ and U^{2+} the dominant ions in the mass spectrum. Gedanken *et al.* (1981) studied the REMPI mass spectrum of dimethyl mercury, however, and found parent-ion formation–fragmentation to be the primary mechanism. Clustering in the supersonic jet was found to broaden the sharp-line two-photon resonances, as expected for a Rydberg-state excitation.

2. Nascent Photoproduct, Free-Radical Detection

One of the most exciting applications of REMPI and R2PI is as a probe of photoproducts, scattering products, and unstable species. In the preceding section REMPI was shown to efficiently ionize photoproducts generated by the multiphoton process itself. Recently, the technique has been extended to the detection of nascent photofragments and reactive species generated by other processes. Mass selectivity is often mandatory in typical low-concentration, high-background environments. These applications follow the pioneering study of Zare and co-workers (Feldman *et al.,* 1977), who used R2PI in the detection of BaCl from the beam–gas reaction of Ba + HCl.

Rockney and Grant (1981) have appled REMPI to the detection of NO_2 fragments from the infrared multiphoton dissociation of CH_3NO_2. Using a CO_2 laser to first photodissociate CH_3NO_2, they detected product fragments at variable time delays with wavelength and mass-spectroscopic resolution. Broad spectra indicated high rotational temperatures of the nascent product. DeGiuseppe *et al.* (1981, 1982) have assigned a number of two-photon transitions of the methyl radical generated by pyrolysis of various molecules via mass-analyzed REMPI. The strong ion signal generated in both of these studies is encouraging for further application of REMPI and R2PI as a photodynamic probe.

E. Two-Laser MPI Spectroscopy

There are a number of advantages to two-laser multiphoton ionization in polyatomic molecules: First, the overall efficiency of the process can be greatly enhanced by direct (one-photon) ionization from the resonant state. Valuable spectroscopic information can also be obtained from two-laser MPI on the ground and resonant excited states and on the dynamics of other processes emanating from these states, as discussed in this section. In addition, two-laser studies interacting with the resonant state or the parent ion can give detailed information on laser ionization–fragmentation mechanisms, as discussed in Sections V and VI.

One possible two-laser scheme was shown in Fig. 1j, where a low-lying excited electronic state is populated by two-photon absorption and directly ionized by a second higher-energy laser. A more common method is to populate the resonant state by one-photon absorption, for reasons discussed later. Four basic experiments are possible with two-laser MPI: (i) the ionizing laser is scanned below the ionization threshold (from the excited state). This is called double-resonance MPI spectroscopy. (ii) The ionizing laser is scanned through the ionization threshold. This gives information on the excited-state ionization potential and other photoabsorption, photoionization processes taking place from the resonant state which is relevant to the following laser ionization–fragmentation discussion. (iii) The ionizing laser is fixed above the ionization threshold and the excitation laser is scanned to trace out the resonant state—a technique very similar to R2PI spectroscopy. (iv) The above processes are studied as a function of the time delay between the excitation and ionization layers. Molecular energy evolution and collisional processes may be probed by this technique.

Before reviewing examples of the above experiments, the drawbacks of two-laser MPI should be mentioned. The apparatus is, of course, more expensive, especially when vacuum-UV radiation is required for the ionizing step. Another problem is that the resonant-state lifetime must be at least on the order of the laser pulse lengths. Two lasers with uncorrelated intensity fluctuations, for example, are not expected to be more efficient than single-laser REMPI (or R2PI) where the ionizing step is one photon. Spectral regions and intensity conditions must also be found where neither laser causes significantly more ionization than the two lasers working together. These can be serious restrictions for a number of molecular MPI systems.

1. Double-Resonance MPI Spectroscopy

Double-resonance experiments have a number of well-known advantages. In terms of the two-laser MPI experiment, a resonant state is pumped and transitions from that state appear as resonances in the second laser ion current spectrum. The resonant level becomes a new "ground state" with only one rotational level populated. Selection rules for transitions from this level are restricted so the excited-state spectrum is simplified. New levels may be accessed by this technique since two separate Franck–Condon envelopes are involved. This method is particularly useful when two structured electronic transitions are in resonance simultaneously in a single-laser REMPI process. Williamson and Compton (1979) have used two dye lasers pumped by a single N_2 laser to populate the $B^3\Pi$ state of iodine with one photon and then trace out higher-order resonances with the second laser. Interference with two- and higher-photon states by the B state was thus avoided.

Infrared–UV double-resonance MPI is also possible. Esherick and Anderson (1980) populated the first vibrational level of NO with a tunable IR laser and studied two-photon resonant four-photon ionization from that level. Again great spectral simplification was obtained. Collisional redistribution in the vibrationally excited ground state by added buffer gas was monitored in this study by changes in the double-resonance spectrum.

Another double-resonance technique introduced by Cooper *et al.* (1981) uses a different approach to give ground-state information. Here, R2PI is monitored at a fixed wavelength while a second laser causes downward transitions from the resonant state. This causes a depletion of R2PI when resonant with ground-state levels, thus tracing out the equivalent of a Raman spectrum but with ionization detection.

2. Excited-State Ionization Thresholds

Figure 18 (Parker and El-Sayed, 1979) shows the excited-state ionization threshold spectrum of DABCO, obtained with the apparatus shown in Fig. 10. The lowest excited state was populated by two-photon absorption and then ionized by a second tunable visible–near-UV laser [see Fig. 1(j)]. Superimposed on this spectrum is the photoelectron energy spectrum for DABCO (dashed line), shifted by the excited-state energy. A

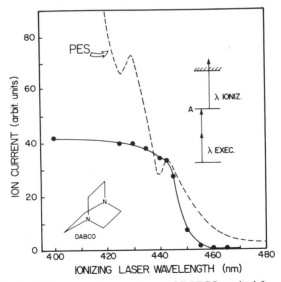

Fig. 18. Threshold photoionization spectrum of DABCO excited from the lowest excited electronic state (solid line). Dashed line shows the photoelectron spectrum shifted by the excited-state energy (Parker and El-Sayed, 1979, courtesy of North-Holland Publishing Company).

much sharper onset is observed from the excited state (a 3s Rydberg state) than from the ground state, which may be attributed to more vertical Franck–Condon factors from the Rydberg level. This study was carried out in the bulk gas phase with less than optimal tuning to the resonant state. Other results of this study are described in a following subsection (IV.E.4).

Duncan *et al.* (1981a) have reported a more refined two-laser R2PI study of benzene and naphthalene at threshold. Their result for naphthalene, rotationally cooled in a supersonic beam and mass selectively detected with the apparatus shown in Fig. 8, is shown in Fig. 19. At zero voltage on the draw-out field an extremely sharp (~ 2 cm^{-1}) excited-state ionization threshold is observed, while at increased draw-out voltages (present during the laser pulse) the ionization threshold shifts to lower energy and broadens considerably. This is a direct observation of field ionization of the large Rydberg orbitals immediately below the IP, an effect which is well documented in atomic physics (Cooke and Gallagher, 1978) and obviously (from this study) also operable in polyatomic molecules. Extremely accurate IPs may be determined by this technique, in the zero-field limit. The sharp resonances in Fig. 19 are due to R2PI by the second laser alone. Step functions are observed in other regions of the excited-state threshold ionization spectrum which indicate the relative Franck–Condon factors for transitions to the ground-state ion. Similar direct ionization was observed in benzene, but broad features assigned to

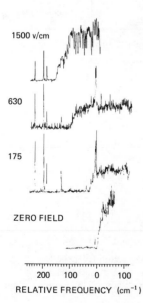

1500 v/cm

630

175

ZERO FIELD

200   100   0   100
RELATIVE FREQUENCY (cm^{-1})

Fig. 19. Voltage dependence of the naphthalene ionization threshold, excited from the $S_1 8^1$ state. Sharp features are due to R2PI from the second laser only (Duncan *et al.*, 1981a).

vibrational autoionization of Rydberg states at the threshold energy are superimposed on the threshold steps. This elegant study demonstrates some of the great potential of two-laser R2PI in polyatomic molecules.

Related R2PI experiments in alkali-metal dimers and clusters have been reported by Mathur *et al.* (1978) and Leutwyler *et al.* (1980, 1981) using cw Ar^+ laser radiation. Isotopic separation, dissociation energies, and extremely accurate ionization potentials (± 1.2 cm^{-1}) were demonstrated in these studies. Ionization potentials in some molecules may be determined by single-laser two-photon ionization and nonresonant multiphoton ionization if sharp steps are observed in the spectral region dividing *n* and *n–1* photon ionization. Williamson *et al.* (1979) have used this method in assigning the photoionization threshold of pyrrole.

3. Excited-State Spectroscopy via Two-Laser MPI

Antonov *et al.* (1978) reported one of the earliest two-laser R2PI resonant-state spectra, for the NO_2 molecule. In this experiment the ionizing laser is a fixed-frequency H_2 laser (7.7 eV) and the excitation laser is scanned through the visible NO_2 absorption regions. The two-laser R2PI spectrum is essentially identical to the corresponding absorption spectrum, indicating that excited-state transitions to the ion continuum have a weak spectral dependence.

An interesting two-color R2PI technique has been recently reported by Miller *et al.* (1982). Third-harmonic generation in rare gases was used to generate tunable radiation in the vacuum UV which, when coupled with the fundamental light, produced two-color R2PI in NO. High-lying Rydberg states were populated by the vacuum-UV photon and a single visible (blue) photon was believed to cause ionization. No ionization was observed without the blue photon. Sharp, single rotational levels were observed in the spectra, consistent with the high (~ 0.2 A) resolution of this technique.

Zacharias *et al.* (1980) have obtained isotope-specific resonance spectra by two-laser R2PI of NO. State selectivity was also demonstrated, and PRE analysis allowed extraction of the photoionization cross section from a single rotational level of the resonant state. Ebata *et al.* (1982) have also reported two-laser double resonance spectra of the NO molecule.

4. Lifetimes via Two-Laser MPI

Excited-state lifetimes may be obtained via two-laser MPI by varying the time delay between the excitation and ionizing lasers. This was first demonstrated by Andreyev *et al.* (1977) in formaldehyde, where a N_2 laser populated the $^1A_2$ excited state, which was then ionized by a H_2

laser. A multiexponential decay was found which is in rough agreement with fluorescence lifetimes observed under similar experimental conditions.

Similar lifetime measurements were performed in the two-laser REMPI study of DABCO mentioned earlier. The two-photon allowed resonant state, being the lowest excited state, is long lived, with a fluorescence lifetime of ~1 μsec and quantum yield near unity. Two separate lasers are thus needed to measure the excited-state decay by MPI, which was found to be identical to the fluorescence decay. Intensity dependences of the fluorescence and total ion count for one- and two-laser REMPI indicated that significant population loss occurs at the third proton level in two-photon resonant four-photon ionization (Parker and El-Sayed, 1979). Two-laser MPI avoids losses at this third level, and thus greatly enhances the overall efficiency. Most two-photon states are not so strongly allowed as this lowest state of DABCO, so further pumping to dissociative levels may be difficult to avoid in other two-laser REMPI studies.

Intersystem crossing and other radiationless transitions may be very efficiently studied by time-delayed two-laser MPI. Duncan *et al.* (1981b) have examined intersystem crossing in ultracold supersonic beams of benzene by populating the $S_1 6^1$ level with a frequency-doubled dye laser and ionizing the excited molecules with an ArF laser. The collision-free triplet lifetime was found to be 470 ± 50 nsec, and the intersystem crossing quantum yield ~80%, in agreement with other measurements.

V. LASER IONIZATION–FRAGMENTATION MECHANISMS

Up to now, measurement of the total ion count has been considered without regard to the charge production mechanism or to the nature of the charges produced. The simple schemes of Fig. 1 assume that ionization results once the requisite number of photons has been absorbed. This may not always be the case. Instead of direct ionization a neutral state lying above the first IP ("superexcited state") may be reached, which then may ionize (autoionization), radiate, dissociate, or absorb more photons. Once an ion is produced one might ask in which rovibronic state it is formed. This type of information can be obtained by photoelectron spectroscopy (see, for example, Eland, 1974), where the photoelectron kinetic-energy distribution gives, via balance of the total excess energy ($nh\nu - IP$), the ion internal energy distribution. Analysis of the ionization mechanism via photoelectron spectroscopy is discussed in the first part of this section.

It may also be asked what happens to the ion after it is formed. This is a particularly important question with laser ionization since the superexcited state or parent ion is produced while still in the presence of the intense laser field. Allowed transitions as well as multiphoton absorption from these species are very probable and can lead to extensive fragmentation of the molecule, as observed in the mass spectrum of laser-produced ions. Spontaneous and laser-induced fragmentation processes are discussed in the second part of this section.

The preceding questions are the same as those posed in single-photon ionization photoelectron and mass spectroscopy. In addition to the photon flux, one-photon and resonant multiphoton ionization differ in the state from which ionization originates. While this is obviously the ground state in the former processes, it is best considered the resonant state in the usual sequential or PRE limit of the MPI process. In this case, molecular photoionization may be carried out from the resonant state using the same detection and theoretical methods of one-photon ionization, but with high-intensity, high-resolution lasers. This is the second major application of laser ionization of polyatomics.

A. Photoelectron Energy Distributions Following Laser Ionization

Since laser ionization is equivalent in energy to near-threshold one-photon ionization, only the highest occupied molecular orbitals are probed, in contrast to HeI (21.2 eV) photoionization, which allows excitation of the complete set of valence orbitals. Just as in near-threshold one-photon ionization, autoionizing states may be populated, greatly affecting the resulting photoelectron kinetic energy. Ignoring these states, the photoelectron energy distribution following nonresonant n-photon ionization is given by

$$KE = nh\nu - IP_1 + (E_{int} - E_{int}^+) \qquad (15)$$

where IP_1 is the first IP, and E_{int} and E_{int}^+ the rotational, vibrational energy of the neutral molecule and resulting ion, respectively. With visible–near-UV lasers, KE can be up to a few electron volts, providing ultrahigh-intensity effects such as field acceleration do not occur.

A single vibronic state is selected by the initial one-photon or multiphoton absorption in a R2PI or REMPI process, and in the PRE regime, photoionization takes place from this new "ground state" at a later time determined by the photon flux and the cross section and Franck–Condon factors for the resonant- to ionic-state transition. High-lying resonant states may exhibit vastly different properties than the real ground state, however, even on the time scale of a 5-nsec pulse. Dissociation, internal conversion, intersystem crossing, and collisional processes may all occur during the pulse length, or time delay in a two-laser experiment, resulting in a complex ionization and fragmentation dependence. If the resonant state requires two or more photons for ionization, the higher resonant or near-resonant levels may also influence the neutral molecular geometry and stability prior to photoionization. The most simple case, direct one-photon ionization from the resonant state, will be treated first, followed by a discussion of autoionization and energy evolution effects on the photoelectron spectrum.

1. Direct Ionization

Rydberg states, which are often selected as the resonant level in a REMPI process, are by definition similar in geometry to the ion. Franck–Condon intensity rules for direct photoionization of Rydberg electronic states give $\Delta v = 0$ where v is a vibrational quantum number. If a valence level is selected, the Franck–Condon factors between the excited molecule and ion will depend on their respective geometries, and when the two geometries differ, a number of quanta in vibrations that most closely correspond to the change in equilibrium geometry will be excited. Direct photoionization of a Rydberg electron correlated with the first IP will produce photoelectrons of energy $KE = h\nu - IP + [E(v) - E^+(v)]$ where $E(v)$ and $E^+(v)$ are total vibronic energies of the excited and ionic states, respectively, with v the same in both states. State-selected ions and monoenergetic electrons useful in subsequent studies may be produced under this condition, and rovibronic state selection may also be possible in favorable spectroscopic cases. Direct photoionization of a valence electron can produce a range of photoelectron peaks, depending again on the relative geometry change, and at room temperature this distribution can become very broad or continuous with larger less symmetric molecules. Two-laser experiments in supersonic beams will provide the cleanest source of state-selected ions and most informative photoelectron spectra.

2. Autoionization

Autoionization is an indirect process where molecules are produced first in a superexcited state and then spontaneously emit electrons. The light must be resonant with the superexcited-state energy in the first step, and the resulting ion plus free electron must have the same symmetry as the autoionizing state. There are two distinct autoionization mechanisms, vibrational autoionization and electric autoionization. In the first process, the Rydberg electron (for example, in REMPI) is excited to a higher n value (electronic origin below the IP) but with vibrational levels in the ion continuum. Breakdown of the Born–Oppenheimer approximation is then invoked to couple the vibrational energy of the core with the distant Rydberg electron, resulting in ionization. A vibrational propensity rule of $\Delta v = -1$ is most probable in this process, and so the ejected electrons are of almost zero energy.

Electronic autoionization produces an ion in a different electronic state from that of the core of the ionizing level. In one-photon excitation this could arise from excitation of an inner-shell electron to a Rydberg state whose ionization limit is consequently an excited state of the ion. Any final state of lower energy than the autoionizing level may be produced with branching ratios dependent of Franck–Condon factors, and a distribution of electron energies is expected. Only direct ionization or vibrational autoionization can result from photoabsorption by a Rydberg electron. However, internal conversion to a nearby valence state or direct excitation to a valence state in R2PI can allow access to Rydberg levels converging on higher IPs. Autoionizing states reached from excited electronic states are, in general, different from those populated by one-photon transitions from the ground state. Another electronic autoionization process has recently been implicated in a number of polyatomics (Baer et al., 1979), whose effect is to produce low-energy electrons and correspondingly vibrationally hot ground-state ions. Although the mechanism is not yet clearly understood, it might be attributed to autoionization during predissociation. Most photoelectron spectrometer transmission functions are strongly peaked at zero energy, so it is experimentally difficult to determine the relative importance of the different autoionization mechanisms.

Superexcited states may also involve two excited electrons, each of energy less than IP_1 but of total energy exceeding IP. Such states, about which very little is known, are directly accessible by resonant-state absorption in R2PI and REMPI processes. Ionic states forbidden by one-electron excitation may also be reached by excited-state ionization. REMPI and R2PI may thus open pathways to more complete knowledge of the manifold of superexcited and ionic states in which a molecule can exist. Progress along these lines is reviewed in Section VI.

One more process may affect the photoelectron energy distribution. Instead of dissociation or autoionization a superexcited state may absorb more photons stepwise to higher-lying superexcited states until dissociation or autoionization becomes more probable. Highly energetic electrons and vibrationally hot ions (which then have a great proclivity to dissociate) will be produced by autoionization from such high-lying superexcited levels. Although most one-excited-electron autoionizing states have broad linewidths indicative of lifetimes on the order of 10^{-13} sec or less, multielectron state lifetimes may be substantially longer since more than the normal two-electron interaction is required for autoionization. Evidence for absorption by superexcited states is also reviewed in Section VI.

3. Molecular Energy Evolution from the Resonant State

Dissociation, internal conversion, intersystem crossing, collisions, and other processes taking place during the laser ionization event can greatly alter ionization rates and photoelectron spectra, even when they precede direct ionization. Dissociation from the resonant state or after ionization may produce neutral fragments which then multiphoton ionize by their own mechanism, producing their own characteristic photoelectrons. Collisional processes both shift excited-state population within its own manifold and increase internal conversion and intersystem crossing rates between manifolds. These latter processes produce vibrationally excited lower-energy electronic states which, owing to Franck–Condon factors, may require substantially higher-energy photons for direct ionization. Photoabsorption by highly vibrational excited states or states with substantially different geometry than the ion can also strongly favor dissociation over ionization from a superexcited state. Since autoionization and the above processes are common in polyatomics, it is not surprising that few laser ionization photoelectron spectra exhibit only the electron energies predicted by Eq. (15).

B. Dissociation Processes Following Laser Ionization

While photoelectron spectroscopy can indicate the internal energy of positive ions formed via resonant MPI, it gives very little information on the fate of those ions after the instant of ionization. Often only a small amount of internal energy is required for an ion to dissociate to a smaller daughter ion plus neutral fragments. At higher internal energies the ion may fragment to several species with branching ratios dependent on the amount of energy provided. Excitation up the "autoionization ladder" will produce excited ions at each level, each type with a different dissociation branching ratio. Fluorescence is another means besides dissociation by which an excited ion may lose energy. Even more important, open-shell ions often have low-lying excited states resonant at the laser wavelength, allowing absorption up the "parent-ion ladder" to levels of greater instability. Internal conversion in such ions is known to be facile, resulting in hot ground-state ions. The laser may also act on the charged (and neutral) fragments produced from parent-ion dissociation. All of these processes result in a distribution of laser-produced ions, rather than only parent ions expected from the schemes of Fig 1. Mass spectroscopy is the standard tool for investigating spontaneous and laser-induced fragmentation mechanisms following photoionization.

Dissociation may also take place after the laser pulse. Excited ions may fragment on the microsecond time scale and thus change mass during the mass analysis procedure. Such metastable ions are commonly seen in electron impact mass spectroscopy and have been observed following laser ionization. These observations can be very informative since they indicate the amount of internal energy provided to the ion by the specific photoionization process. An example of this is described in Section VI.

1. Fragmentation Mechanisms

Figure 20 shows the four basic mechanisms for laser-induced ionization–fragmentation of a neutral molecule ABC. Two-photon resonance through ABC* is followed by one-photon ionization to ABC⁺ or one-photon absorption to a superexcited state ABC**. In the neutral-fragment photoionization mechanism (a) the resonant state or superexcited state may dissociate (jagged lines) to form a vibrationally excited product AB† (plus C) which is then two-photon ionized to form AB⁺. The autoionization ladder (b) involves stepwise absorption through superexcited states, each of which may dissociate with different branching ratios to energetically allowed products ABC⁺, AB⁺, and A⁺. The parent-ion ladder (c) requires either direct ionization or autoionization (wavy line) to ABC⁺, which then successively absorbs photon through its excited states ABC⁺*, each of which may dissociate with different branching ratios. Finally, in the ionic fragment dissociation staircase (d), ABC⁺ is formed

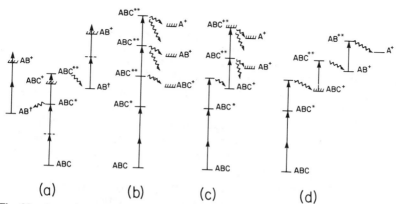

Fig. 20. Laser ionization–fragmentation mechanisms: (a) neutral-fragment ionization, (b) autoionization ladder, (c) parent-ion ladder, (d) ionic-fragment dissociation staircase.

and then is excited to ABC^{+*}, which dissociates to AB^+. This species absorbs a photon to AB^{+*}, which then dissociates to A^+, etc.

The oversimplification of these schemes is obvious. Intensity-dependent competition between autoionization and further absorption (schemes (b) and (c), respectively) can occur and other combinations are possible. Multiphoton processes can affect photoproducts as well as the parent molecule, and the sequential nature of these schemes (i.e., rate equations) will not be valid for all conditions. Even with these qualifications such schemes have proven useful in characterizing laser fragmentation for a number of molecules (see Section VI).

In theory, all of the above processes can be assigned cross sections, predissociation rates, and branching ratios and then inserted into a population rate-equation model along with the laser intensity to predict the intensity-dependent ion yield and fragmentation pattern. While such efforts have been attempted, this is, in practice, almost impossible for even the simplest molecule owing to the scarcity of input data. Besides the uncertainty in laser intensity, relaxation rates, and excited-state ionization cross sections mentioned already, not enough is known about photodissociation of ions. Direct experiments on laser-induced ion fragmentation are becoming available (see, for example, Kim and Dunbar, 1980), but few are performed with the same photon flux as REMPI or R2PI. A major problem with single-laser studies is that the photoproducts are formed with varying amounts of internal energy—which greatly affects their absorption probabilities (Franck–Condon factors), radiationless transition rates, and fragmentation branching ratios. These complications are so severe that even qualitative success with rate-equation modeling of REMPI fragmentation (i.e., distinguishing a unique scheme) is at present unlikely.

2. Statistical Approach to Laser Ionization–Fragmentation

Considering the number of pathways available, especially with high laser intensity (typical REMPI experiment) it is not surprising that a statistical, rather than dynamical, approach to predicting fragmentation patterns has proven successful (Silberstein and Levine, 1981). The goal of this method is not to determine the absorption/dissociation mechanism for laser fragmentation, but rather to determine how much (if any) these specific mechanisms affect the resulting fragmentation pattern. In the statistical limit all such dynamic effects are absent and the fragment distribution is the one of maximal entropy subject only to the conservation of energy, atoms, and charge. Calculation of this statistical distribution for an average energy absorbed is quite sensitive to the heats of formation of each fragment from the parent molecule, many of which are unknown.

There are several computation-free methods of testing the statistical limit, however: (i) Since there is no specification of structure, isomers should yield the same pattern at the same energy input. (ii) All fragments should also have thermal internal-state distributions. (iii) The same pattern should be recovered through different resonant states, although the intensity needed to accomplish this may differ. This reflects the amount of energy put into the molecule.

There are, of course, cases where the statistical approach fails because of mechanistic constraints. Most obvious are the molecules where the parent ion at certain wavelengths does not absorb and thus cannot fragment. At higher laser intensities two-photon absorption can be expected, however, leading to fragmentation possibly comparable to mass spectra without the parent absorption bottleneck. Several examples of nonstatistical behavior using current laser intensities and a number of examples of purely statistical fragmentation using the above criteria will be described in Section VI.

VI. EXPERIMENTAL ANALYSIS OF LASER IONIZATION–FRAGMENTATION MECHANISMS

Mass analysis of ions produced by R2PI and REMPI of polyatomic molecules is a new and very active extension of MPI research, with the first mass spectra reported in 1978. Kinetic-energy analysis of laser-produced ions and electrons is even more recent, dating back to only 1980. These types of investigations are directed more at determining the relative importance of the different ionization and fragmentation mechanisms outlined in Section V than in exploiting or perfecting the technique. An exceptional amount of attention has been paid to benzene as a prototype polyatomic molecule. In this section the systematics of laser ionization and fragmentation of benzene will be presented, followed by a survey of other representative molecules which exhibit different response than benzene to intense laser fields. It should be emphasized that this synthesis is quite preliminary, and the variations are expected as wider-ranging and more refined studies are reported.

As anticipated from Section V, complex photoelectron energies and extensive fragmentation is usually observed at high laser intensities, while at low intensity often no fragmentation is found. This greatly favors R2PI as the superior detection scheme since fragmentation compromises the quality of a mass-selective technique. R2PI and REMPI are not separated in the following discussion since evidence thus far suggests that intensity is much more influential than wavelength in determining the fragmentation pattern. There are several notable exceptions to this rule, however, and these are also strong examples of nonstatistical fragmentation behavior.

A. Benzene.

Benzene is a popular target for resonant MPI study because of its intermediate size, convenient vapor pressure, and well-understood one- and two-photon spectrum. REMPI, R2PI, and two-laser MPI of benzene have been reported for a wide range of wavelengths, focusing conditions, and sample environments (high pressure to supersonic beams), making this molecule the best-characterized if not the best-understood example of resonant multiphoton ionization.

1. Ionization Mechanism

The first multiphoton ionization photoelectron spectrum was reported by Meek et al. (1980) for R2PI of benzene. At the ArF and KrF wavelengths the electron energy distribution reflected only the excess energy available past the IP (i.e., $2h\nu$ − IP). A small, broader peak to lower energies was interpreted as the result of the competition between intramolecular relaxation and photoionization at the ArF energy. Miller and Compton (1981b) reported the first REMPI photoelectron spectrum, in this case for two-photon resonant three-photon ionization of benzene. Measurements again showed no excess energy than $3h\nu$ − IP. Benzene is too large a molecule, especially at room temperature, for photoelectron spectra to distinguish between autoionization and direct ionization, although these studies do rule out further absorption by a superexcited state.

Duncan et al. (1981a) have obtained the threshold ionization spectrum of supersonic cooled benzene originating from the $S_1 6^1$ state by two-laser R2PI. Autoionizing states were observed in this region and contributions from autoionization and direct ionization were believed to be nearly equal. The width of the autoionizing features indicated a lifetime on the order of 10^{-14} sec, and the mechanism was assigned as electronic autoionization from the $n = 3$ member of a Rydberg state converging on the second IP.

2. Fragmentation Mechanism

The first REMPI mass spectrum of a polyatomic was reported by Zandee and Bernstein (1979a) for two-photon resonant (3s Rydberg state) three-photon ionization of benzene. A focused dye laser was used ($\sim 10^{10}$ W/cm^2) and extensive fragmentation resulted with C$^+$ the most abundant fragment under the tightest focus (Zandee and Bernstein, 1979b). From appearance potentials of the fragment ions as many as nine 3.2-eV photons were estimated to have absorbed during the laser pulse length. Boesl *et al.* (1978) reported mass-analyzed R2PI of benzene at $\sim 10^6$ W/cm^2 and found only the parent ion. Antonov *et al.* (1980) followed R2PI over the 10^4–10^9 intensity range with a KrF laser and found a smooth transition through the above extremes. In the intensity ranges where fragmentation is observed, two-laser experiments have greatly assisted in determining the fragmentation mechanism.

Both the autoionization ladder, Fig 20(b), and the neutral fragment photoionization, Fig 20(a), were eliminated by a two-color experiment of Boesl *et al.* (1980). R2PI through the S_1 state (259 nm) produced only parent ions but the addition of a visible pulse at 520 nm 15 nsec later caused extensive fragmentation. No increase in the total charge was found on addition of the second laser, ruling out ionization of neutral fragments formed after ionization.

A similar two-laser experiment has been described by Pandolfi *et al.* (1981) with the variation of higher draw-out fields and a 50 nsec time delay between the UV and visible pulses. Benzene parent ion was produced at 266 nm and accelerated out of the focal region before the second pulse at 510 nm could cause fragmentation. The fragmentation pattern was also studied as a function of the draw-out field and Stevenson's rule, that on dissociation the positive charge remains on the larger species, was confirmed for multiphoton fragmentation.

The relative importance of the parent-ion ladder, Fig. 20(c), and the ionic fragmentation staircase Fig. 20(d) was investigated by Hering *et al.* (1981) using 20-psec-long pulses at 266 nm. With one pulse, fragmentation was found to proceed up the parent-ion ladder only to the energy range required for C_4^+ and C_3^+ fragment formation. The addition of another pulse 5 nsec later allowed C_1^+ formation, presumably via the ion fragmentation staircase. Since the staircase takes longer to climb than the parent-ion ladder, the stepwise process is interrupted when only 20-psec pulses are available. Interestingly, Hetherington *et al.* (1981) detected C_2 formed in the dissociation $C_6^+ \rightarrow C_4^+ + C_2$ using picosecond CARS under similar conditions.

Newton and Bernstein (1982) have studied ionic fragmentation by forming ions via electron impact and dissociating them with wavelengths used in REMPI experiments. Some two dozen intensity-dependent branching ratios were established for the fragmentation of these benzene derived ions, allowing a quantitative "pathway diagram" for benzene REMPI dissociation. Other authors have argued against ionic fragmentation since smaller ions (e.g., $C_3H_x^+$, $C_2H_x^+$) are very unlikely to absorb over wide spectral ranges, yet many fragmentation patterns are identical at diverse wavelengths. A combination of both mechanisms—the parent-ion ladder and fragment-to-fragment dissociation—is very likely, with the weighting dependent on the particular molecule.

Miller and Compton (1981b) have reported ion kinetic-energy distributions following single-laser REMPI of benzene. Using the apparatus shown in Fig. 9, they found near thermal kinetic energies for the H^+ and C^+ fragments, indicating that these ions are not produced from photodissociation of ground-state C_2^+ or CH^+. They also suggested that excess photon energy remains most often as internal rather than kinetic energy in the fragments.

Carney and Baer (1982) derived kinetic-energy releases for the various fragment ions of benzene from TOF peak shapes. They found their data in best agreement with the statistical theory of Rebentrost and Ben-Shaul (1981), which predicts that a superexcited parent-ion undergoes several consecutive dissociations, Fig. 20(c). The fragment-ion absorption—dissociation mechanism also was in accord with their kinetic-energy releases but was not found realistic on the basis of the wavelength independence of the fragmentation pattern.

B. Other Molecules

1.   Ionization Mechanism—Photoelectron Spectroscopy

Photoelectron kinetic-energy analysis is much more informative when applied to the laser ionization of smaller molecules. As discussed in Section V, there is a wealth of information available from the study of photoionization of excited states, both for the ionization mechanism and the relaxation dynamics of the excited state. Photoelectron spectra of the excited-state molecules formed by other means have been reported previously [see, for example, Jonathan et al. (1971) on the $^1\Delta_g$ state of O_2].

Miller and Compton (1981a) have studied two-photon resonant four-photon ionization of NO using photoelectron energy analysis. Four separate ionization processes were observed for different vibrational levels of the $\tilde{A}$ state of NO: (i) direct ionization producing state-selected ions and monoenergetic electrons with energies predicted by Eq. (15), (ii) an unassigned autoionization process which produced zero-energy electrons, (iii) vibrational ionization, and (iv) "quasi-free-free" absorption by a superexcited state which then autoionized. Two-photon resonant ($\tilde{C}$ state) three-photon ionization resulting in state-selective direct ionization was also observed.

Glownia *et al.* (1982) have recorded photoelectron spectra following four-photon ionization in NH_3, resonantly enhanced by a variety of two- and three-photon vibronic Rydberg states. Zero-energy electrons were observed exclusively following resonance enhancement by all of the resonant states but the $\tilde{C}$ state, where state-selective direct ionization was also present. Vibrational ionization was assigned as the process producing the zero-energy electrons from the $\tilde{B}$ state. Internal conversion from three-photon resonances to the $\tilde{B}$ state, which then vibrationally autoionizes, was also observed. No spectral evidence for internal conversion mechanism was found for the $\tilde{C}$ state, explaining its stability and resulting direct ionization.

Photoelectron energy analysis by Fisanick *et al.* (1981) for REMPI of $Cr(CO)_6$ suggested that at shorter wavelengths electrons are released by both Cr (neutral-fragment photoionization) and $Cr(CO)_6$, which is then totally fragmented to Cr^+. The autoionization mechanism is again not indicated by this study, or by REMPI photoelectron spectroscopy of acetaldehyde by the same group (Robin, 1980).

2. Fragmentation Mechanisms

Most molecules including benzene undergo significant and usually wavelength-independent fragmentation with typical REMPI or focused R2PI fluxes. Separate fragment ions rarely show an integral intensity dependence or an integer difference between successive ions. This suggests that combinations of the fragmentation mechanisms, Fig. 20(a)–(d), are likely, or, equivalently, that the least expensive energy route is seldom followed. Multiple fragmentation pathways are assumed in the statistical approach to laser dissociation–ionization, and several experimental examples of this case are discussed in this section.

Not all molecules show intensity-dependent fragmentation. Several examples have become available where only the parent ion appears over a span of wavelengths, and over the available laser intensity range. Such behavior is certainly easier to approach than multiple fragment spectra. The arguments so far lend support to the parent-ion-dissociation ladder.

Figure 21 shows one of the most direct confirmations of parent-ion dissociations for two-photon resonant four-photon ionization of 2,4-hexadiyne, a benzene isomer (Carney and Baer, 1981a). The two-dimensional mass spectrum in Fig. 21(a) shows a dramatic change in fragmentation as the laser wavelength (constant intensity) is increased from 540 to 510 nm. At 538 nm only the parent ion is observed, which correlates with the lack of one-photon absorption, as predicted by the threshold photoelectron spectrum in Fig. 21(b). Intermediate wavelengths show mixtures of the

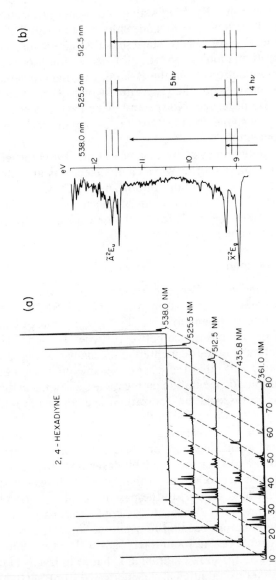

Fig. 21. REMPI fragmentation of 2,4-hexadiyne (Carney and Baer, 1981a): (a) REMPI mass spectra at several laser wavelengths, constant intensity; (b) threshold photoelectron spectrum for the wavelengths shown in part (a).

two extremes—parent-ion and C^+ formation. Similar behavior was observed in REMPI fragmentation of toluene.

In the open tertiary amines, trimethylamine and triethylamine, neither parent nor daughter ion was found to absorb between 520 and 380 nm (Parker *et al.*, 1981). Daughter ions [the parent-H for $N(CH_3)_3$ and parent-CH_3 for $N(C_2H_5)_3$] can become accessible when the excess energy above the *n*-photon ionization exceeds the daughter-ion appearance potential. The fraction F of signal due to daughter ion in trimethylamine is plotted in Fig. 22(a) versus the laser wavelength, and is shown to become large in the proper excess-energy region of both three- and four-photon ionization processes (two-photon resonance enhanced). Above 380 nm only the parent and daughter appear, regardless of laser intensity, while at 355 nm extensive intensity-dependent fragmentation was observed. This latter

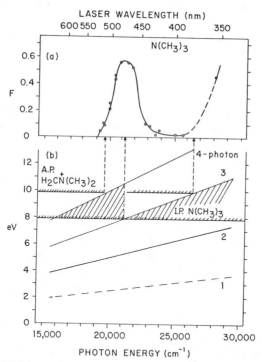

Fig. 22. REMPI fragmentation of trimethylamine (Parker *et al.*, 1981): (a) Fraction F of trimethylamine daughter ion in the total ion count vs laser wavelength; constant intensity. (b) Energy-level diagram for two-photon resonant three- and four-photon ionization of trimethylamine. Shaded areas indicate excess energy available to the molecule. Note that the daughter ion appears in the regions where the excess energy exceeds the daughter-ion appearance potential.

behavior is also found in REMPI mass spectra of many other molecules (besides benzene), e.g., t-butyl benzene, methyl iodide, DABCO and ABCO (Lichtin et al., 1981b), acetaldehyde (Fisanick et al., 1981), and pyrrole (Cooper et al., 1980), to name just a few examples, and can be ascribed to efficient absorption to low-lying resonant states of the parent ion (and beyond) over all of the incident laser wavelengths. In other words, where the neutral molecule is two-photon resonant, the parent ion (created after n-photon ionization) is very often one-photon resonant; thus there is no "bottleneck" to limit further absorption and fragmentation.

Carney and Baer (1981b) in a study of REMPI fragmentation of H_2S found that the autoionization mechanism best accounts for the observed intensity dependence of the fragments S^+, HS^+, and H_2S^+, and for the appearance of these fragments at all resonance wavelengths investigated. H_2S and other small molecules such as NO have limited pathways for ionization and fragmentation because of their less probable resonance conditions, possibly making autoionization the observable absorption-fragmentation mechanism.

3. Two-Laser Studies

Letokhov and co-workers at Moscow have reported pioneering studies of two-laser R2PI fragmentation of polyatomics using N_2 laser excitation and H_2 laser ionization. In their study of benzaldehyde (Antonov et al., 1978) they observed only the parent ion and parent-minus-hydrogen daughter ion over long time delays (microsecond range), consistent with rapid intersystem crossing from S_1 to long-lived T_1 states. Applying the same technique to benzophenone, $(C_6H_5)_2CO$, they also found fast conversion to long-lived T_1 states at the parent-ion mass. Two components were observed for the $C_6H_5CO^+$ daughter ion, one attributed to conversion to the T_1, ionization and then dissociation, and the other due to absorption by T_1 of another N_2 laser photon resulting in dissociation and then ionization of the $C_6H_5CO\cdot$ radical by the H_2 laser. A sharp dropoff in the second component with longer time delays was attributed to a 14-nsec radiative lifetime for the radical species. Other molecules were also investigated with similar results.

The two-laser REMPI study of DABCO described earlier was repeated with mass analysis by Newton et al. (1981). DABCO undergoes extensive fragmentation, including bifurcation, in two-photon resonant four-photon ionization at 560 nm. One-photon ionization of the Ã state by an unfocused 266-nm laser beam produced only DABCO$^+$ ion, however, indicating that fragmentation due to high flux can be avoided by bypassing the three-photon-level bottleneck. Higher intensities at 266 nm caused extensive dissociation, as expected.

DeVries *et al.* (1981) measured the first ion kinetic-energy distributions following R2PI and two-laser REMPI of I_2, using the TOF drift technique. A flash-lamp-pumped dye laser populated the two-photon state, which was ionized by an ArF laser photon. Among their several findings, the authors observed negative-ion formation, dissociative autoionization, and sufficient ion kinetic energy to invoke a parent-ion dissociation mechanism.

4. Experimental Tests of Statistical Fragmentation

Lubman (1981) has used the isomer method to test the statistical limit for laser ionization in several isomer pairs including naphthalene and azulene, and found that the same patterns can be obtained at different wavelengths by adjusting the laser intensity. Hudgens *et al.* (1981) also found identical fragmentation patterns for the *trans* 1,2-dichloroethane isomers. Lichtin *et al.* (1981b) tested the maximal entropy limit using the alternate-doorway method and found it to hold reasonably well for a diverse group of large and small molecules. Silberstein and Levine (1981) were able to compute very similar patterns for all of the experimental test cases of this study. Numerous examples of low-energy electrons and low ion kinetic energies (the second test of the statistical model) have already been described. In short, above $\sim 10^9$ W/cm^2 a surprising range of molecules appear to have little if any dynamic constraints to ionization–dissociation.

5. Metastable Ions

Metastable ions, when detected, can provide detailed insight into laser ionization–fragmentation mechanisms. The first observation of a metastable ion produced by R2PI was reported by Proch *et al.* (1981), who assigned $C_6H_5^+$ as a product of long–lived (~ 2 μsec) vibrationally hot aniline ions. Broad asymmetric peaks were observed at mass 66, which grew broader to longer wavelengths as shown in Fig. 23. The internal energy of the parent-ion precursor was not determined (by photoelectron spectroscopy) in this study Baer and Carney (1982) studied aniline ion dissociation at the same energy using one- and five-photon excitation, compared to the three-photon absorption processes studied by Proch *et al.* Competition between absorption and internal conversion to a vibrationally hot ion (which presumably does not absorb owing to poor Franck–Condon factors) was found to take place with three-photon excitation. Five-photon excitation with its attendant high intensity showed no metastable species

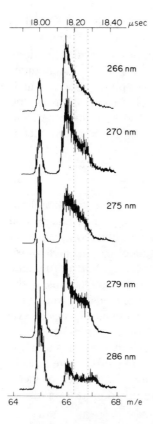

Fig. 23. Portion of the REMPI TOF mass spectrum of aniline at $m/e = 66$ as a function of laser wavelength, constant laser intensity (Proch *et al.*, 1981, courtesy of North-Holland Publishing Company).

because further absorption dominated over internal conversion. One-photon excitation placed the internal energy of metastable ground-state aniline ion at roughly 5 eV. Metastable ions were also observed but not analyzed, in the REMPI spectrum of triethylamine (Parker *et al.*, 1981).

VII. LASER IONIZATION AS AN ANALYTICAL TOOL

The third area of activity in resonant multiphoton ionization of polyatomics involves the pursuit of what may be called the "perfect" detector—a device which is molecule specific, state selective, and ultrasensitive to the point of single-molecule sensitivity. In the related field of atomic resonance ionization spectroscopy (Hurst *et al.*, 1979), laser schemes have been worked out which, with present-day lasers, will allow state-selective detection of virtually every element on the periodic table, and with single-atom sensitivity. Molecules are more challenging targets than atoms, however, as can be appreciated from the discussions of the preceding sections. In this section the prospects and limitations of laser ionization as the perfect detector for molecules will be considered, on an absolute scale and especially on a practical scale, by comparison with other analytical techniques. The reader should also refer to chapter 7 by Harris and Lytle for a more general review of laser analytical spectroscopy.

A few conclusions from the preceding sections are worth repeating. First, any molecule can be ionized with unit probability, given the proper laser. Many devices can detect single ions, thus there is no theoretical limitation to single-molecule detection. Almost any molecule with an IP less than ~11 eV should be ionizable by R2PI or REMPI with present lasers. Two (or more) laser techniques should offer an efficient scheme for even the most difficult molecule, again given the proper lasers. Specificity, however, is not guaranteed by simply increasing the laser flux—some compromise between ultimate sensitivity and specificity will often be necessary in common sample environments. Two related problems, overlapping spectra and too-rapid energy evolution from the resonant state, limit the selectivity of resonant MPI.

While supersonic beams help tremendously in relieving spectral congestion and in concentrating the sample within the laser bandwidth, other processes such as rapid intramolecular relaxation and predissociation can still cause continuous absorption profiles in some spectral regions. Molecule-specific detection in this case could be limited to samples where background molecules with similar spectra are not present.

A more serious drawback is dissociation itself, since it introduces new species to the system. Some molecules are not efficiently resonance ionized with present-day lasers, owing to rapid dissociation from all of the excited states. As an example, R2PI could be used to detect KI from the crossed-beam reaction of $K + CH_3I$. KI dissociates faster than it ionizes, however, producing $K + I$. Atomic RIS can very efficiently detect K or other alkali metals from dissociation of alkali halides (Grossman *et al.,* 1977), unfortunately, K is already present in the scattering region. Care must also be taken that the technique used for the sample does not also ionize background through higher-order effects. A procedure used to resonance ionize K in the above example could inadvertently photodissociate KI producing a new source of K. Ion pair formation in other systems can produce easily photodetached anions and thus another source of confusing neutral species.

Sensitivity is also affected by energy evolution. Rapid internal conversion and intersystem crossing trade electronic for vibrational energy, and the new Franck–Condon factors can be very unfavorable for further absorption. Absorption followed by dissociation from a superexcited state is also more likely with these conditons. Collisional process can have the same effect, either through redistribution within the excited-state manifold or by promoting crossing rates between manifolds.

Although mass sensitivity may be used to regain lost spectral selectivity, laser-induced fragmentation of the parent ion may lower the quality of the mass spectrum. Working at low intensity to decrease fragmentation also decreases the probability of ionization. Fragmentation processes can be exceedingly complex, making the intensity dependence a much less useful additional dimension for two-dimensional mass spectroscopy. These limitations apply even more directly to resonant multiphoton ionization photoelectron spectroscopy. It may thus be concluded that laser ionization cannot be a general (all molecule) perfect detector. For many applications, however, laser ionization is perhaps the best substitute, especially when compared to other highly successful techniques.

A. Comparison with Laser-Induced Fluorescence and Electron Impact Mass Spectroscopy

The proper comparison of laser ionization with laser-induced fluorescence (LIF) for detection of large molecules is in terms of sensitivity since both methods offer spectral selectivity. As mentioned previously, ionization allows complete measurement of the ground state, rather than the 50% saturation limit of LIF, and 100% collection efficiency compared to at best ~5% for LIF. This factor of sensitivity grows much larger when practical considerations such as scattered light, detector background, and signal conversion efficiency are included. A reasonable sensitivity limit for LIF of large polyatomics is $\sim 10^8$ molecules/cm^3, while for laser ionization 10^4 molecules/cm^3 sensitivity has been demonstrated and single-molecule sensitivity is possible. LIF is limited by the fluorescence quantum yield while in laser ionization up-pumping takes an active role in overcoming nonradiative (and radiative) processes.

Laser ionization does not compete as effectively with electron impact (EI) as the most efficient ionization source for mass spectroscopy. (The aspects of specificity versus generality are ignored here, although it could be pointed out that in practice the on-resonance MPI yield varies greatly compared to EI, and that at present laser ionization is not applicable to anywhere near the molecular range of EI.) A fair comparison is between EI TOF and laser-ionization TOF since the complete mass spectrum can be obtained by both on short time scales. EI-TOF sources typically operate at 10 kHz, compared to 10 Hz for most pulsed lasers, and the fraction of sample ionized is usually $\sim 10^{-4}$ for EI compared to 25% for favorable R2PI conditions. In terms of efficiency, laser ionization is at best comparable, at a significantly higher price. For these reasons laser ionization has the best potential as a supplement rather than a replacement of conventional mass spectrometers.

Zacharias *et al.* (1980) have provided a beautiful example of the two-dimensional assets of laser ionization in a two-laser R2PI study of NO with mass-spectroscopic and fluorescence detection. By scanning a high-resolution excitation laser in the region of 226 nm (ionizing laser 266 nm) resonances due to ^{14}NO and ^{15}NO were observed completely overlapped, of course, in the LIF spectrum. The excitation spectra of either isotope could be obtained separately through laser ionization simply by monitoring the appropriate mass, thus illustrating state selectivity. Electron impact of the same sample revealed a strong background signal, which laser ionization discriminated against by better than 1000:1. Charge transfer processes taking place at higher pressures were easily investigated with the same apparatus.

Isomeric selectivity is also possible with laser ionization, even for very challenging isomer pairs. Figure 24 compares the two-photon resonant three-photon ionization mass spectra of *t*-butyl and *n*-butyl iodide (~370 nm) with the 70-eV EI spectra (Parker and Bernstein, 1982). Significant differences are seen in the REMPI spectra, while the EI spectra are very similar. Although spectral selectivity between the two neutral species is poor owing to resonance overlap, the ionic spectra of their parent-minus-I daughter ions must be quite different since *t*-butyl$^+$ does not fragment over the intensity range shown. Isomerization during the 5-nsec pulse length is also ruled out. Similar REMPI results were found for isopropyl and *n*-propyl iodide, whose EI spectra are indistinguishable. The isomer pairs azulene and naphthalene show very different R2PI mass spectra at the same laser intensity (Lubman *et al.*, 1980), although the intensities can be experimentally adjusted separately to give the same patterns, in accord with the statistical prediction.

B. Absolute Sensitivity Estimates

The most direct measure of the absolute sensitivity of laser ionization has been reported by Wessel *et al.* (1981) for R2PI of naphthalene. In this study a temperature-controlled large-volume ultraclean proportional counter was used to directly detect 5×10^4 naphthalene molecules/cm^3, as measured by temperature extrapolation from the vapor-pressure curve. The authors estimated that as few as 10 molecules/cm^3 could be detected with improved counting electronics and a higher-power laser with a longer pulse length or wider spectral bandwidth. They also pointed out that the TEM$_{00}$ power from the laser most affects the actual intensity since higher-order modes are focused much less effectively.

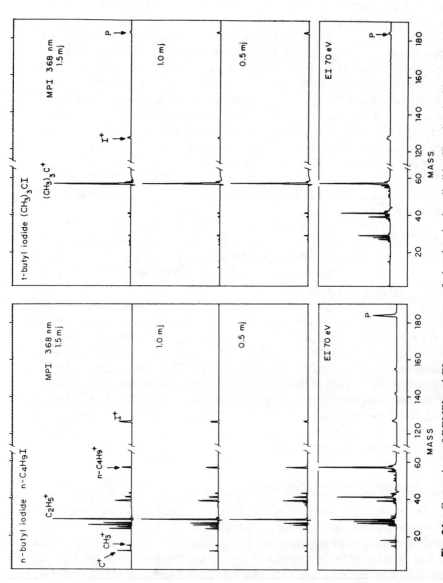

Fig. 24. Comparison of REMPI and EI mass spectra of *t*-butyl and *n*-butyliodide. [Reprinted with permission from Parker and Bernstein, 1982. Copyright (1982) American Chemical Society.]

Boesl *et al.* (1981) have made careful intensity-dependent studies of mass-resolved R2PI in several organic molecules. They found 25% of the ground-state population of aniline which was resonant with the laser bandwidth was ionized at 10^7 W/cm^2, an easily obtained intensity. By extrapolating their observed signal they estimated 7×10^5 molecules/cm^3 could be detected (with mass analysis) under their conditions, or roughly parts per billion sensitivity with resonance at the aniline S_1 origin. In addition, at 10^7 W/cm^2 essentially no fragmentation was observed in the aniline mass spectrum.

Antonov and Letokhov (1981) have estimated the ultimate sensitivity of R2PI for detection of air impurities, assuming cooling in a supersonic beam for spectral selectivity. Using a transmission factor of 0.5 for TOF detection, an ion yield of 0.5, and a photoionization volume of 0.1 cm^3, they estimate parts per trillion sensitivity. For many molecules an ion yield of 0.5 requires intensities so high that extensive fragmentation will take place. These authors also discuss other aspects of two-laser ionization detection including vibrational state resonance enhancement.

VIII. CONCLUDING REMARKS

A number of applications of laser ionization in polyatomic molecules have now been outlined. REMPI has already proven to be an extremely productive technique for detection and characterization, via polarization analysis, of new excited electronic states. R2PI is a more sensitive and general method than LIF for the study of one-photon absorption in large molecules. Two-laser experiments such as double-resonance, excited-state ionization threshold and time-delayed two-step ionization offer an even greater potential for understanding the spectroscopy and dynamics of excited vibrational and electronic levels.

Laser ionization is also an efficient technique for the investigation of ionization, autoionization, and fragmentation from high-lying excited states. Great efforts are currently underway towards characterizing the systematics of laser ionization–fragmentation of polyatomics. These studies have a practical goal of developing an ultrasensitive, molecule-specific, state-selective tool which couples spectroscopic and mass-spectrometric detection. Although the ideal case has not yet been realized significant progress has already been made along these lines, the best example so far being direct, molecule-specific detection of concentrations down to 10^4 molecules/cm^3.

The laser-ionization technique as described has, of course, several inherent limitations, the most important of which is vapor pressure. At least a few millitorr of sample is needed for mass-spectrometric or gas ionization chamber studies, although some information is available in the liquid phase. Selective excitation in the presence of similar background will require supersonic-beam environments, which complicate the otherwise simple apparatus. Another drawback is fragmentation, which often accompanies multiphoton ionization and compromises mass selectivity. Careful choice of the laser(s) wavelength and intensity will be needed to minimize fragmentation while providing a high ionization yield. Any ultrasensitive technique has restrictions such as the ones described in this chapter. Laser ionization is a remarkably general tool, however, which can be expected to make many more contributions to the understanding of molecular systems.

REFERENCES

Ackerhalt, J. R., and Eberly, J. H. (1976). *Phys, Rev. A* **14,** 1705.

Ackerhalt, J. R., and Shore, B. W. (1977). *Phys. Rev. A* **16,** 277.

Amirav, A., Even, U., and Jortner, J. (1980). *Chem. Phys.* **51,** 31.

Andreyev, S. V., Antonov, V. S., Knyazev, I. N., and Letokhov, V. S. (1977). *Chem. Phys. Lett.* **45,** 166.

Antonov, V. S., and Letokhov, V. S. (1981). *Appl. Phys.* **24,** 89.

Antonov, V. S., Knyazev, I. N., Letokhov, V. S., Matiuk, V. M., Movshev, V. G., and Potapov, V. K. (1978). *Opt. Lett.* **3,** 37.

Antonov, V. S., Letokhov, V. S., and Shibanov, A. N. (1980). *Appl. Phys.* **22,** 293.

Aron, K., Otis, C., Demeray, R. E., and Johnson, P. (1980). *J. Chem. Phys.* **73,** 4167.

Baer, T., and Carney, T. (1982). *J. Chem. Phys.* **76,** 1304.

Baer, T., Guyon, P. M., Nenner, I., Tabche-Fouhaille, A., Botter, R., Ferreira, L., and Govers, T. (1979). *J. Chem. Phys.* **70,** 1585.

Bakos, J., Kiss, A., Szabo, L., and Tender, M. (1972). *Phys. Lett. A* **39A,** 317.

Berg, J. O., Parker, D. H., and El-Sayed, M. A. (1978). *J. Chem. Phys.* **68,** 5661.

Berkowitz, J. (1979). "Photoabsorption, Photoionization, and Photoelectron Spectroscopy." Academic Press, New York.

Bernstein, R. B. (1982). *J. Phys. Chem.* **86,** 2271.

Boesl, V., Neusser, H. J., and Schlag, E. W. (1978). *Z. Zaturforsch. A* **33A,** 1546.

Boesl, V., Neusser, H. J., and Schlag, E. W. (1980). *J. Chem. Phys.* **72,** 4327.

Boesl, V., Neusser, H. J., and Schlag, E. W. (1981). *Chem. Phys.* **55,** 193.

Braun, C. L., Scott, T. W., and Albrecht, A. C. (1981). *Chem. Phys. Lett.* **84,** 248.

Brossel, J., Cagnac, B., and Kaslter, A. (1953). *C. R. Hebd. Seances Acad. Sci.* **237,** 984.

Brophy, J. H., and Rettner, C. T. (1979). *Chem. Phys. Lett.* **67,** 351.

Bunkin, F. V., and Prokhorov, A. M. (1964). *Sov. Phys.—JETP (Engl. Transl.)* **19,** 739.

Carney, T. (1981). Ph.D. Thesis, University of North Carolina, Chapel Hill.

Carney, T., and Baer, T. (1981a). *J. Chem. Phys.* **75,** 477.

Carney, T., and Baer, T. (1981b). *J. Chem. Phys.* **75,** 4422.

Carney, T., and Baer, T. (1982). *J. Chem. Phys.* **76,** 5968.

Cervenan, M. R., and Isenor, N. R. (1975). *Opt. Commun.* **13,** 175.

Collins, C. B., Johnson, B. W., Popescu, D., Musa, G., Pascu, M. L., and Popescu, I. (1973). *Phys. Rev. A* **8,** 2197.

Cooke, W. E., and Gallagher, T. F. (1978). *Phys. Rev. A* **17**, 1226.
Cooper, C. D., Williamson, A. D., Miller, J. C., and Compton, R. N. (1980). *J. Chem. Phys.* **73**, 1527.
Cooper, D. E., Klimcak, C. M., and Wessel, J. E. (1981). *Phys. Rev. Lett.* **46**, 324.
Cremaschi, P., Johnson, P. M., and Whitten, J. L. (1978). *J. Chem Phys.* **69**, 4341.
Dalby, F. W., Petty-Sil, G., Pryce, M. H. L., and Tai, C. (1977). *Can. J. Phys.* **55**, 1033.
Delone, G. A., Delone, N. B., and Piskova, G. K. (1972). *Sov. Phys. JETP (Engl. Transl.)* **35**, 672.
deVries, M. S., Van Veen, N. J. A., Baller, T., and DeVries, A. E. (1981). *Chem. Phys.* **56**, 157.
Dietz, T. G., Duncan, M. A., Liverman, M. G., and Smalley, R. E. (1980). *J. Chem. Phys.* **73**, 4816.
DiGiuseppe, T. G., Hudgens, J. W., and Lin, M. C. (1981). *Chem. Phys. Lett.* **82**, 267.
DiGiuseppe, T. G., Hudgens, J. W., and Lin, M. C. (1982), *J. Phys. Chem.* **86**, 36.
Duncan, M. A., Dietz, T. G., and Smalley, R. E. (1979). *Chem. Phys.* **44**, 415.
Duncan, M. A., Dietz, T. G., and Smalley, R. E. (1981a). *J. Chem. Phys.* **75**, 2118.
Duncan, M. A., Dietz, T. G., Liverman, M. G., and Smalley, R. E. (1981b). *J. Phys. Chem.* **75**, 7.
Ebata, T., Imajo, T., Mikami, N., and Ito, M. (1982). *Chem. Phys. Lett.* **89**, 45.
Eberly, J. H., and O'Neil, S. V. (1979). *Phys. Rev. A* **19**, 1161.
Eland, J. H. D. (1974). "Photoelectron Spectroscopy." Oxford Univ. Press, London and New York.
Esherick, P., and Anderson, R. J. M. (1980). *Chem. Phys. Lett.* **70**, 621.
Feldman, D. L., Lengel, R. K., and Zare, R. N. (1977). *Chem. Phys. Lett.* **52**, 413.
Fisanick, G. J., Eichelberger, T. S., Heath, B. A., and Robin, M. B. (1980). *J. Chem. Phys.* **72**, 5571.
Fisanick, G. J., Gedanken, A., Eichelberger, T. S., Kuebler, N. A., and Robin, M. B. (1981). *J. Chem. Phys.* **75**, 5215.
Franken, P. A., Hill, A. E., Peters, C. W., and Weinreich, G. (1961). *Phys. Rev. Lett.* **7**, 118.
Gedanken, A., Robin, M. B., and Kuebler, N. A. (1981). *Inorg. Chem.* **20**, 3340.
Gedanken, A., Kuebler, N. A., and Robin, M. B. (1982a). *J. Chem. Phys.* **76**, 46.
Gedanken, A., Robin, M. B., and Yafet, Y. (1982b). *J. Chem. Phys.* **76**, 4798.
Gerrity, D. P., Rothberg, L. J., and Vaida, V. (1980). *Chem. Phys. Lett.* **74**, 1.
Glownia, J. H., Riley, S. J., Colson, S. D., and Nieman, G. C. (1980). *J. Chem. Phys.* **73**, 4296.
Glownia, J. H., Riley, S. J., Colson, S. D., Miller, J. C., and Compton, R. N. (1982). *J. Chem. Phys.* **77**, 68.
Göppert-Mayer, M. (1931). *Ann. Phys. Leipzig* [5] **9**, 273.
Granneman, E. H. H., Klewer, M., Nienhus, G., and Vander Wiel, M. J. (1977). *J. Phys. B* **10**, 1625.
Grossman, L. W., Hurst, G. S., Kramer, M. G., and Young, J. P. (1977). *Chem. Phys. Lett.* **50**, 207.
Halpern, A. M., Gerrity, D. P., Rothberg, L. J., and Vaida, V. (1982). *J. Chem. Phys.* **76**, 102.
Heath, B. A., Fisanick, G. J., Robin, M. B., and Eichelberger, T. S. (1980). *J. Chem. Phys.* **72**, 5991.
Held, B., Mainfray, G., and Morellec, J. (1972). *Phys. Lett. A* **39**, 57.
Hering, P., Maaswinkel, A. G. M., and Kompa, K. L. (1981). *Chem. Phys. Lett.* **83**, 222.
Hetherington, W. M., Korenowski, G. M., and Eisenthal, K. B. (1981). *Chem. Phys. Lett.* **77**, 275.

Hudgens, J. W., Seaver, M., and DeCorpo, J. J. (1981). *J. Phys. Chem.* **85**, 761.

Hurst, G. S., Payne, M. G., Kramer, S. D., and Young, J. P. (1979), *Rev. Mod. Phys.* **51**, 767.

Huxley, L. G. H., and Crompton, R. W. (1974). "The Diffusion and Drift of Electrons in Gases." Wiley, New York.

Johnson, P. M. (1975). *J. Chem. Phys.* **62**, 4562.

Johnson, P. M. (1976). *J. Chem. Phys.* **64**, 4638.

Johnson, P. M. (1980a). *Acc. Chem. Res.* **13**, 20.

Johnson, P. M. (1980b). *Appl. Opt.* **19**, 3920.

Johnson, P. M., and Otis, C. E. (1981). *Annu. Rev. Phys. Chem.* **32**, 139.

Johnson, P. M., Berman, M. R., and Zakheim, D. (1975). *J. Chem. Phys.* **62**, 2500.

Jonathan, N., Morris, A., Ross, K. J., and Smith, D. J. (1971). *J. Chem. Phys.* **54**, 4954.

Keldish, L. V. (1964). *Zh. Eksp. Teor. Fiz.* **47**, 1945.

Kim, M. S., and Dunbar, R. C. (1980). *J. Chem. Phys.* **72**, 4405.

King, D. S., Schenck, P. K., Smyth, K. C., and Travis, J. C. (1977). *Appl. Opt.* **16**, 2617.

Kramers, H. A., and Heisenberg, W. (1925). *Z. Phys.* **31**, 681.

Lahmani, F., and Srinivasan, R. (1976). *Chem. Phys. Lett.* **42**, 111.

Lambropoulos, P. (1976). *Adv. At. Mol. Phys.* **12**, 87.

Lambropoulos, P. (1980). *Appl. Opt.* **19**, 3926.

Lecompte, C., Mainfray, G., Manus, C., and Sanchez, F. (1975). *Phys. Rev. A* **11**, 1009.

Leopold, D. G., Hemley, R. J., Vaida, V., and Roebber, J. L. (1981). *J. Chem. Phys.* **75**, 4758.

Letokhov, V. S., Mishin, V. I., and Puretzky, A. A. (1977). *Prog. Quantum Electron.* **5**, 139.

Leutwyler, S., Hermann, A., Wöste, L., and Schumacher, E. (1980). *Chem. Phys.* **48**, 253.

Leutwyler, S., Hofmann, M., Harri, H., and Schumacher, E. (1981). *Chem. Phys. Lett.* **77**, 257.

Lichtin, D. A., Zandee, L., and Bernstein, R. B. (1981a). *In* "Lasers in Chemical Analysis" (G. M. Hieftje, J. C. Travis, and F. E. Lytle, eds.), p. 125. Humana Press, Clifton, New Jersey.

Lichtin, D. A., Bernstein, R. B., and Newton, K. R. (1981b). *J. Chem. Phys.* **75**, 5728.

Lineberger, W. C., and Patterson, T. A. (1972). *Chem. Phys. Lett.* **13**, 40.

Lompre, L. A., Mainfray, G., Manus, C., Repoux, S., and Thebault, J. (1976). *Phys. Rev. Lett.* **36**, 949.

Lompre, L. A., Mainfray, G., Manus, C., and Thebault, J. (1978). *J. Phys. (Paris)* **39**, 610.

Lubman, D. M. (1981). *J. Phys. Chem.* **85**, 3752.

Lubman, D., and Kronick, M. (1982). *Anal. Chem.* **54**, 660.

Lubman, D. M., Naaman, R., and Zare, R. N. (1980). *J. Chem. Phys.* **72**, 3034.

McClain, W. M. (1971). *J. Chem. Phys.* **55**, 2264.

Mainfray, G., and Manus, C. (1980). *Appl. Opt.* **19**, 3934.

Maker, P. D., Terhune, R. W., and Savage, C. M. (1964). "Quantum Electronics," p. 1559. Columbia Univ. Press, New York.

Mallard, W. G., Miller, J. H., and Smyth, K. C. (1982). *J. Chem. Phys.* **76**, 3483.

Marinero, E. E., Rettner, C. T., and Zare, R. N. (1982). *Phys. Rev. Lett.* **48**, 1323.

Mathur, B. P., Rothe, E. W., Reck, G. P., and Lightman, A. J. (1978). *Chem. Phys. Lett.* **56**, 336.

Meek, J. T., Jones, R. K., and Reilly, J. P. (1980). *J. Chem. Phys.* **73**, 3503.

Miller, J. C., and Compton, R. N. (1981a). *J. Chem. Phys.* **75**, 22.

Miller, J. C., and Compton, R. N. (1981b). *J. Chem. Phys.* **75**, 2020.

Miller, J. C., Compton, R. N., and Cooper, C. D. (1982). *J. Chem. Phys.* **76**, 3967.

Morrison, R. J. S., Rockney, B. H., and Grant, E. R. (1981). *J. Chem. Phys.* **75**, 2643.

Murakami, J., Ito, M., and Kaya, K. (1980). *J. Chem. Phys.* **72**, 3263.

Murakami, J., Ito, M., and Kaya, K. (1981). *Chem Phys. Lett.* **80**, 203.

Newton, K. R., and Bernstein, R. B. (1982). To be published.

Newton, K. R., Lichtin, D. A., and Bernstein, R. B. (1981). *J. Phys. Chem.* **85**, 15.

Nieman, G. C., and Colson, S. D. (1979). *J. Chem. Phys.* **71**, 571.

Otis, C. E., and Johnson, P. M. (1980). *Rev. Sci., Instrum.* **51**, 1128.

Otis, C. E., and Johnson, P. M. (1981). *Chem. Phys. Lett.* **83**, 73,

Pandolfi, R. S., Gobeli, D. A., and El-Sayed, M. A. (1981). *J. Chem. Phys.* **85**, 1779.

Parker, D. H., and Avouris, P. (1979). *J. Chem. Phys.* **71**, 1241.

Parker, D. H., and Bernstein, R. B. (1982). *J. Phys. Chem.* **86**, 60.

Parker, D. H., and El-Sayed, M. A. (1979). *Chem. Phys.* **42**, 379.

Parker, D. H., Berg, J. O., and El-Sayed, M. A. (1978). *Springer Ser. Chem. Phys.* **3**, 320.

Parker, D. H., Pandolfi, R., Stannard, P. R., and El-Sayed, M. A. (1980). *Chem. Phys.* **45**, 27.

Parker, D. H., Bernstein, R. B., and Lichtin, D. A. (1981). *J. Chem. Phys.* **75**, 2577.

Petty, G., Tai, C., and Dalby, F. W. (1975). *Phys. Rev. Lett.* **34**, 1207.

Proch, D., Rider, D. M., and Zare, R. N. (1981). *Chem. Phys. Lett.* **81**, 430.

Raman, C. V., and Krishnan, K. S. (1928). *Nature (London)* **121**, 501.

Rava, G. P., Goodman, L., and Krogh-Jespersen, K. (1981). *J. Chem. Phys.* **74**, 273.

Rebentrost, F., and Ben-Shaul, A. (1981). *J. Chem. Phys.* **74**, 3255.

Reilly, J. P., and Kompa, K. L. (1980). *J. Chem. Phys.* **73**, 5468.

Rianda, R., Moll, D. J., and Kupperman, A. (1980). *Chem. Phys. Lett.* **73**, 469.

Robin, M. B. (1980). *Appl. Opt.* **19**, 3941.

Rockney, B. H., and Grant, E. R. (1981). *Chem. Phys. Lett.* **79**, 15.

Rockney, B. H., Cool, T. A., and Grant, E. R. (1982). *Chem. Phys. Lett.* **87**, 141.

Rothberg, L. J., Gerrity, D. P., and Vaida, V. (1981). *J. Chem. Phys.* **75**, 4403.

Scott, T. W., and Albrecht, A. C. (1981). *J. Chem. Phys.* **74**, 3807.

Scott, T. W., Twarowski, A. J., and Albrecht, A. C. (1979). *Chem. Phys. Lett.* **66**, 1.

Silberstein, J., and Levine, R. D. (1981). *J. Chem. Phys.* **75**, 5735.

Speiser, S., and Jortner, J. (1976). *Chem. Phys. Lett.* **44**, 399.

Stuke, M., and Wittig, C. (1981). *Chem. Phys. Lett.* **81**, 168.

Tai, C., and Dalby, F. W. (1978). *Can. J. Phys.* **56**, 183.

Turner, R. E., Vaida, V., Molini, C., Berg, J. O., and Parker, D. H. (1978). *Chem. Phys.* **28**, 47.

Vaida, V., Robin, M. B., and Kuebler, N. A. (1978). *Chem. Phys. Lett.* **58**, 557.

Voronov, G. S., and Delone, N. B. (1965). *Pis'ma Zh. Eksp. Teor. Fiz.* **1**, 42.

Wessel, J. E., Cooper, D. E., and Klimcak, C. M. (1981). *In* "Laser Spectroscopy for Sensitive Detection" (J. Gelbwachs, ed.), SPIE 286, p. 48. Springer-Verlag, Berlin and New York.

Wiley, W. C., and McLaren, I. H. (1955). *Rev. Sci. Instrum.* **26**, 1150.

Williamson, A. D., and Compton, R. N. (1979). *Chem. Phys. Lett.* **62**, 295.

Williamson, A. D., Compton, R. N., and Eland, J. H. D. (1979). *J. Chem. Phys.* **70**, 590.

Zacharias, H., Schmiedl, R., and Welge, K. H. (1980). *Appl. Phys.* **21**, 127.

Zakheim, D. S., and Johnson, P. M. (1978). *J. Chem. Phys.* **68**, 3644.

Zakheim, D. S., and Johnson, P. M. (1980). *Chem. Phys.* **46**, 263.

Zandee, L., and Bernstein, R. B. (1979a). *J. Chem. Phys.* **70**, 2574.

Zandee, L., and Bernstein, R. B. (1979b). *J. Chem. Phys.* **71**, 1359.

Zandee, L., Bernstein, R. B., and Lichtin, D. A. (1978). *J. Chem. Phys.* **69**, 3427.

Zoller, P., and Lambropoulos, P. (1980). *J. Phys. B* **13**, 69.

5 OPTICAL-PHASE-SHIFT METHODS FOR ABSORPTION SPECTROSCOPY

Donald M. Friedrich

Department of Chemistry
Hope College
Holland, Michigan

I. INTRODUCTION

The art and science of interferometry are well known to the practitioners of optics. Application and analysis of optical phase shifts have been central to optical testing and physical measurements since the discovery of interference fringes by Young (1802), Hershel (1809), and Brewster (1817). The experiments and interferometers of Fizeau, Rayleigh, and Michelson are also familiar from the lore of nineteenth-century physics. Maxwell's brilliant synthesis of optics and electromagnetism then established interferometry as a technique of nearly unsurpassed precision and theoretical completeness. Multiple interference of standing waves was studied by Fabry and Perot (1899) in the parallel-plate interferometer, or "etalon," which now bears their names (Born and Wolf, 1975, Chap. VII; Tolanski, 1973). Originally this device served for resolution of closely spaced atomic spectral lines (atomic fine structure) and for determination of spectral linewidths. All laboratory applications of interferometry utilized devices of reasonably simple design which were supposed to be used under reasonably simple optical conditions (monochromatic, quasicoherent, point-source light beams). However, between 1948 and 1951, Gabor published his synthesis of diffraction and imaging theory which provided the theoretical foundation for recording complex images via an

ULTRASENSITIVE LASER SPECTROSCOPY

optical-phase-shift method called holography (Gabor, 1948, 1949, 1951; Vest, 1979).

All of these earlier interferometric methods and analyses awaited recognition of the possibility of sustained stimulated emission of coherent light. With the solution in the 1960s of certain materials problems, a variety of laser light sources have encouraged the application of interferometry in ways hitherto unused in spectroscopy and chemistry. Because it is constructed of a Fabry–Perot cavity, the laser itself proved, in a sense, to be one of the earliest studies in atomic physics by laser interferometry. In analyzing the spectral gain characteristics of the active laser medium, Lamb (1964) demonstrated the utility of optical saturation methods for determining the widths and positions of narrow spectral lines (Siegman, 1971). Such intracavity techniques have been extended to the determination of very small absorption coefficients, and as such form a whole new method of absorption spectroscopy which is described by Harris in Chapter 6 of this book. However, the high spectral power density and the coherence of laser light are also exactly the source characteristics needed to apply true phase-shift detection methods to physical and chemical measurements at places remote from the laser cavity and uncoupled from the gain of the active medium. The motivation for this comes from the high sensitivity of interferometric methods to quite small changes in the refractive index of the transparent dielectric material contained between the mirrors of an interferometer.

Of course many previous applications of interferometry benefited greatly from the introduction of a laser light source. Even the lowly He–Ne laser has had a profound impact upon such traditional interferometric areas as optical testing and alignment, metrology, production of diffraction gratings, refractive index determinations, Schlieren photography, and, of course, holography. Here, however, we wish to limit our review to the much smaller subset of applications of optical-phase-shift methods to chemistry and spectroscopy which seem to be truly new, having been introduced recently to these fields principally because of the laser and its associated technologies.

We conclude this introduction with a brief survey of and comments on the relation of optical-phase-shift methods to other, related techniques. This theme of comparing techniques will be taken up in detail in the following sections. These applications rely on detection of a change in refractive index which is the consequence of some earlier, usually optically initiated, event, such as the absorption of light (which need not be from a laser). Refractive index changes may be created by thermal, chemical, or other relaxation processes subsequent to the excitation step. Nonradiative decay of excited states causes a small temperature rise which leads to a small transient decrease in the refractive index within the irradiated region of the sample. The transient then decays or relaxes toward the ambient-temperature value at a rate governed by the bulk thermal diffusivity in the sample. For sufficiently large thermal transients in fluids, mass diffusion may lead to uncontrolled convective transients, which are difficult to interpret. In detecting thermal transients, interferometry provides the same kind of information as that obtained from thermal lensing methods, which are described in Chapter 3 by Fang and Swofford. However, optical phase shifts are determined by the magnitude of the refractive index change, in contrast to thermal lensing, which detects a change in the curvature of the optical wave front of a single laser beam as modified by the spatial gradient of the refractive index in the irradiated sample. Both methods extract similar information about the magnitude of the excitation process and the course of the subsequent chemical and thermal relaxation processes. It was recognized quite early (Hu and Whinnery, 1973) that thermal lensing was inherently more sensitive to the spatial distribution of the excitation and, hence, that the interferometric method ought to be quantitatively useful even under conditions of excitation by laser beams of lower quality than TEM_{00}. This apparent advantage has, until recently, been demonstrated only once, by Stone (1973), who used an incoherent Xe arc-lamp excitation source to induce refractive index transients within a modified Jamin (Born and Wolf, 1975, p. 310) interferometer which was monitored by a He–Ne laser of low power. Stone's experiment (Stone, 1972), which will be discussed below, is one of a class of measurements which detect the spatial shift of an interference fringe imaged on a detector located at the output of the interferometer. This type of measurement is in fact sensitive to thermal lensing, i.e., curvature of the temperature field, because the beam is partially masked at the detector in order to convert a spatial fringe shift into an intensity shift at the photocathode of the detector. Very recently this problem has been solved elegantly by Davis (1980), who designed an optical-phase-sensitive detector to measure directly the optical phase shifts of a single-frequency cw low-power monitoring laser. By detecting all of the light emerging from the interferometer, rather than a spatially selected component, the consequences of thermal lensing are greatly reduced. Davis (1980) has called this technique "phase fluctuation optical heterodyne" (PFLOH) spectroscopy.

If the sample is a solid, thermal transients will provide two phase-shift transients, usually of comparable magnitude but of opposite sign. The thermally induced refractive index transient will be negative owing to the dependence of the index n on sample density, which decreases with increasing temperature. This will effectively shorten the optical path of the light in the irradiated arm of the interferometer. However, the attendant thermal expansion of the solid will mechanically increase the optical path, tending to cancel the effect of the thermal refractive index transient. Thermal lensing spectroscopy is not sensitive to this problem, by comparison.

Besides thermally and mechanically induced phase shifts, there is also the possibility of chemically induced refractive index changes. These changes may be either transient or permanent, depending on whether the sample has suffered irreversible photochemistry during the excitation step. The simplest kind of refractive index transient of molecular origin arises while the irradiated molecules are in some excited state. In general, the polarizabilities of excited electronic states are different and usually larger (Mathies and Albrecht, 1974) than ground-state polarizabilities. Hence electronic excitation of solute molecules in some transparent solvent will produce both thermal and electronic refractive index transients. The thermal transient will decay (nonexponentially) with a relaxation time characteristic of the solvent thermal diffusivity and the size of the irradiated region of sample. However, the transient due to the production of excited-state molecules of greater polarizability will decay at the rate determined by the lifetime of the electronic excited state. In addition, the polarizability transient will be positive, opposite to the thermal transient, and the thermal transient will grow as the excited-state population decays. Unlike the isotropic thermal transient, the polarizability transient may be anisotropic if the excited-state population is generated by irradiation with polarized light. Since the monitor laser can also be polarized, the three diagonal elements of the polarizability tensor are measurable. Friedrich and Klem (1979) have provided an analysis of this effect which will be discussed below.

If permanent photoproducts are generated by the excitation step, then the new compounds may have polarizabilities different from that of the parent compound. Some chemical intuition can be applied here to estimate the direction of the shift. For example, the loss of polarizable groups, such as halogens during photolysis, should lead to a decreased polarizability, while photochemical additions are expected to lead to increased polarizability for the same reason. Simple rearrangements will lead to smaller polarizability changes unless the geometry or conjugation of the active compound changes significantly.

II. SOURCES OF PHASE SHIFT

It is necessary to distinguish between the different sources of the refractive index transient. As pointed out in Chapter 3 on thermal lensing

(Fang and Swofford) and Chapter 1 on photoacoustic spectroscopy (Tam), by far the most extensive analyses and measurements have been made of spectroscopic excitations which yield thermal transients. This is natural, since most excitations decay efficiently through rapid nonradiative, thermal channels. Production of the thermal transient by nonradiative decay has generally been treated either as a steady-state process under cw irradiation or as an infinitely short ("delta function") heat source under pulsed laser irradiation. However, the analyses of Friedrich and Klem (1979) and more recently by Davis and Petuchowski (1981) treat cases in which the heat is evolved more slowly, on the time scale of the bulk thermal relaxation of the fluid (liquid or vapor sample).

There are two rather obvious experimental conditions which require such analysis. In solid solutions, optical excitation energy can be trapped in metastable excited states, which may have lifetimes exceeding the solvent thermal relaxation (Friedrich and Klem, 1979). The most notable example is that of the lowest triplet state of stable "closed-shell" compounds which display phosphorescence. Depending on molecular environment, triplet-state phosphorescences can range from several seconds to less than milliseconds (McGlynn et al., 1969). Likewise, in vapor-phase studies of energy transfer and relaxation via the familiar $V \to R \to T$ (vibration to rotation to translation) mechanism, the rate limiting step may frequently be $V \to R$ followed by rapid $R \to T$ relaxation. Thus, vibrational pumping can lead to a significant population of molecules in slowly (ca. 10^{-5} sec) decaying excited vibrational states. Whether or not these relaxation rates are fast depends primarily on the value of the thermal relaxation time of the sample t_c, which is given by

$$t_c = w_p^2/4D \tag{1}$$

where D is the bulk thermal diffusivity in square centimeters per second. Since the size of the irradiated region is under experimental control (w_p is the e^{-2} radius of a Gaussian excitation beam), the relaxation time can be varied over a few orders of magnitude, although not with impunity for successful interferometry, as we shall discuss later. In the case of rapidly decaying metastable molecular sources of heat, one expects to see thermal transients with the general shape of Fig. 1, in which there is a growth to a maximum followed by decay. The rise time is determined by the molecular relaxation mechanisms operating, while the decay is the thermal decay familiar from thermal lensing experiments. However, if the lifetime of the metastable state is longer than the thermal relaxation time, then the sample reaches the condition of "secular equilibrium," in which the thermal power is delivered very slowly to the medium, and the temperature rise is small.

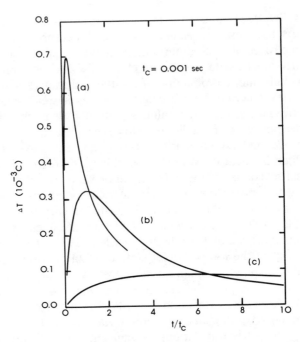

Fig. 1.   Temperature change ΔT at the beam center ($r = 0$) vs time scaled by the characteristic thermal relaxation time. These curves are given for a thermal input function value of $\phi_e HUN\sigma/2\pi kJ = 10^{-6}$ deg sec with $t_c = 0.001$ sec. Radiationless decay of the excited state of exponential lifetime τ_e provides the heat source. (a) $\tau_e = 0.1t_c$, (b) $\tau_e = t_c$, (c) $\tau_e = 10t_c$. [Adapted from Fig. 3 of Friedrich and Klem (1979). Courtesy of North-Holland Publishing Company.]

Thus the analysis has two principal components. First, the time dependence of the thermal transient must be related to the relaxation mechanisms in the sample. Then the time dependence and polarization characteristics of any nonthermal refractive index transients must be related to the molecular electronic anisotropy and mechanical rotational diffusion in solution. The sources of refractive index change are easiest to connect for the case of excitation by short optical pulses so that convolution of the laser excitation pulse will be unnecessary. The temporal and spatial dependence of the thermal transient generated by metastable excited states is essentially an extension of the Green's-function problem worked out by Twarowski and Kliger (1977) for impulse ("delta function") heat generation in solution. Adopting their notation, and using a Gaussian TEM_{00} distribution of excitation energy, we write the heat generated in an optically thin sample as

$$Q(r, t)dr\, dt = \frac{4\phi_e HUN\sigma}{J\tau_e w_p^2} \exp\left(-\frac{2r^2}{w_p^2} - \frac{t}{\tau_e}\right) r\, dr\, dt \qquad (2)$$

where H is the total number of photons in the laser pulse, $N\sigma$ (cm^{-1}) is the absorption coefficient, J is 4.184 J/cal, and w_p is the excitation beam radius in the sample. (We will frequently use $N\sigma$ for the absorption coefficient in order to avoid confusion with the polarizability α, which is also discussed in this chapter.) The integrated photon power H is multiplied by the yield ϕ_e of metastable excited states per photon absorbed and by U, which is the thermal energy released by nonradiative decay per metastable excited state created in the excitation process. U is the sum of the nonradiative electronic decay together with the decay of lower vibrational levels into which the metastable electronic state relaxes (Friedrich and Klem, 1979). The heat source $Q(r, t)$ has units of calories per centimeter second and for its time behavior depends only on the rate of the nonradiative relaxation mechanism. Thus, only the exponential lifetime τ_e appears in $Q(r, t)$.

The Green's function for the temperature rise in this problem of cylindrical symmetry has been given many times (Carslaw and Jaeger, 1963):

$$G(r, r', t, t') = (4\pi kt)^{-1} \exp\left(-\frac{r^2 + r'^2}{4D(t - t')}\right) I_0\left(\frac{rr'}{2D(t - t')}\right) \qquad (3)$$

where I_0 is the Bessel function of imaginary argument. The thermal diffusivity D is related to the thermal conductivity k (cal cm^{-1} sec^{-1} deg^{-1}) by

$$D = k/\rho c_p \qquad (4)$$

where ρ is the mass density (g cm^{-3}) and c_p is the specific heat (cal g^{-1} deg^{-1}). Since the heat source $Q(r, t)$ is factorable into time- and radial-dependent functions, $Q(r, t) = q(r)f(t)$, the temperature change is given by the integral

$$\Delta T(r, t) = \int_0^\infty dr' \int_0^t dt'\, q(r')f(t')G(r, r', t, t'). \qquad (5)$$

If $f(t) = 1$, we recover the results derived for cw excitation (Stone, 1972; Hu and Whinnery, 1973), while if $f(t) = \delta(t)$, we obtain the results for impulse heat sources (Twarowski and Kliger, 1977). The time dependence of the temperature rise at the beam center is shown in Fig. 1 for three cases of interest. In curve (a), the nonradiative decay is ten times faster than the thermal relaxation. Already one can see that this is effectively a case of impulse heat source. In curve (b) the thermal relaxation time is matched to the lifetime of the metastable state. Notice that the rise is quite rapid followed by a long tail. This is due to the very different relaxation mechanisms at work. The molecular nonradiative decay is exponential, while the thermal relaxation is essentially hyperbolic. Finally, the condition of "secular" equilibrium is demonstrated in curve (c) where the exponential molecular decay is only ten times longer than the thermal relaxation. The clear lesson from curves (a) and (b) is that the thermal relaxation time should be adjusted to be less than or equal to the molecular relaxation time if one is to extract information about the molecular processes from analysis of the time dependence of thermal transients.

In an extensive appendix to their description of the PFLOH method, Davis and Petuchowski (1981) present solutions to the heat conduction problem for several different conditions. These are in the form of analytical or integral expressions for the temporal and radial temperature dependence $T(r, t)$ following pulsed, cw, and sinusoidal excitation. They consider both uniform ("flat-top") and Gaussian radial distributions of the excitation intensity. Although the motivating model for the calculations is that of vibrational relaxation in the vapor phase (V $\rightarrow$ T rates), the analytical and integral results can be applied to heat transfer problems in solution and solids.

A refractive index transient will also be generated if the excited states have electronic polarizabilities different from the ground state. In order to connect the microscopic, molecular polarizability to the bulk refractive index, some model of the local field is required. For a low-power He–Ne cw monitoring laser, the light is usually sufficiently intense that both fluctuations (at too low intensity) and nonlinear effects (at too high intensity) can be neglected (Huang, 1967). In this situation, the Lorentz–Lorenz model (Born and Wolf, 1975, pp. 98ff) may be used to relate the refractive index to the molecular polarizability:

$$[(n^2 - 1)/(n^2 + 2)]M = \tfrac{4}{3}\pi N_A \rho \alpha \tag{6}$$

where M is the molecular weight and N_A is Avogadro's number. The polarizability has units of volume, and in this equation α represents the orientation average of the diagonal elements of the molecular polarizability tensor.

If the sample is a dilute solution in which the solute is excited by a laser pulse of Gaussian cross section, then the number density of the excited states so prepared is given by

$$N_e(r, z, t) = 2\phi_e H N \sigma \, \exp\left(\frac{-2r^2}{w_p^2}\right) \exp(-N\sigma z) \exp\left(\frac{-t}{\tau_e}\right). \tag{7}$$

The first exponential in Eq. (7) expresses the Gaussian nature of the distribution; the second exponential accounts for the Beer's-law decrease of excitation intensity as the beam travels through the sample; the last exponential term describes the time dependence of the excited-state population with a net lifetime of τ_e. We assume no depletion so that $N_e \ll N$.

The refractive index change depends upon the change in the polarizability upon excitation and upon the population of the excited state. Assuming that the change $\Delta n \ll 1$, we note

$$\Delta n_e = \tfrac{2}{9}\pi \left[\frac{(n_0^2 + 2)^2}{n_0} \right] N_e \Delta\alpha. \tag{8}$$

Substituting Eq. (7) into Eq. (8), we obtain finally

$$\Delta n_e = \frac{4(n_0^2 + 2)^2}{9n_0 w_p^2} \, \overline{\Delta\alpha}\phi_e HN\sigma \, \exp\!\left(\frac{-2r^2}{w_p^2} - N\sigma z - \frac{t}{\tau_e} \right). \tag{9}$$

Note that this relation uses the orientationally averaged polarizability change $\overline{\Delta\alpha}$. This is valid if the excited-state solute molecules become orientationally randomized on a time scale shorter than the excited-state lifetime. However, if the solute molecules are locked in a viscous solution so that rotational randomization is slow relative to the other relaxation mechanisms, then Eq. (9) must be modified to account for the anisotropy of the excitation (since even unpolarized light is anisotropic), the polarization of the molecular electronic transition, the anisotropy of the excited-state polarizability, and the polarization of the cw monitoring laser in the interferometer. This exercise in photoselection has been worked (Friedrich and Klem, 1979) for the case of collinear excitation and monitoring laser beams. We find that by setting the monitor polarization at an angle of 54.7° relative to the polarization vector of the linearly polarized excitation beam, the $\Delta\alpha$ then extracted from Eq. (9) is one-third the average polarizability change (isotropic, or orientationally averaged value) or one-ninth the trace of the tensor, $\mathrm{Tr}(\Delta\boldsymbol{\alpha})$.

The diagonal elements of the polarizability tensor can in principle be resolved by observing the "electronic" part of the refractive index transient (Eq. (9)) with the monitor polarization first parallel ($\Delta\alpha_\parallel$) and then perpendicular ($\Delta\alpha_\perp$) to the excitation polarization vector. If the molecular transition moment is polarized along the x axis of the molecules, which are randomly and isotropically distributed in solution, then the two anisotropic experimental polarizability changes are given by

$$\Delta\alpha_\parallel = \tfrac{1}{15}(3\Delta\alpha_{xx} + \Delta\alpha_{yy} + \Delta\alpha_{zz}), \quad \Delta\alpha_\perp = \tfrac{1}{15}(\Delta\alpha_{xx} + 2\Delta\alpha_{yy} + 2\Delta\alpha_{zz}). \tag{10}$$

Note that $\Delta\alpha_{yy}$ and $\Delta\alpha_{zz}$ have the same coefficients. $\Delta\alpha_{xx}$ can be resolved, since $\Delta\alpha_{xx} = 6\Delta\alpha_\parallel - 3\Delta\alpha_\perp$. In order to resolve $\Delta\alpha_{yy}$ and $\Delta\alpha_{zz}$, a y-axis- or z-axis-polarized transition must be excited to create the same metastable lower excited state. This is possible in principle, since there are usually orthogonally polarized transitions available in the spectra of molecules. However, the problem of thermal interferences from higher, nonradiative internal conversion steps may become severe if the excess photon-excitation energy is too large.

Finally, we note that the polarization dependence of the refractive index transient is uniquely measured by optical-phase-shift methods. This offers the possibility of determining solvent–solute interactions and rotational relaxation mechanisms for small populations of metastable excited states or photoproducts by measuring the time-dependent decay of the anisotropy of the refractive index transient. This anisotropy may be expressed either as polarization ratio $I_\parallel/I_\perp$ or as a degree of polarization ($I_\parallel - I_\perp)/(I_\parallel + I_\perp)$.

Photochemical products constitute a third source of the optical phase shift. Usually in photochemistry, such as flash photolysis, small changes in the concentration of photochemical products are monitored by direct absorption or emission techniques. These methods are ubiquitous, but have their limitations, since in many photochemical problems the concentrations are very low and the products may not fluoresce or phosphoresce. Nonradiative detection methods such as thermal lensing, optoacoustic, or optical-phase-shift techniques are welcome additions to the detection methods available to the photochemist. The photoproduct may be monitored in a subsequent excitation of thermal transient spectroscopy, although background absorption from the much greater concentration of parent compound must be absent for this method to be successful. If the photoproduct is not substantially excited by the *monitoring* laser, the thermal contribution of the photoproduct will be negligible if sufficient time is allowed for the sample to cool after the photochemical actinic pulse. Then the photoproduct will contribute two new perturbations to the interferometer, namely a shift in the refractive index and a change in the absorption, both due to conversion of parent compound to photoproduct.

III. EFFECT OF INDEX AND ABSORPTION CHANGES

Comparison of these two effects provides an instructive evaluation of the optical-phase-shift method in general. While specific experimental realizations of optical-phase-shift methods will be described in the following sections, it will be helpful now to work through a simplified analysis of what may be expected from a typical interferometric monitoring of a transparent sample in which a small change is induced in both the refractive index and in the absorption cross section.

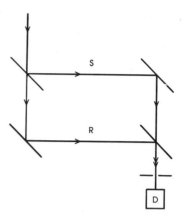

Fig. 2. Basic double-beam interferometer design considered in this chapter. The sample arm is S and the reference arm is R. The detector D monitors intensity changes caused by time-dependent phase changes in the S arm relative to the R arm. The intensity changes may be observed by monitoring spatial fringe shifts, as in the Stone–Jamin interferometer (Stone, 1972) or by monitoring the time-dependent phase shift as in the phase demodulation Mach–Zender interferometer developed by Davis and Petuchowski (1981).

Consider a simple split-beam interferometer of the type shown in Fig. 2. The optical beams in the sample arm (S) and the reference arm (R) are supposed to be of nearly equal amplitude. The beams recombine at the output mirror. The electric field of the reference beam at the detector may be simply written in complex analytic representation

$$E_R = E_R^0 \exp[i(\omega t + \phi_R)]\exp[-\tfrac{1}{2}N\sigma l]. \tag{11}$$

The sample beam suffers both phase and absorption shifts due to the photochemical changes induced in the sample by an actinic pulse. The electric field of the sample beam at the detector is

$$E_S = E_S^0 \exp[i(\omega t + \phi_S + \Delta\phi)]\exp[-\tfrac{1}{2}N^*(\sigma^* - \sigma)l]\exp[-\tfrac{1}{2}N\sigma l] \tag{12}$$

where $\omega = 2\pi c/\lambda$, ϕ_R and ϕ_S are fixed delays of the beams, and E_R^0 and E_S^0 are the field amplitudes at the detector prior to excitation of the sample. N^* is the concentration (molecules cm^{-3}) of photoproduct B generated by the simplest reaction A $\rightarrow$ B; absorption cross sections of parent A and photoproduct B are σ and σ^*, respectively, given at the monitoring wavelength λ. After excitation, both absorption and refractive index of the sample change, introducing attenuation changes and phase shifts into the sample beam. The phase shift due to changing refractive index is

$$\Delta\phi = 2\pi l\Delta n/\lambda. \tag{13}$$

The sample beam intensity is additionally attenuated according to Beer's law by $\exp[-N^*(\sigma^* - \sigma)]$. The field amplitude is attenuated by the square root of this factor. The output of the detector is then proportional to the intensity given by

$$I = (E_S + E_R)(E_S + E_R)^*. \tag{14}$$

Without loss of generality, we simplify the result by setting $E_R^0 = E_S^0$ so that $(E_S^0)^2 = \frac{1}{2}$, and also by adjusting the relative fixed phases so that $\cos(\phi_S - \phi_R) = 0$ (i.e., $\phi_S - \phi_R = \frac{1}{2}m\pi$ where m is an odd integer). Then substituting Eqs. (11) and (12) into Eq. (14), we obtain the relative change of intensity at the detector:

$$\Delta I/I = -\tfrac{1}{2}\{1 - \exp[-N^*(\sigma^* - \sigma)l]\} + \Delta\phi \exp[-\tfrac{1}{2}N^*(\sigma^* - \sigma)l]. \quad (15)$$

Since we suppose that the absorption change is small, $N^*(\sigma^* - \sigma) \ll 1$, the exponentials are expanded, giving finally

$$\Delta I/I \sim -\tfrac{1}{2}N^*(\sigma^* - \sigma)l + \Delta\phi. \quad (16)$$

Thus Eq. (16) becomes an evaluative tool for deciding whether to choose a phase-shift method or direct absorption to monitor the accumulation and fate of the photoproduct (or, in fact, any other weak absorber). In order to use Eq. (16) in a nonphotochemical context, it is only necessary to replace $N^*(\sigma^* - \sigma)$ by the absorption coefficient $N\sigma$ of the sample. The phase shift from all sources must exceed the optical density change in order for the optical-phase-shift method to be competitive with simple absorbance. Since both methods require measurement of a relative intensity change $\Delta I/I$, the same technical problems apply to both, and hence in principle the two methods should have similar limits of sensitivity. The great advantage of optical-phase-shift methods, like that of the other non-radiative spectroscopies, lies in the possibility of enhancing refractive index changes.

IV. DETECTION LIMITS

Since an important goal of a sensitive spectroscopic technique is to provide a measure of very low concentrations and/or very small absorptivities, the phase shift $\Delta\phi$ in Eq. (16) for each source should be expressed in these terms in order to evaluate the sensitivity of the method. We now proceed to give numerical estimates for the phase shift to be expected from each source for typical limiting experimental conditions. (Further comparison of different optical-phase-shift methods and related techniques are given in the next section.)

We first consider the "photochemical phase shift" due to the generation of a photoproduct with photophysical parameters different from the parent compound. The change in the refractive index Δn due to the change of chemical composition of the solution is given by Eq. (8) where N is replaced by N^*. The photoproduct concentration will also possess the spatial distribution representative of the cross section of the actinic laser beam. For an n-photon photochemical reaction generated by an excitation cylindrical laser beam with radial dependence $L(r)$, the spatial distribution of the photoproduct will also be cylindrical, with radial dependence $L^n(r)$. At the beam center (BC) the refractive index change will be [see Eq. (9)]

$$\Delta N_{BC} \sim [2\pi(n_0^2 + 2)^2/9n_0]N^*\overline{\Delta\alpha}^* \tag{17}$$

where $\overline{\Delta\alpha}^*$ is the difference between parent and photoproduct polarizabilities. Here we have neglected the Beer's-law factor for excitation of an optically thin sample of absorption coefficient $N\sigma$ at the excitation wavelength λ_p.

The "photochemical phase shift" in an interferometer, such as the one illustrated in Fig. 2, will be

$$\Delta\phi_{BC} = 2\pi l\Delta n_{BC}/\lambda = (4\pi^2/9n_0\lambda)(n_0^2 + 2)^2 lN^*\Delta\alpha^*. \tag{18}$$

Since, typically, $n = 1.5$, $\lambda = 6 \times 10^{-5}$ cm, and $l = 1$ cm, then $\Delta\phi_{BC} \sim 7.3 \times 10^4 N^*\overline{\Delta\alpha}^*$. For a 10-Å^3 change of the polarizability, $\Delta\phi_{BC} = 7.3 \times 10^{-19}N^*$. Thus, in order to be observed at a sensitivity limit of 10^{-4} rad, the concentration of photoproduct N^* must be at least 1.4×10^{14} molecules cm^{-3}. In the vapor phase this corresponds to a photoproduct partial pressure of 0.004 Torr or a concentration of 2×10^{-7} mol L^{-1} in solution.

In order to compare the phase-shift term with the direct absorption term of Eq. (16), we must know the change in absorption coefficient caused by the photochemistry. A wide range of possibilities may be encountered: the concentration change may be large, but the change in absorption coefficient may remain small; or change in the extinction coefficient may be large, but the photochemical quantum yield may be very small. If the change in absorption coefficient is relatively large, as when the photoproduct absorption band lies in a spectral region where the parent sample is transparent and the photochemical quantum yield is large, then direct absorption is the preferred method of photoproduct detection. However, if the photoproduct has no absorption bands in the accessible spectrum (usually visible or UV), or if the photoproduct absorption bands lie under more intense parent absorption bands, then observation of the photochemical process may be quite difficult by direct absorption photometry. In these and related situations, sensitive indirect methods are required to detect small changes caused by the photochemistry. Let us pose here the problem of detecting a photoproduct in which the extinction coefficient of the photoproduct is only 1% larger than that of the parent, which may have a typical value of 10^3. The relation between absorption cross section (cm^2 molecule^{-1}) and extinction coefficient ε (L mol^{-1} cm^{-1}) is $\sigma = \varepsilon \times 2.303 \times 10^3/6.022 \times 10^{23}$. Thus the absorption cross section of the parent compound is $\sigma = 3.82 \times 10^{-18}$ cm^2 molecule^{-1}. If the cross section difference is 1% of this, then the absorbance change is 2×10^{-6} or about fifty times smaller than our estimated sensitivity limit of $\Delta I/I = 10^{-4}$. This simple comparison is probably overly optimistic for the limits of sensitivity of a typical tunable spectrophotometric apparatus employing an incoherent light source and a monochromator. The ratio of the shot-noise-limited signal-to-noise ratios for detecting a 1-mW He–Ne source compared to a 1-μW incoherent source is about 30 (mW/μW)$^{1/2}$, giving the bright coherent source a clear advantage in lowering the limit of observable $\Delta I/I$. Furthermore, there are other considerations which may make the direct absorption method less useful. For example, in elucidating photochemical reaction mechanisms, one wishes to follow the generation of photoproducts directly in addition to monitoring the depletion of parent concentration. Simply observing the combined effects of depletion of N and accumulation of N^* a single absorption measurement is frequently not adequate.

The real problem here, of course, is with measurement of $\Delta I/I$ itself. It is well known that such "high-background" measurements are inherently less sensitive than "zero-background" methods, which can employ photon counting at low light levels. A classic example lies in the comparison of direct absorption and fluorescence excitation spectroscopies (see Chapter 2 by Birge). The photochemical holographic diffraction method developed by the IBM group (Burland *et al.*, 1980; Bjorklund *et al.*, 1980, 1981) provides an optical-phase-shift method of much improved sensitivity. It is a zero-background method, since there is no diffracted monitoring beam unless a holographic grating is photochemically generated in the sample. This method of considerable simplicity and promise is reviewed in more detail in the next section.

We turn now to an evaluation of the interferometric technique (i.e., $\Delta I/I$ measurement) for thermally generated phase shifts. In the limit of impulse heat generation, the refractive index transient near the beam center given by Twarowski and Kliger (1977) may be written for an m-photon process as

$$\Delta N_{BC}(\text{thermal}) = S_m F_m(t)[1 - 2mr^2 w_p^{-2} F_m(t)] \tag{19}$$

where we have separated the contribution of a thermal "input function" S_m and a thermal "decay function" F_m, viz.

$$S_m = \frac{dn}{dT} \frac{h\nu_p H_m N\sigma D}{kJ} \left(\frac{2}{\pi w_p^2}\right)^m, \tag{20}$$

$$F_m(t) = \left(1 + \frac{2mt}{t_c}\right)^{-1}. \tag{21}$$

All of these terms have been defined previously, except that σ represents an m-photon absorptivity, rather than the absorption cross section (which depends on the laser intensity as I^{m-1} for an m-photon excitation). The excitation energy is $h\nu_p H$ at wavelength λ_p. A Gaussian profile of the excitation beam is assumed, with w_p as the e^{-2} beam radius in the sample. The thermal relaxation time t_c is given in Eq. (1). Note that the "input function" S_m is negative if $dn/dT < 0$, as is usual.

The first term in the brackets of Eq. (19) gives a phase shift of $\Delta\phi$, and the second term, being quadratic in r, generates a thermal lens of negative focal length ("concave lens") given by

$$f_0^{-1} = 4mlS_m w_p^{-2} \tag{22}$$

where l is the thickness of the lens. Near the beam center, $(r/w_p)^2 \ll 1$ so that only the first term contributes significantly to the phase shift $\Delta\phi$. However, as we shall describe later, any measurement of ΔI which involves spatial masking of the interferometer output will detect contributions to the intensity change from both the phase-shift and the thermal lensing terms. The thermal contribution to the phase shift is

$$\Delta\phi_{BC}(t) = (2\pi l/\lambda)S_m F_m(t). \tag{23}$$

At time $t = 0$ the maximum phase shift is obtained as $\Delta\phi_{BC}(0) = 2\pi l S_m \lambda^{-1}$.

To evaluate the sensitivity in terms of optical absorption, we now regroup the terms in Eq. (23):

$$\Delta\phi_{BC}(0) = N\sigma \frac{h\nu_p H}{\pi w_p^2} \frac{4\pi lD}{\lambda k J} \frac{dn}{dT}. \tag{24}$$

The first term $N\sigma$ is the sample absorption coefficient at λ_p. The second term is a two-dimensional energy density, with units of joules per square centimeter, which we denote as $\mathcal{J} = h\nu_p H/\pi w_p^2$. The sample length l may depend on the radius of the excitation beam w_p. For a visible excitation beam ($\lambda_p \sim 5 \times 10^{-5}$ cm) focused to 0.01 cm radius, the confocal parameter b is 12.6 cm. If we restrict the physical length of the samples to shorter lengths, e.g., 1 cm, maintaining excitation focusing no tighter than $w_p > 0.01$ cm, then the effective length l is independent of the confocal parameter and is equal to the physical sample length. It may be desirable to disregard this restriction for multiphoton excitation studies. For example, in two-photon excitation spectroscopy, Swofford and McClain (1975) have shown that the optimum detection conditions obtain when the entire confocal parameter region about the focal point is monitored by the detecting apparatus. For three-photon and higher-order spectroscopies, the effective sample length depends strongly on the beam radius.

Typical thermo-optical parameters are $dn/dT \sim -5 \times 10^{-4}$ deg^{-1}; $D \sim 10^{-3}$ cm^2 sec^{-1}; $k \sim 4 \times 10^{-4}$ cal cm^{-1} sec^{-1} deg^{-1}; $\lambda \sim 6.3 \times 10^{-5}$ cm. Then Eq. (24) gives an estimate for the phase shift of $\Delta\phi \sim 60(N\sigma)\mathcal{J}$. Again setting a practical sensitivity limit of about 10^{-4} for following $\Delta\phi$ transients by wide-band dc monitoring of $\Delta I/I$, we find a limit for the energy density product of $(N\sigma)\mathcal{J} > 1.7 \times 10^{-6}$ J cm^{-3} in order to provide an observable heat source in the sample. Three examples are shown in Table I for typical excitation pulses from nitrogen, Nd–YAG, and flash-pumped dye lasers. For these estimations, the beam radius was increased with increasing peak laser power in order to maintain a common power density of 1.6×10^8 W cm^{-2} for each example. Increasing power densities much higher than this can lead to dielectric breakdown problems in solutions.

In conclusion, it is important to recall that the limit of detection sensitivity is determined by technical limits imposed on the measurement of intensity changes $\Delta I/I$. Our intention in this section was to provide comparison of the measurement of small changes in refractive index and absorption coefficient in the two-beam interferometer. We chose the common wide-band, or "dc" detection condition, which has a practical limit of about $(\Delta I/I)_{min} \sim 10^{-4}$, in order to provide this comparison for pulsed excitation. In specific experiments considerable improvement over this technical limit may be achieved by abandoning the pulsed mode of excitation and using more complex signal processing. We shall see examples in the next section.

From Table I, the limit of detectable absorption coefficient $N\sigma$ scales with excitation energy for one-photon absorption. Under excitation by the 500-mJ flash-pumped dye laser, using the "typical" solution parameters quoted above, a minimum absorption coefficient of 3.5×10^{-8} cm^{-1} means that a typical solute with extinction coefficient of 10^4 L mol^{-1} cm^{-1}

TABLE I

Laser	Pulse energy (mJ)	Pulse duration (nsec)	Radius, w_p (cm)	Absorption Coefficient Limit for $\Delta\phi > 10^{-4}$ rad, $N\sigma$ (cm^{-1})
Nitrogen-pumped dye laser	0.1	2	0.01	1.6×10^{-4}
Neodymium-pumped dye laser	10	10	0.04	1.0×10^{-6}
Flash-pumped dye laser	500	300	0.06	3.5×10^{-8}

may be detected at concentrations greater than or equal to $3.5 \times 10^{-12}M$. Weak molecular transitions with extinction coefficients as small as 3.5×10^{-4} L mol cm^{-1} may be measured for compounds as dilute as $10^{-4}M$. In very pure solvents, the method may be applied to recording the absorption spectra of forbidden electronic and vibrational transitions. As an example we calculate the minimum concentration of benzene (in a pure, "inert" solvent) for detection of a two-photon electronic excitation ($^1B_{2u} \leftarrow {}^1A_{1g}$ band, ca. 260 nm by one-photon spectroscopy). This is necessarily a pulsed experiment. The absorptivity σ for excitation of this two-photon forbidden (u $\leftarrow$ g) electronic transition is about 10^{-52} cm^4 sec molecule^{-1} photon^{-1}. The absorption coefficient is $N\sigma$ cm^{-1}, and given a limit of 3.5×10^{-8} cm^{-1}, we find a lower limit $[N] = 1.6 \times 10^{-3}M$ for two-photon excitation by the 500-mJ, 300-nsec, flash-pumped dye laser.

V. EXPERIMENTAL METHODS

Following the observations of intracavity thermal lensing in a cw laser by Gordon *et al.* (1965), the phenomenon was rapidly developed to measure very small absorption coefficients. Early thermo-optical methods were single-beam experiments in which one monitored the change in the divergence of a cw excitation laser beam due to thermal lensing. By the early 1970s it became apparent that the simplest experimental conditions and interpretive analysis would be obtained if the sample were positioned outside the laser cavity (Hu and Whinnery, 1973). In order to avoid the complications of intracavity methods, Stone (1972) introduced a phase-shift method using the modified Jamin interferometer shown in Fig. 3. In this device two 6-mm-thick quartz optical flats serve as the mirrors (actu-

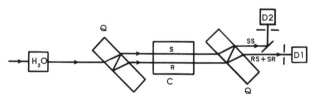

Fig. 3. Jamin-type interferometer as modified by Stone for optical-fringe-shift measurements of very small absorption coefficients. This device uses a single TEM$_{00}$ cw laser to simultaneously excite and monitor the refraction changes in liquid samples. The S beam is much more intense than the weaker R beam. It is assumed that the temperature rise in the reference R beam is negligible relative to the sample S beam. Water (H$_2$O) filters the 3.39-μm radiation from the cw He–Ne laser which might otherwise heat the sample. The sample cell C is placed between quartz flats Q. The fringes are masked and detected at the photomultiplier D1, while the laser intensity is monitored at detector D2. [Adapted from Fig. 1 of Stone (1972).]

ally, beam splitters) of the interferometer. A strong cw beam ("S" arm of the interferometer in Fig. 3) provides the excitation power for the sample arm. Two reflections within the first quartz flat provide a weak reference beam ("R" arm of the interferometer). As the sample absorbs energy from the cw S beam, the temperature rises in that arm, causing a monotonic decrease in the refractive index of the sample. The resultant phase shift of the S beam relative to the unperturbed R beam gives a fringe shift of the interference pattern formed when the S beam is recombined, after two reflections, with the R beam at the second quartz flat. The fringe pattern is masked so that the single detector views the steeply rising or falling edge of one fringe. This is the point of maximum sensitivity of the intensity measurement to changes of the relative phase $\Delta\phi$. In the previous section, this point was defined where $\cos(\phi_S - \phi_R) = 0$. In order to cancel fluctuations in the laser intensity, a portion of the transmitted S beam (SS in Fig. 3) is sent to a second reference detector for ratio recording of the $\Delta I/I$. A significant advantage of this arrangement is the dual-beam nature of the interferometer, which automatically cancels large-scale bulk variations in the refractive index of the sample due to ambient temperature changes.

As with all cw thermo-optical transient measurements, the temperature increases (and the refractive index decreases) monotonically with time. If the sample is exposed to a cw laser beam at time $t = 0$, initially the thermal transient grows linearly with time, while at long times the temperature rises logarithmically with time [cf. Eqs. (19)–(21)]:

$$\Delta n(r, t) = \frac{dn}{dT}\left[\frac{N\sigma P}{4\pi Jk}\ln\left(1 + \frac{2t}{t_c}\right) + \frac{2r^2 w_p^{-2}}{1 + t_c/2t}\right] \qquad (25)$$

where P is the cw power in watts. Notice that the denominator in the second term is *different* from $F_1(t)$ for the pulsed experiment. This analytical result depends on an infinitely wide sample cell, so that thermal equilibrium is never reached: the temperature rises indefinitely. In a finite sample, the heat conduction problem must account for constant-temperature finite-boundary conditions (Gordon *et al.*, 1965). This is in contrast to the behavior of a thermal lens formed by cw laser excitation. The cw thermal lensing term contains an extra time-dependent "decay function" $(1 + t_c/2t)^{-1}$ which leads to an asymptotic value of the r^2-dependent temperature term of $2(r^2/w_p^2)$ and hence to a limiting focal length at long times, even though the temperature continues to rise in the "infinite" solution model. Thus for quantitative fitting of the cw phase-shift data, it is advantageous to use a wide sample cell. Stone (1972) recommends a cell radius at least ten times larger than the beam radius w_p so that the "infinite" solution model will be valid for many thermal relaxation times $t <$

$100 t_c$. In the practical recovery of quantitative thermo-optical parameters from the time-dependent $\Delta I/I$ data, analysis of thermal lensing (outside of the cavity) makes convenient use of $I(0)$ and $I(\infty)$ values to fit the data. The monotonic logarithmic behavior of the phase-shift data at long times does not have the advantage of an easily extracted value of $I(\infty)$.

It is difficult to compare the sensitivities of the pulsed and cw methods of measuring thermally generated phase shifts. The cw method has the advantage of accumulating thermal energy over time from the cw laser beam, while observation of the competition between molecular kinetic and thermal relaxation rates is easier with the pulsed technique. We choose the same parameters as for the Nd–YAG pumped dye laser example (Table I), and estimate the absorption coefficient limit for depositing 10 mJ from a cw laser beam. If the sample is exposed to a 1-mW cw laser for 10 sec (10 mJ) then to achieve a phase shift of 10^{-4} rad, the absorption coefficient will be 1×10^{-5} cm^{-1}. If a 10-mW cw laser is used for 1 sec, the absorption coefficient will be 2×10^{-6} cm^{-1}. The two are not identical, because the temperature change is logarithmic in time.

Some care must be exercised in construction of the sample cell. As in photoacoustic spectroscopy, absorption of the laser excitation energy by the cell windows can contribute to the limiting background signal. The use of high-quality quartz optically contacted to the cell body without cement permitted Stone (1972) to realize absorption coefficient limits of 10^{-5} cm^{-1}.

Stone also demonstrated the use of the interferometer for spectroscopy by modifying the apparatus as shown here in Fig. 4 (Stone, 1973). The tunable source was incoherent light from a monochromator illuminated with a Xe arc lamp. The monitoring laser beams now were both weak in the interferometer, while the incoherent tunable excitation light was directed only along the S arm of the interferometer. The absorption spectrum of chlorobenzene recorded by this method is reproduced here in Fig. 5. It is one of the earlier spectra of the higher overtones of the CH stretching mode in benzene.

Because this is a spatial-fringe-shift method, significant complications from intensity redistribution due to thermal lensing and/or thermal deflection (Jackson *et al.*, 1981) effects can be encountered when using the Jamin configuration with relatively tightly focused beams either in the pulsed or cw mode. We will analyze this problem in the next section.

By employing a single-frequency cw monitoring laser, Davis and Petuchowski (1981) demonstrated that the phase shift could be detected directly in the time domain, rather than by the more indirect spatial-fringe-shift method. This makes spatial redistribution of the light intensity due to thermal lensing effects insignificant, since all of the light emerging from

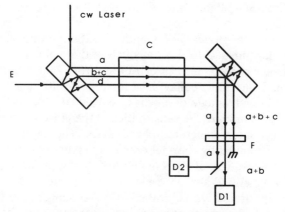

Fig. 4. Stone's modification of the fringe-shift interferometer shown in Fig. 3 to permit sample excitation by an external source E. This was light filtered from an intense incoherent Xe arc in Stone's original work. The cw laser monitoring beams R and S are both weak in this device, and are presumed to give no contribution to the observed temperature increase. The b + c path carries the excitation beam E and the signal beam S, while the reference beam R is directed along the a path. F is a filter to pass only the 632.8-nm monitor light to the detectors D1 and D2. [Adapted from Fig. 2 of Stone (1973).]

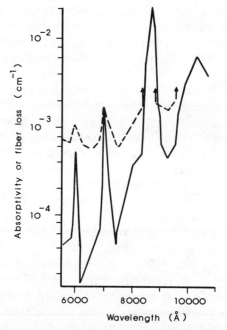

Fig. 5. Absorptivity of chlorobenzene (solid line) determined by the fringe-shift method of Fig. 4 compared with the transmission loss spectrum (dotted line) of a glass fiber filled with bromobenzene. The CH overtone absorptions of the benzene ring are clearly evident [From Fig. 5 of Stone (1973).]

the interferometer is imaged on the detector. Furthermore, absorption by the cell windows has little effect on the phase shift in vapors. The phase shift associated with heating a solid is small because of the compensation of the negative $(dn/dT)\Delta T$ term by the positive thermal expansion of the window material $\frac{1}{3}n\beta\Delta T$ (Stone, 1972). The phase shift due to window absorption is caused by secondary heating of the sample. In the liquid this can be a problem, as indicated by Stone (1972), but in vapor samples the effect of window heating is to generate a longitudinal acoustic wave, which to first order is not detected by the phase-shift method (Davis, 1980). This is an improvement over acousto-optical detection, which is affected by window heating. Davis and Petuchowski (1981) were specifically interested in detecting the time dependence and absolute magnitude of the deposition of thermal energy in vapor-phase mixtures due to relaxation of vibrationally excited molecules. The direct phase demodulation (PM) technique of PFLOH spectroscopy has theoretical sensitivity limits of better than 10^{-10} cm^{-1}.

We shall review here only the salient features of the PFLOH experiment, and refer the reader to the recent paper by Davis and Petuchowski (1981). Of particular interest is their extensive appendix, which solves the heat conduction and convection problem for several practical situations likely to be encountered in thermo-optical spectroscopy.

The PFLOH apparatus is composed of a Mach–Zehnder interferometer in a configuration very similar to the modified Jamin interferometer used by Stone. Essentially, the quartz plates in Fig. 4 are replaced by individual beam splitters in Fig. 6. The reference arm R and the sample arm S may both contain the sample, as in the Stone apparatus. However, now R serves as a local oscillator which beats in the square-law detector with the sample beam S which consists of the carrier frequency ν plus PM sidebands which contain the signal information. Mixing the reference carrier with the signal in the detector demodulates the PM signal producing the output given by Eq. (14). (Davis correctly describes this process as heterodyning with zero offset, rather than homodyning, which is a self-mixing process and does not lead to demodulation of the PM sidebands). Davis and Petuchowski have used the interferometer with gas samples only in the S arm and also with the apparatus filled with the sample gas so that both arms pass through the sample as in the Stone interferometer. For liquid samples in which ambient temperature variations have considerably more effect on the sample refractive index, the dual-beam configuration is surely to be preferred.

Again, by properly setting the relative phases and intensities of the two arms, the relative intensity change is just equal to the phase shift as given by Eq. (16). Recall that we had argued that a reasonably high sensitivity

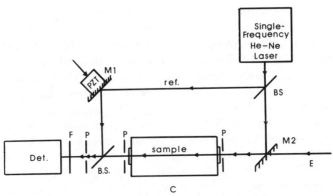

Fig. 6. Schematic diagram of the Mach–Zender interferometer arrangement used by Davis and Petuchowski for optical heterodyne detection of laser-induced refractive index changes. The cell windows and other components of the system are tilted at a small angle to prevent back reflections entering the single-frequency He–Ne laser. Ml is mounted on a PZT which compensates for slow phase and intensity shifts; M2 is a germanium flat which couples CO_2 laser excitation into the sample cell C. Aperture P and filter F define and filter the excitation beam. [Adapted from Fig. 1 of Davis and Petuchowski (1981).]

for dc measurement of $\Delta I/I = \Delta \phi$ is about 10^{-4}. This is practical for wideband dc detection of transients. However, it is well known that considerable improvement in sensitivity can be obtained for very slowly changing signals or for sinusoidally modulated signals if narrow-band detection is used. As pointed out by Davis (1980), the phase-shift method can be used in the wide-band detection mode to record time-dependent transients for kinetic studies, or the method can be used in a narrow-band ac detection mode of much greater sensitivity to detect trace amounts of gases in mixtures. Since coherent detection is possible with a single-frequency monitoring laser, the result of signal-to-noise ratio theory for coherent signals may be applied to this optical technique. For unity signal to noise Davis (1980) shows that the minimum detectable refractive index change is

$$\Delta n(\text{min}) = \frac{\lambda}{2\pi l}\left(\frac{2h\nu\Delta f}{\eta_Q P}\right)^{1/2} \tag{26}$$

where Δf is the signal processor bandwidth and η_Q is the detector quantum efficiency. In terms of our continuing example, with $l = 1$ cm and monitor laser power of 1 mW, we obtain $\Delta n = 3 \times 10^{-13}(\Delta f)^{1/2}$ $\text{Hz}^{-1/2}$ given $\eta_Q \sim 0.7$ (Davis, 1980). The ac signal processing techniques (e.g., lock-in detection) have been highly refined in the last decade and practical processing bandwidths on the order of one hertz are not uncommon.

[Informative discussions of lock-in detection amplifiers and their applications may be obtained from manufacturers of these instruments. See also Magrab and Blomquist (1971); Malmstadt *et al.* (1974), and Horowitz and Hill (1980) for background information.] The PFLOH signal decreases as the reciprocal of the modulation frequency f_M for $f_M > 50$ Hz owing to the time lag introduced by the thermal relaxation t_c. This regime corresponds to the $\Delta T = t/t_c$ behavior for short irradiation times with a cw laser. At very low modulation frequencies ($f^{-1} > t_c$) logarithmic behavior is expected [Eq. (25)], but is defeated by a servo-PZT [Fig. 6)] with a bandwidth of 9 Hz. The servo does not respond to high-frequency modulation, but maintains a constant relative phase ($\phi_S - \phi_R$) at the detector. Davis and Petuchowski (1981) report an optimum modulation frequency of 23 Hz for their system. With a 2.8% modulation depth superimposed on the cw CO_2 excitation laser, they demonstrated a sensitivity of 1.8×10^{-12} Hz$^{-1/2}$ for refractive index changes and 1.8×10^{-9} cm^{-1} W^{-1} for detection of small absorption coefficients.

These interferometers are very sensitive to acoustical noise. A mechanical vibration of one of the mirrors by 1 Å corresponds to a phase shift of 10^{-3} rad. Acoustic waves in air between the mirrors are easily detectable. Thus careful vibration isolation and acoustic damping are necessary. For pulsed measurements at the 10^{-4}-rad sensitivity level, we have found it adequate to mount the interferometer comonents on a heavy, air-supported optical table of honeycomb construction. The interferometer arms and mirrors are covered with a Styrofoam enclosure which is surrounded by a second enclosure of acoustic fiber glass and wood. The optical entrance and exits are kept as small as practicable and are sealed with quartz windows. For more sensitive measurements, Davis and Petuchowski (1981) recommend acoustic enclosures of heavy aluminum lined with lead and foam.

We turn now to a diffraction type of phase-shift experiment which has been reported by Burland *et al.* (1980) at IBM. Their holographic method was developed to detect small photochemical changes in solid samples, where the photoproduct does not diffuse on the time scale of the measurement. Although not an interferometric method, the analysis closely resembles the results described above. The method requires a cw monitoring laser of good spatial quality, but since the phase-shift measurement is in the spatial domain, a single-frequency laser is not required. However, unlike the spatial-fringe-shift measurements, the detected beam is not apertured, and so the method avoids complications associated with intensity redistribution within the beam. The holographic method is clearly described by Bjorklund *et al.* (1980). We briefly review their analysis here. A schematic of the experimental arrangement is shown in Fig. 7.

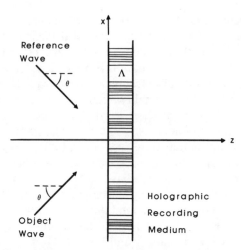

Fig. 7. Formation of a photochemical hologram in a sample of thickness l by interference of an object wave E^{obj} and a reference wave E^{ref}. The darkened areas indicate the regions where the two waves interfere destructively and little photochemistry is produced. The lighter areas are regions of constructive interference. The fringe spacing is Λ and the angle of incidence of each beam is θ. [Adapted from Fig. 1 of Bjorklund *et al.* (1980).]

The reference and object beams are derived from a single cw laser and crossed at the angle θ in the sample. An interference pattern is formed, and sample excitation and photochemistry take place at the maxima. The excitation and photochemistry produce an image of the interference pattern, recorded as periodic refractive index and absorbance changes in the sample. This diffraction grating then diffracts part of the reference beam in first order into the detector as shown. If there is no excitation or photochemistry, there is no signal at the detector, and thus this is a zero-background measurement with extremely high potential sensitivity.

As with the interferometric methods, the electric fields of the object and reference light waves are represented in their complex analytic forms. However, the temporal phase difference between the beams is not of interest here, but rather the relevant phase difference is contained in the different directions of the propagation wave vectors $\mathbf{K}_0$ and $\mathbf{K}_r$:

$$\mathbf{E}^{obj} = E_0^{obj}\hat{y}\,\exp[i(\mathbf{K}_0 \cdot \mathbf{r} - \omega t)], \quad \mathbf{E}^{ref} = E_0^{ref}\hat{y}\,\exp[i(\mathbf{K}_r \cdot \mathbf{r} - \omega t)]. \quad (27)$$

Superposing the electric fields at the sample ($z = 0$), an intensity interference pattern is formed with a periodicity

$$\Lambda = \lambda/2n \sin \theta, \quad (28)$$

$$I = B[E_0^{obj^2} + E_0^{ref^2} + 2E_0^{obj}E_0^{ref} \cos(2\pi x/\Lambda)]. \quad (29)$$

If the intensity is given in Einsteins per centimeter squared–second, the proportionality constant B in cgs units is

$$B = (\lambda/8\pi h N_A)(\varepsilon/\mu)^{1/2} \tag{30}$$

where N_A is Avogadro's number, μ is the magnetic permeability, and ε is the permittivity or dielectric constant of the sample medium. [The equation for B given in Bjorklund *et al.* (1980, Eq. (5)) contains a typographical error. The correct relationship is given here in Eq. (30).]

For simple one-photon ("two-step") photochemistry in optically thin samples, where cw excitation of the ground-state concentration of solute molecules [A] produces a steady-state population of intermediate excited states [A*], the photoproduct concentration [P] initially increases linearly with time (before ground-state depletion sets in):

$$[P] = 2303\phi^*\varepsilon I[A]t \tag{31}$$

where ϕ^* is the quantum yield for photochemistry (i.e., for production of sample changes which lead to shifts in the refractive index and absorption coefficient), and ε is the molar decadic extinction coefficient. The photoproduct concentration will be spatially modulated in the x direction as an image of the sinusoidal intensity pattern

$$[P] = [P_0] + [P_1]\cos(2\pi x/\Lambda) \tag{32}$$

obtained by substituting Eq. (29) into Eq. (31).

For visible excitation light in ordinary dielectric samples, the periodicity is about 2×10^{-5} cm, and so a "thick" hologram is produced where the sample thickness $l\phi \gg \Lambda$. When the object beam is blocked the hologram will diffract a portion of the reference beam into the zeroth order, producing a duplicate wave front of the object beam. Beams are also diffracted into ± 1 order, which may be used to continuously monitor the progress of the hologram formation (see Fig. 8). Two types of hologram are formed. Periodic variations of the refractive index of amplitude n_1 produce a phase hologram, while periodic variations in the absorption coefficient of amplitude α_1 produce an absorption hologram. Kogelnik (1969) shows that the diffraction efficiency η_{diff} will be

$$\eta_{\text{diff}} = \left[\sin^2\left(\frac{\pi n_1 l}{\lambda \cos \phi}\right) + \sinh^2\left(\frac{\alpha_1 l}{2 \cos \theta}\right)\right] \exp\left(\frac{-2\bar{\alpha}l}{\cos \theta}\right) \tag{33}$$

where the average absorption coefficient of the irradiated sample is $\bar{\alpha}$. Initially the changes in refractive index and absorption coefficient are small enough that

$$\eta_{\text{diff}} \approx \left[\left(\frac{\pi n_1 l}{\lambda \cos \theta}\right)^2 + \left(\frac{\alpha_1 l}{2 \cos \theta}\right)^2\right] \exp\left(\frac{-2\bar{\alpha}l}{\cos \theta}\right). \tag{34}$$

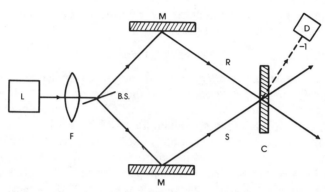

Fig. 8. Experimental arrangement used by the IBM group to record the temporal growth of a photochemical hologram. The object beam (S) and reference beam (R) are of nearly equal intensities. The cw argon or dye-laser light L is focused into the sample C by lens F. The beams are separated at beam splitter BS and directed toward the sample at the crossing angle $2\Theta = 0.013$ rad by mirrors M. The -1-order diffracted reference beam is continuously monitored at detector D. [Adapted from Fig. 3 of Bjorklund *et al.* (1980).]

Since the changes n_1 and α_1 are proportional to the photoproduct concentration $[P_1]$ [Eq. (32)], the diffraction efficiency increases quadratically with time, $\eta_{\text{diff}} \propto I_L^2 t^2$ where I_L is the laser intensity. For photochemistry involving m photons (sequential or simultaneous) the diffraction efficiency varies as the $2m$ power of the laser intensity, $\eta_{\text{diff}} \propto I^{2m} t^2$, assuming "initial" conditions for the measurement of constant $\bar{\alpha}$.

To evaluate the sensitivity of this technique, the intensity of diffracted light falling on the detector is $I_{\text{det}} = \eta_{\text{diff}} I_{\text{ref}}$. For highest sensitivity, photon counting may be employed for the diffracted beam. The limit or sensitivity will then depend upon the maximum integrating time period consistent with a given experiment. For example, the IBM group used the holographic method to determine photochemical kinetics, in which the time-dependent growth of the holographic efficiency over several minutes constituted the primary data.

We consider first application of photon counting to the phase hologram, where $\pi n_1 \lambda^{-1} \gg \alpha_1 2^{-1}$ [Eq. (34)]. The number of counts from the detector is

$$N_{\text{det}} = \frac{\Delta t \lambda P_{\text{ref}} \eta_Q}{hc} \left(\frac{\pi n_1 l}{\lambda \cos \theta} \right)^2 \exp\left(\frac{-2\bar{\alpha} l}{\cos \theta} \right) \tag{35}$$

where P_{ref} is the cw power in the reference beam and η_Q is the photomultiplier quantum efficiency, which we now take to be about 0.3. For optically thin samples we now neglect the Beer's-law exponential in this estimation. It is practical to use angles θ near $1°$. With a film sample of 0.02 cm (still a thick hologram, since $l \gg \Lambda$), the count per sampling period Δt is then

$$N_{\text{det}} \sim 1.2 \times 10^{24} n_1^2 P_{\text{ref}} \Delta t \quad \text{counts/sec W.} \tag{36}$$

Now, the rms fluctuations of the signal N_{det} are proportional to the square root of the signal, assuming shot-noise limit. This means we suppose that light scattered from inhomogeneities in the sample is not the limiting noise source. Thus for a signal-to-noise ratio of unity the minimum detectable change n_1 of the refractive index is

$$n_1(\text{min}) \sim 9.2 \times 10^{-13} (P_{\text{ref}} \Delta t)^{-1/2} \quad \text{sec}^{1/2} \, W^{1/2}. \tag{37}$$

Now if we let $P_{\text{ref}} = 1$ mW, and count for $\Delta t = 1$ sec, then n_1 (min) $\sim 3 \times 10^{-11}$. Thus, in principle, the holographic method has an index resolution comparable to that of the PFLOH technique quoted by Davis and Petuchowski (1981).

In a similar manner, neglecting the phase hologram, the resolution of changes in the absorption coefficient is given by

$$\alpha_1(\text{min}) \sim 6.3 \times 10^{-8} (P_{\text{ref}} \Delta t)^{-1/2} \quad \text{cm}^{-1} \, \text{sec}^{1/2} \, W^{-1/2}. \tag{38}$$

Again, for cw power of 1 mW and 1 sec counting time, $\alpha_1(\text{min}) = 6 \times 10^{-5}$ cm^{-1}. This cannot be compared directly to the detection limits for sample absorption in PFLOH even though the numerical value in Eq. (38) is essentially the same as that quoted by Davis and Petuchowski (1981) for the PFLOH ac detection made. This is because Eq. (38) shows resolution of a *change* in the sample absorption, while the PFLOH method is quoted for limits of detection of sample absorption. The magnitude of the absorption and index changes depend, as with other methods discussed in this text, upon the integrated intensity of the primary excitation laser which is responsible for inducing the physical or chemical change in the sample. However, it is important to note that in order for the absorption hologram to be observed at all ($\Delta t = 1$ sec), the absorption coefficient must change by at least 6×10^{-5} cm^{-1}. This is just ten times better sensitivity than the fringe-shift method described above (ca. 5×10^{-4} cm^{-1}).

If the changes n_1 or α_1 relax according to some temporal relaxation function $F(t)$, then upon cessation of the excitation laser beam (which may be much more intense than a monitoring reference beam), the detector signal will relax according to $F^2(t)$. In this way the holographic method may be applied to the study of chemical and thermal relaxation processes. For a linear grating of the sort shown in Fig. 7, the characteristic relaxation time t_c is given by (Sacedo *et al.*, 1978)

$$t_c^{-1} = (\Lambda^2/4\pi^2 D)^{-1} + \tau_e^{-1} \tag{39}$$

where D is either the thermal or mass diffusivity and τ_e is the exponential chemical lifetime of the excited or photochemical state produced by the excitation. If τ_e is very long, then the relaxation time has a form very similar to Eq. (1) above for thermal relaxation of a Gaussian cylindrical heat source. If the object–reference beam crossing angle 2θ is approximately 2°, then the grating spacing $\Lambda = 10^{-3}$ cm at 500 nm gives a thermal relaxation time $t = 8 \times 10^{-5}$ sec for typical solvents with $D = 10^{-3}$ cm^2 sec^{-1}. Thus, if the grating is produced by a short coherent excitation pulse, the thermal relaxation will be considerably more rapid than observed in conventional thermal lensing experiments or in the interferometric methods described above. The use of transient grating techniques to determine excitation decay kinetics is well known from the pioneering work of Siegman and colleagues (Sacedo et al., 1978) and has been treated by Eichler (1977). Application of this idea to thermal diffraction for sensitive calorimetric absorption measurement has recently been demonstrated in Harris's lab (Pelletier et al., 1982). They determined a minimum absorption coefficient of 7×10^{-4} cm^{-1} in 0.17 μL sample volume, using 0.44 W of Ar$^+$ ion laser excitation. Relaxation times were very short (32 μsec), in agreement with Eq. (39).

Finally, we reproduce here the experimental holographic growth curves obtained by the IBM group (Bjorklund et al., 1980). In Fig. 9 is shown the efficiency I_{det}/I_L of the photochemical holograms generated by photolysis of dimethyl-s-tetrazine [DMST] in 200–300 μm films of polyvinylcarbazole (PVK). The quadratic time dependence of the effi-

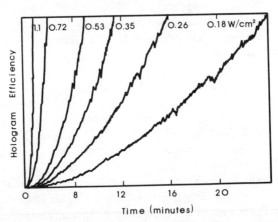

Fig. 9. Holographic growth curves for DMST in PVK using the 514.5-nm line from an argon ion laser. Next to each curve is indicated the laser power density used to produce the hologram. [From Fig. 4 of Bjorklund et al. (1980).]

ciency is clearly evident. An analysis of the laser intensity dependence of the holographic signal showed the value of m (in I^{2m}) to be 1.82 ± 0.07. This demonstrated that the DMST photolysis proceeded by absorption of two photons per photolyzed solute molecule.

VI. EFFECTS OF THERMAL GRADIENTS

In this last section we comment on the effects of thermal lensing and photothermal deflection on spatial-fringe-shift measurements. It is not difficult to operate the fringe-shift apparatus with the laser beams focused into the interferometer with confocal parameters as short as or shorter than the spacing between the mirrors (quartz flats). Fringe shifts are readily observable by expanding the recombined output beam through a short-focal-length lens and projecting the magnified fringes on a screen several centimeters away. A slit in the screen is placed in front of a photomultiplier, and the rising or falling edge ($\pm\frac{1}{2}\pi$) of a fringe is positioned on the slit by adjusting the quartz flats, the expansion lens, or a phase compensator plate in the R arm of the interferometer. We have operated the interferometer in two modes. In the cw mode, as shown in Fig. 10, the arms S and R are weak beams, while the strong beam is directed around the interferometer and sent backward through the S arm to provide excitation. In this way the excitation beam can be turned on

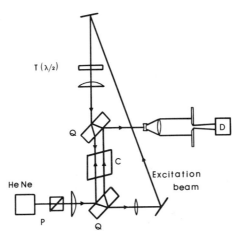

Fig. 10. Modified fringe-shift apparatus used in our laboratory to test the sensitivity of the fringe-shift method to interference from thermal lensing and photothermal deflection effects. The component notation is the same as in Fig. 2, except that prism P and retarder T are used to uncouple the excitation beam E from the He–Ne laser cavity.

and off without affecting the continuous operation of the monitor beams. The retardation plate ensures that no excitation light reenters the He–Ne laser, which could lead to instabilities. In the pulsed mode, the cw excitation beam is blocked, and pulses from a nitrogen-pumped dye laser are directed back along the path of S. Our finding is that when the excitation beam radius in the sample cell becomes small enough (~ 0.01 cm), the observed intensity change at the detector becomes a mixture of fringe shifting and intensity redistribution due to thermal gradients, i.e., thermal lensing and photothermal deflection. That these derive from thermal gradients in the excitation region can be verified by blocking the reference arm R in the interferometer and observing the time-dependent signal from the detector.

When the intensity shift due to thermal lensing or deflection is in the same direction as the intensity change due to the fringe shift, the mixture of effects is not readily apparent in the transient response of the detector. However, it is possible to adjust the fringe pattern at the slit so that the intensity changes for fringe-shift and thermal gradients are in opposite directions. This effect is seen in Fig. 11, where the large fringe-shift transient (positive) is preceded by a significant negative thermal gradient transient.

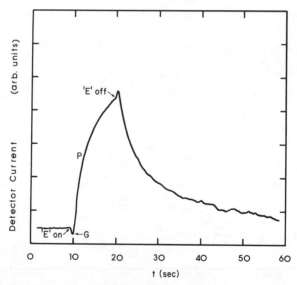

Fig. 11. Typical thermal growth and decay curves in 10 cm ethanol samples, showing interference from thermal lensing and photothermal deflection in the fringe-shift measurement using the apparatus of Fig. 10. The intensity change due to phase shift is labeled P and the change due to thermal gradient effects is labeled G.

VII. CONCLUSION

Three phase-shift methods were reviewed. The spatial-fringe-shift method was the earliest application of the optical-phase-shift concept to determination of very small absorption coefficients. However, the interaction of fringe-shift and thermal gradient effects makes the other two more recently developed methods more attractive. By detecting the phase shift in the time domain, the PFLOH method avoids interference from thermal gradient effects. The PFLOH method offers considerable sensitivity and versatility, since it may be operated in the pulse mode for relaxation studies or in the ac mode for detection of very small absorption coefficients. The PFLOH method does require the use of a single-frequency cw monitoring laser and careful acoustic isolation. The method has certain advantages over optoacoustic techniques. The holographic method is not an interferometric technique. Instead it detects reference-beam light diffracted by a phase or absorption hologram in the form of a diffraction grating generated by exciting the sample with an interference pattern formed by the intersection of two phase-coherent laser beams in the sample. The method is especially well suited to photochemical studies in viscous or solid media. It has sensitivities which are comparable to the PFLOH method, and it is also relatively insensitive to transient gradient effects.

Except for the preliminary studies of Friedrich and Klem (1979), none of these methods has been analyzed or exploited for the application of photoselection concepts to the anisotropies produced by polarized-light excitation and detection.

ACKNOWLEDGMENTS

Acknowledgment is made to the Donors of the Petroleum Research Fund, administered by the American Chemical Society, for the support of this research.

REFERENCES

Bjorklund, G. C., Burland, D. M., and Alvarez, D. C. (1980). *J. Chem. Phys.* **73**, 4321–4328.
Bjorklund, G. C., Braeuchle, C., Burland, D. M., and Alvarez, D. C. (1981). *Opt. Lett.* **6**, 159–161.
Born, M., and Wolf, E. (1975). "Principles of Optics," 5th ed., Chapter VII. Pergamon, Oxford.
Braeuchle, C., Burland, D. M., and Bjorklund, G. C. (1981). *J. Phys. Chem.* **85**, 123–127, 618.

Burland, D. M., Bjorklund, G. C., and Alvarez, D. C. (1980). *J. Am. Chem. Soc.* **102**, 7117–7119.

Carslaw, H. S., and Jaeger, J. C. (1963). "Operational Methods in Applied Mathematics," p. 109. Dover, New York.

Davis, C. C. (1980). *Appl. Phys. Lett.* **36**, 515–518.

Davis, C. C., and Petuchowski, S. J. (1981). *Appl. Opt.* **20**, 2539–2554, 4151.

Eichler, H. J. (1977). *Opt. Acta* **24**, 631–642.

Friedrich, D. M., and Klem, S. A. (1979). *Chem. Phys.* **41**, 153–162.

Gabor, D. (1948). *Nature (London)* **161**, 777.

Gabor, D. (1949). *Proc. R. Soc. London, Ser. A* **197**, 454.

Gabor, D. (1951). *Proc. Phys. Soc., London, Sect. B* **64**, 449.

Gordon, J. P., Leite, R. C. C., Moore, R. S., Porto, S. P., and Whinnery, J. R. (1965). *J. Appl. Phys.* **36**, 3.

Horowitz, P., and Hill, W. (1980). "The Art of Electronics," pp. 628–631. Cambridge Univ. Press, London and New York.

Hu, C., and Whinnery, J. R. (1973). *Appl. Opt.* **12**, 72.

Huang, K. (1967). "Statistical Mechanics," p. 82. Wiley, New York.

Jackson, J. D. (1962). "Classical Electrodynamics," p. 205. Wiley, New York.

Jackson, W. B., Amer, N. M., Boccara, A. C., and Fournier, D. (1981). *Appl. Opt.* **20**, 1333–1344, and references contained therein.

Kogelnik, H. (1969). *Bell Syst. Tech. J.* **48**, 2909.

Lamb, W. E., Jr. (1964). *Phys. Rev. A* **134A**, 1429.

McGlynn, S. P., Azumi, T., and Kinoshita, M. (1969). "The Triplet State." Prentice-Hall, Englewood Cliffs, New Jersey.

Magrab, E. B., and Blomquist, D. S. (1971). "The Measurement of Time-Varying Phenomena," pp. 171–178. Wiley, New York.

Malmstadt, H. V., Enke, C. G., Crouch, S. R., and Horlick, G. (1974). "Optimization of Electronic Measurements," Module 4, Instrum. Scientists Ser., pp. 98–104. Benjamin/Cummings, Menlo Park, California.

Mathies, R., and Albrecht, A. C. (1974). *J. Chem. Phys.* **60**, 2500.

Pelletier, M. J., Thorsheim, H. R., and Harris, J. M. (1982). *Anal. Chem.* **54**, 239–242.

Sacedo, J. R., Siegman, A. E., Dlott, D. D., and Fayer, M. D. (1978). *Phys. Rev. Lett.* **41**, 131–134.

Siegman, A. E. (1971). "An Introduction to Lasers and Masers." McGraw-Hill, New York.

Stone, J. (1972). *J. Opt. Soc. Am.* **62**, 327–333.

Stone, J. (1973). *Appl. Opt.* **12**, 1828–1830.

Swofford, R. L., and McClain, W. M. (1975). *Chem. Phys. Lett.* **34**, 455–460.

Tolansky, S. (1973). "An Introduction to Inteferometry," 2nd ed. Wiley, New York.

Twarowski, A. J., and Kliger, D. S. (1977). *Chem. Phys.* **20**, 253.

Vest, C. M. (1979). "Holographic Interferometry." Wiley, New York.

6 LASER INTRACAVITY-ENHANCED SPECTROSCOPY

T. D. Harris

Bell Laboratories
Murray Hill, New Jersey

I. INTRODUCTION

In this chapter a method of measurement known as intracavity absorption-enhancement spectroscopy will be detailed. Developments from the first reports through December 1981 are included. It should be emphasized that intracavity spectroscopy is far from a mature field. Many questions remain to be resolved, and this chapter is best characterized as a progress report. The emphasis will be on experimental methods with some theoretical considerations as they reflect directly on measurement. First an overview of the origin of the intracavity effect will be presented. This is followed by a brief historical account of the early development of intracavity spectroscopy. The remainder of the chapter is divided into three sections based on the nature of the absorber. In the first section those methods which apply when the absorber bandwidth is wider than the laser linewidth are discussed. In the second methods for which absorber bandwidths are less than the laser linewidth are considered. Finally, procedures which have been reported for transients are reviewed.

There is considerable disagreement as to the proper theoretical interpretation of this measurement method and no consensus seems to be forthcoming. As a result, discussion of theory at this time would be of limited utility. This should not diminish the importance of continued work

343

ULTRASENSITIVE LASER SPECTROSCOPY

Copyright © 1983 by Academic Press, Inc.
All rights of reproduction in any form reserved.
ISBN 0-12-414980-4

to fundamentally understand the method. Finally, it should be noted that the references do not comprise a comprehensive survey of the intracavity-enhancement literature.

A. Origin of the Effect

A general understanding of the origin of the intracavity enhancement is necessary for the successful utilization of the method. Since the principle purpose of this chapter is to be an introductory guide for users, these origins will be outlined as they are currently understood. This treatment is not unanimously accepted and should be used with a view toward further developments. The major sources of controversy arise from three factors. The first of these is the large variety of instrumental configurations. There is no doubt that the experimental parameters which effect the enhancement will change with these configurations. As a result, comparisons between configurations are ambiguous. The second factor is the difficulty of making spectral measurements at sufficiently high resolution to give quantitative data on the parameter under study. Atomic and molecular lines a few hundred megahertz wide are studied with a laser several angstroms wide. The third factor is the difficulty of controlling those parameters which seem to be principally responsible for the enhancement. Foremost among these are the number of modes and the proximity to threshold. Not only is control of these parameters difficult but quantitative measurement of them is equally difficult.

1. Multipass Effect

The most obvious source of enhancement is the multipass effect. Since the laser cavity is a resonator, there are multiple passes of the light through the sample. The multipass effect must be distinguished from the cavity lifetime. The latter quantity is determined by the average number of round trips that photons make before leaving the cavity. The average number of passes is equivalent to the advantage of a conventional multipass cell, and is the reciprocal of the fraction round-trip loss. Since a laser has gain, there is also a restoration effect for both broadband and narrowband absorbers. The light that is fed back into the gain has been altered by the loss. This is most easily imagined for the case of the two-mode homogeneous laser. Assume a perfect cavity so the mode with no loss has a transmission of 1.0. Assume also that the other mode is subject to a single-pass loss of 0.001, resulting in a cavity transmission of 0.998. The intensity ratio of light in the two modes arriving back at the gain after one round trip is $1.0/0.998$. After two round trips the ratio is $(1.0)^2/(0.998)^2$, assuming both modes see the same gain. It is obvious that even if the output mirror is 50% transmissive and the gain is 2, the ratio will continue to grow until perturbed in some way. Therefore the multipass effect is both a multiple round-trip effect and a "memory" effect.

2. Threshold Effect

The threshold effect is the second of the three sources of intracavity enhancement and the least controversial. Since a laser is a threshold device a certain gain must be present to initiate oscillation. For the case of a broadband absorber, the effect of loss on output power follows directly from the treatment of the output power versus mirror transmission calculation. The only difference is that lost light is not present as output. This calculation is given by several laser textbooks and need not be recounted here (Siegman, 1971). The governing principle was expressed by Hansch et al. (1972),

$$\text{Enhancement} = \frac{\alpha/\delta}{\alpha - \delta} \qquad (1)$$

where α and δ are, respectively, the gain and loss of the resonator. At very high gain $\alpha - \delta = \alpha$, and the threshold enhancement becomes $1/\delta$ or the cavity Q. As threshold is approached $\{\alpha \cong \delta\}$ and the threshold enhancement approaches $1/(\alpha - \delta)$ or rises as the reciprocal of the proximity to threshold. This is the source of "infinite" enhancement near threshold. Many but not all authors have presented this as a source of unlimited sensitivity, which it is not. Unfortunately, the unavoidable noise is enhanced along with the absorption and no improvement in signal-to-noise ratio is seen as threshold is approached. Some complication may result from applying this single-mode threshold treatment to narrow-band absorbers because each mode's proximity to threshold is now independently variable. However, it does serve to illustrate the important principles.

3. Mode Competition Effect

Mode competition is the third source of enhancement and is a term coined to describe homogeneous gain. It means that all the oscillating modes of the laser compete for the same gain centers. As a result, modes with low loss or high gain grow in intensity at the expense of modes with higher loss or lower gain. If the loss to be measured is spectrally broad compared to the laser, mode competition plays no role in intracavity enhancement. The laser gain has a natural spectral distribution and the mode at the center of the gain spectrum sees the highest gain. If losses are assumed to be spectrally flat, the mode with the highest gain grows in intensity at the expense of all other modes. Consequently, at equilibrium a homogeneously broadened laser should have only one mode. In practice a cw laser may have many hundred to several thousand modes. Although this is a general laser property, it is of particular significance to intracavity-absorption theory. Two different proposals have been advanced in the intracavity-enhancement literature in response to the observations. The first proposal, which is dominant in the Soviet literature, is that equilibrium is never reached. Some perturbation of the laser results in a regression to the initial buildup of laser action and the presence of many modes is explained. The alternative explanation is to assume an equilibrium laser but introduce some inhomogeneity to account for the existence of many modes.

The type of inhomogeneity proposed is called spatial hole burning, illustrated in Fig. 1. If only one mode were present, those gain centers at the nodes of the electric field of this mode will not be stimulated to emit. Other modes will then develop which have electric field peaks at the nodes of the highest-gain mode, resulting in a multimode laser. The correct explanation probably falls somewhere between the two extremes.

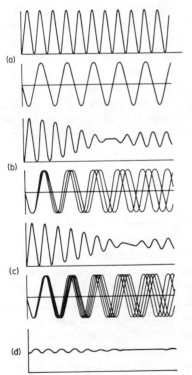

(a)

(b)

(c)

(d)

Fig. 1. Illustration of spatial hole burning: (a) A single mode with electric field and intensity at one move away from one mirror. The stimulated emission is proportional to the intensity. Deviation from peak to valley is 100%. (b) Fields and intensity for three modes. After initial damping, peak-to-valley deviation is 30%. (c) As in b for five modes. A general flattening with a deviation of 20%. (d) Intensity only well away from the mirror for 70 modes. Deviation less than 1%.

Recent work on a numerical solution to laser rate equations suggests that equilibrium is not reached even after 1 sec of lasing action, giving some credence to the nonequilibrium proposal. This problem will be addressed briefly in the section on narrow absorbers. It is central to the argument about the ability to make quantitative intracavity measurements.

B. Historical Development

1. Initial Observations

The high sensitivity of lasers to internal losses has been known since before the first demonstration of laser action. One of the principle concerns of initial laser design was the development of a feedback system with losses smaller than the gain. In addition the use of etalons for mode selection takes advantage of frequency-selective loss addition, with the redistribution of the energy into the modes with lowest loss. Etalons were used very soon after the invention of the laser.

The introduction of laser intracavity-enhanced absorption is generally credited to a group at the Lebedev Institute in Moscow in 1970 (Pakhomycheva *et al.*, 1970). The principle aim of the paper was to explain the spectral channeling of room-temperature Nd^{3+}–glass lasers. It was proposed that the channeling was a result of frequency-selective losses within the homogeneous width of the gain. The laser then oscillated at the lowest-loss wavelengths within this 20-cm^{-1} width. The hypothesis was tested by inserting an etalon into the resonator and recording the spectrum of the dispersed laser emission on film. The etalon was a liquid-filled cell, and its reflectivity was varied by changing the composition and consequently the refractive index of the liquid in the cell. The etalon introduced a frequency-selective loss with a period of 1.1 cm^{-1}. The laser emission exhibited an output spectrum modulated at this period, induced by an etalon reflection coefficient of 2×10^{-7}. Although it was also stated that the dependence of the modulation depth on the reflection coefficient was obtained, no data were presented. The authors concluded that this laser could be used as a high-sensitivity spectrograph. To demonstrate the application, samples of ammonia and methane were separately introduced into the laser. As expected, severe channeling of the emission spectrum resulted. No attempt was made to relate absorber concentration to the degree or depth of channeling. The enhanced sensitivity of the laser to frequency-dependent loss was correlated to the relative spectral widths of the loss and the laser homogeneous broadening.

Shortly after this report, and apparently independently, came a publication from a group at the National Bureau of Standards (NBS) in the

United States (Peterson *et al.*, 1971). The NBS group used a flash-lamp-pumped dye laser operated from 570 to 590 nm expressly for the purpose of investigating the enhanced detection of absorbers placed inside the laser resonator. Both atomic sodium and molecular iodine were used as absorbers. The concentration of sodium could be varied by adjustment of the cell temperature. Detection was also on film with a spectrograph. The enhancement was determined by comparison of intracavity and extracavity sample placement of the lowest concentration of sodium vapor which was "just detectable" on the photograph positive print. An enhancement factor of 70 was obtained. This value was subject to uncertainty because of the limited spectral resolution and the subjectivity of the detection system. Pump energy from threshold to 1.5 times threshold had no measurable effect on the enhancement. The NBS group also determined the dependence of enhancement on laser cavity lifetime, by changing the output mirror from 90 to 99% reflectivity. Again no change was found. The model used to explain the enhanced sensitivity assumed a brief surge above threshold with oscillation at all wavelengths followed by an end to lasing at wavelengths of absorption. Calculated enhancements for the various "switching-parameter" estimates were presented and were in reasonable agreement with the measured enhancement. This parameter was proposed as the absorber loss to switch off laser action.

2. Early Development

The second report of the Soviet group was not published until mid 1972 (Belikova *et al.*, 1972). It consisted primarily of a quantitative treatment of the etalon reflectivity discussed in the initial report. Data were presented which showed a measured etalon reflectivity range from 2×10^{-7} to 2×10^{-6}. This tenfold change in loss resulted in a change in modulation depth from 10% at the lower reflectivity to 20% at the higher reflectivity. The extremely small change in modulation depth is uncharacteristic of later reports from other laboratories. Since the etalon-induced loss is present across the laser spectrum, comparison with narrow-line atomic absorption is difficult. Agreement with a proposed model was claimed although no calculated values were presented.

The follow up by the NBS group was more extensive (Keller *et al.*, 1972). They again used a pulsed dye laser. The uncertainty of the measured enhancement was reduced by using $Eu(NO_3)_3$ in methanol as the absorber. The weak f-f transitions of this ion provided a reliable source of known and variable absorption. In addition it has a linewidth significantly narrower than the laser emission but easily resolvable with the spectrograph. Photographic detection was again employed for broadband emission. A second configuration in which the laser emission spectrum was narrowed with a five-element prism tuning assembly was also studied. This arrangement yielded a laser linewidth of less than 0.1 nm compared to the untuned width of ~6 nm. The enhancement with the broadband system was ~400. The enhancement for the narrow-band case was not given except to say that it was much lower. Again no dependence of enhancement on pump power or cavity lifetime was detected. However, it was noted that the change from a 90% reflector to a 60% reflector was not large, because the cavity was naturally lossy. A revised model based on a set of coupled differential rate equations was proposed. The model gave two important predictions. The first was that detection sensitivity should depend on cavity lifetime and that sensitivity would be greatest for high-quality (low-loss) cavities. Second, the sensitivity should also depend on pump energy with higher sensitivity near threshold. Neither dependence was strong and the model was considered to be in agreement with measurement.

Two other papers appeared before 1973. A brief note describing the detection of Ba^+ and Sr in a flame demonstrated the ability to detect transients (Thrash *et al.*, 1971). The other contribution was notable in several respects. A continuous laser was used for the first time (Hansch *et al.*, 1972), and an enhancement of 10^5 was reported. This value was not matched until very recently despite considerable effort in many quarters. Finally, the detection scheme was ideally suited to intracavity spectroscopy. Evacuated cells containing I_2 were placed both inside and outside the cavity. The intensity of the light at the absorbed wavelength was monitored by detection of fluorescence from the external cell. Since the absorber and the detector are spectrally matched a true representation of the quenching could be determined. These workers also proposed a model based on coupled differential rate equations. However, since a cw laser was used a steady-state approximation could be made, which made the equations analytically solvable. The major predictions of the model were threefold: The depth of quenching should be linearly proportional to absorbance, the sensitivity should be greatest for near-threshold operation, and sensitivity should increase with the number of modes. With these five papers, the field was launched. By this time work was under way in at least a dozen laboratories around the world.

II. EXPERIMENTAL METHODS

We now turn to the mechanics that enable the successful application of the method. This discussion is divided according to the relative linewidths of the laser and absorber. Substantially different problems and character-istics are encountered depending on whether the laser is narrow or broad compared to the absorber spectrum. In addition we include a discussion of those procedures which apply to the detection of transient species. We begin with absorber linewidths broader than the laser emission.

A. Broadband Absorbers

Intracavity enhancement spectrometry for broadband absorbers has both a major advantage and a major disadvantage. The advantage is that high-resolution recording of the laser emission is not required, thus solv-ing the principle problem encountered for narrow absorbers. In order to gain this advantage, the enhancement from mode competition has been sacrificed. These two factors coupled with the generally nonlinear charac-ter of the output versus loss data have directed the useful applications for this configuration. Very few reports using pulsed lasers have been made. Pulsed lasers are less stable than continuous lasers, and as a result, the sensitivity achievable without mode competition is not high enough to make intracavity sample placement an effective configuration. This is graphically demonstrated by an early report of the inability to detect an optical density of 0.4 external to the laser, indicating a noise in excess of 50% of the signal (Keller et al., 1972). Those reports which did involve pulsed lasers used an indirect measurement of sample loss and will be discussed below.

1. Direct-Loss Methods

The principle problems of direct-loss measurement with broad absorb-ers are the nonlinear dependence of laser power on loss, and the necessity of making separate measurements with and without the sample present. There is some dispute on the first of these. Two separate papers from the same institution have claimed a linear dependence of laser output power on loss. The first was for solutions of a Co complex placed in a He–Ne laser and claimed a linear dependence of $\log(I_0/I)$ on concentration for quenching up to 80% (Konjevic and Kokovic, 1974). The second was for several gas-phase organic species inside a line-tuned CO_2 laser (Konjevic et al., 1977). Again a linear dependence of $\log(I_0/I)$ on concentration was claimed, up to a 50% quenching. No other reports of linear dependence for broadband losses could be found. There are at least five other cases in which distinctively nonlinear response was shown. In addition, the theory of brandband loss in cw lasers does not predict a linear dependence. This apparent conflict between theory and experiment cannot yet be resolved.

An early example of the utility of making direct intracavity measurements despite the difficulty is shown in Fig. 2 (Chackerian and Weisbach, 1973). These authors operated a continuous, single-mode, CO laser with an intracavity gas cell. Power-loss measurements were made on successive NO–He mixtures of increasing dilution but constant total pressure. The results were in good agreement with the theory for the output of a single-mode laser. An enhancement of 250 was estimated. Accurate operation of a multipass cell of comparable sensitivity would be difficult at best.

The most successful application of direct output measurements in intracavity spectroscopy has received little recognition among other practitioners in this field. A technique called Laser Magnetic Resonance (LMR) spectroscopy has been used to record far-infrared spectra of many free radicals (Davies and Evenson, 1975). For this application the radical is generated inside the laser and between the poles of a magnet equipped with modulating coils. Variation of the field is used to Zeeman-tune the absorption into resonance with the laser. This conveniently allows the sample to be "removed" without disturbing any of the components. The modulation of the field results in two important advantages. First the signal is moved toward higher frequency, where there is much less noise. Second, the entire measurement is performed over a very narrow range of laser output. This overcomes the problem of long-range linearity so that the signal is proportional to concentration. An absorbance detection limit of 5×10^{-10} cm^{-1} was estimated. This certainly places LMR among the most sensitive absorbance measurements ever reported.

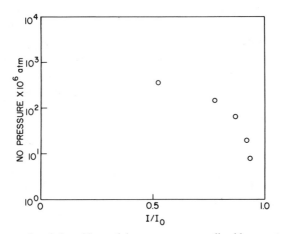

Fig. 2. Intracavity nitric oxide partial pressure vs normalized laser output for the nitric oxide line at 1935.48 cm^{-1}. [Adapted from Chackerian and Weisbach (1973).]

2. Indirect Methods

The difficulty of trying to relate laser output to loss has promoted a number of attempts to make indirect measurements of the loss. The first of these was reported in 1971 in an effort to measure gain in potential chemical lasers (Djeu et al., 1971). A CS_2-O_2 flame was mounted inside a continuous CO laser. Rather than measure the increased emission, or decreased in the case of loss, at the peak center, the wavelength range of oscillation of the CO laser was measured. In a later paper this width was shown to vary linearly with absorber concentration for loss measurement (Djeu, 1974). The second advantage of this method is that the measurement is always done at threshold, thus maximizing the available sensitivity.

A second approach to the problem was taken by Cresenzi and Shirk (1979) using a flash-lamp-pumped dye laser. An electro-optic Kerr cell was placed in the resonator with the sample. The purpose of the Kerr cell was to act as a reference loss. Any increase in loss by the sample was exactly compensated by the Kerr cell. Since the loss induced by the Kerr cell as a function of voltage can be accurately and reliably predicted, there is no need to calibrate the output-loss behavior of the laser. A type of nulling technique results. The sensitivity of this device suffered the usual problems of pulsed dye lasers, but the reference loss principle is important and greatly simplifies quantification.

The principle of reference loss has also been applied to a continuous dye laser (Shirk et al., 1980). In this report the reference loss was an electro-optic Pockels cell. Good results were achieved and two analytical measurements were made which have verified the utility of the approach (Harris and Mitchell, 1980; Harris and Williams, 1981). The principle advantage was not only a linear response but that the measured absorbance was independent of those factors which change the absolute sensitivity. A considerable improvement in sample throughput and the ability to scan wavelength without recalibration were realized. The application of this configuration is discussed in the chapter on Analytical Applications (Chapter 7 of this volume).

Another indirect method of intracavity absorbance measurement has also appeared (Ramsey and Whitten, 1980). These authors took advantage of the relaxation kinetics of the $Nd^{3+}-YAG$ laser to relate time delay before lasing to intracavity loss. A linear dependence between delay and absorbance was found. A detection limit of 10^{-3} cm^{-1} was reported but improvement was projected.

A proposal to recover some sensitivity through mode competition was made in 1976 by Batishche et al. A cavity was proposed which contained one gain medium but two separate oscillators, one of which contains the sample. Separation based on polarization, direction, and wavelength was discussed. An example cavity based on the separation by polarization is shown in Fig. 3. Sensitivity should be improved because the energy loss caused by the sample could be redistributed into the second cavity. The polarization scheme was tested using a ruby laser, but comparison with a single cavity is difficult because of differing requirements for optimization. The data presented were of generally poor quality but an increase of 30 in sensitivity over a comparable single resonator was seen.

A similar device was proposed and tested independently by Cresenzi and Shirk in 1979. This design also incorporated the reference-loss device, so it is difficult to access the advantage of each separately. If the earlier derivations, that enhancement was improved linearly with the number of modes, are to be trusted, the improvement of two modes over one would certainly not be worth the added complexity. The authors claim that the two-channel device was distinctly better at measuring sample loss, but no further applications have appeared.

3.   Recommendations

This concludes the discussion of broadband absorbers. As was seen, the absorber linewidths may in fact be quite narrow but wider than a single-mode laser. Samples can vary from a free-burning flame to liquids. Pulsed-laser applications have not shown promise unless some indirect measurement is used. Consequently cw lasers are recommended despite the greater expense and complexity. A reference-loss method greatly simplifies quantification and wavelength tuning. The detection limit is directly proportional to laser noise, so every effort should be made in this area. To this end a good prototype system is shown in Fig. 2 in Chapter 7 of this book. A well-regulated cw pump laser with a carefully controlled TEM_{00}

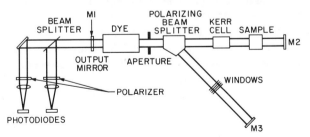

Fig. 3.   Example of a dual-beam intracavity laser for broadband absorbers. Mode competition is provided by one absorbed beam and one unperturbed beam. [From Crescenzi and Shirk (1979). Courtesy of North-Holland Publishing Company.]

transverse mode should be employed. This pump should be focused in a jet-stream dye laser. If possible an electro-optic noise reduction device should be inserted between the pump laser and dye laser but this arrangement has not been proven in practice. All optical elements should be attached to the Invar resonator of the dye laser. Every optical element should be wedged at least 3° and no two surfaces should be parallel. Inadvertent etalons result in large intensity fluctuations. A reference-loss device should be employed where possible. The most important specification for optical elements is surface finish. Optics which are as defect free as possible will result in lower noise and much improved results. Finally, the sample cell must be evacuable for gases or it must be a flow cell for liquids to allow for reliable reference measurement.

B. Narrow-Band Absorbers

The enhanced detection of narrow-band absorption with broadband lasers constitutes the majority of the intracavity literature. An enormous variety of experimental configurations has been reported. This variety is probably the principal source of the controversy associated with the method. An assessment of which reports are more or less reliable is impossible. In an attempt to provide a coherent picture of the area, the discussion has been divided into pulsed and continuous lasers. These two sections are further divided based on the methods of dispersion and detection.

Many different methods of reporting data have been used. To avoid confusion, terminology will be defined. The absorbance of a sample is $\log(I_0/I)$ where I_0 is the intensity which would have been measured in the absence of the sample and I is the intensity at the same wavelength with the sample present. The transmittance T is I/I_0. These terms apply to extracavity, conventional measurements. The terms to be used for intracavity measurements are illustrated in Fig. 4. Again I and I_0 are the emission of the laser with and without the sample, respectively. The value I/I_0 will be labeled the apparent transmittance. Similarly $\log(1/T)$ or $\log(I_0/I)$ will be called apparent absorbance. These terms are equally applicable to broad absorbers if I and I_0 are the total laser output with and without the sample, respectively.

1. Pulsed Lasers

a. Photographic Detection

The intracavity effect with narrow absorbers was first studied using pulsed lasers. The principal attraction of pulsed dye lasers is the relative

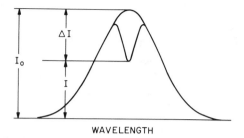

WAVELENGTH

Fig. 4. Depiction of the terms to be used in describing narrow-band absorbers in broadband lasers. Note that ΔI cannot be determined without knowledge of the unperturbed output for the $5:1$ laser to absorber width.

ease with which they can be operated throughout the visible region. In addition their lasing bandwidth is extremely large, several nanometers in some cases. This condition provides the potential for very large enhancement from mode competition. The problem has been the inherent instability of the devices. The results of each configuration will be summarized using two factors. First, the minimum extinction which could be detected and second, the relationship between intracavity loss and change in laser output are considered. The enhancement alone is not a useful property for comparison, as has been shown (Shirk *et al.*, 1980). If noise is enhanced equally with signal, no advantage from increased enhancement is realized.

The majority of reports for pulsed lasers have used photographic detection. This is no doubt due to its ease and availability. Photographic detection has two properties which are undesirable from the standpoint of pulsed-laser intracavity spectroscopy. The worse is that the linearity of the film response is always suspect. Second, unless some sort of gating scheme is used, the detected emission is integrated over the duration of the pulse. This integration severely complicates the interpretation of data.

The first reports using pulsed lasers and photographic detection which attempted to correlate the laser–absorber response were in 1974. Five separate laboratories reported results in that year. First von Weyssenhoff and Rehling (1974) reported on a flash-lamp-pumped dye laser with sodium in a flame as the absorber. A linear dependence of apparent absorbance on concentration for a range of apparent absorbance from 0.02 to 1.0 was found. The resolution the detection system was much too poor to faithfully reproduce the line shape. An estimate of optical extinction in a flame is difficult and was not made, so enhancement could not be determined. Spiker and Shirk (1974) reported results with a flash-lamp-pumped laser on solutions of Ho^{3+} and Pr^{3+}. Resolution was more than adequate to faithfully reproduce the line shape. A nonlinear concentration dependence of apparent I_0/I was found from values 1.0 to 6.0. However, the same data showed linearity when plotted as $\Delta I/I_0$ versus concentration up to a value of 0.5. Both photographic and photoelectric detection were employed. Photoelectric detection was accomplished by installing a phototube at the exit slit of a monochromator and scanning the monochromator after each laser pulse. A spectral profile was generated which was an average of many pulses. Photoelectric detection was found to give better detection limits than photographic detection. An estimated enhancement of 100 was quoted but no minimum extinction was reported. Green and Latz (1974) also detected atomic sodium in a flame using a flash-lamp-pumped dye laser. An empirical approach to photographic detection was used. The plate transmission was measured on a densitometer. The transmission at the absorption peak was plotted against the solution sodium concentration. These plots were found to be linear and reproducible under proper operating conditions. While analytically useful this method does not allow comment on the relationship between laser response and sample extinction.

Schroder et al. in 1975 reported on a flash-lamp-pumped, Q-switched dye laser with I_2 vapor as the absorber and time-gated detection. The principal purpose of the study was to determine the time development of the quenching profile. Photographic detection was used with an overall resolution of 0.2 Å. This was apparently sufficient to resolve the absorber structure. A plot was given showing a linear relationship of apparent absorbance to the logarithm of concentration over the range of apparent absorbance from 0.08 to 0.4. Minimum extinction was near 2×10^{-4}. No additional quenching appeared after 100 nsec of lasing, even though the pulse duration was ~350 nsec. These results are in direct conflict with the assertions that the intracavity effect is simply predicted by the duration of lasing.

It should be apparent to the reader that no agreement is at hand. The probable cause is that those parameters which determine the functionality of the response are nearly impossible to reproduce from laboratory to laboratory. The result is that if the available equipment necessitates using pulsed lasers and photographic detection, some sort of known absorber as a comparison standard is an absolute requirement. Some of the uncertainty for pulsed systems may be eliminated if a clearly linear detection scheme is used in place of film. One case of this was mentioned above in the work of Spiker and Shirk (1974) using photoelectric detection. This was possible only because the spectral resolution greatly exceeded the absorber linewidth.

b. Nonphotographic Detection

The first report of a pulsed laser with broadband, nonphotographic detection was by Horlick and Codding in 1974. Solutions of Pr^{3+} and Eu^{3+} were placed inside a Q-switched, frequency-doubled ruby, pumped dye laser. Detection was by a self-scanning linear diode array coupled to a monochromator. Resolution was stated to be 2.5 Å, which is sufficient for the bandwidth of the absorber. Apparent absorbance was linear with concentration for both species up to an apparent absorbance of 1.0. It should be noted that this pumping scheme gives operation far above threshold and for short time periods, so behavior may differ substantially from a flash-lamp-pumped system. No minimum extinction was reported.

Maeda *et al.* in 1975 reported on a flame seeded with sodium and lithium inside a flash-lamp-pumped dye laser. The laser spectrum was channeled by insertion of an 8.7-Å free-spectral-range etalon. In this configuration the laser operates with a few widely spaced longitudinal modes and these modes can be easily resolved with a spectrograph. The measured spectral width of the mode was comparable to the atomic linewidth. In this way the detection system could clearly distinguish the absorbed from the nearest unperturbed channels. The results are shown in Fig. 5. Plots of apparent absorbance versus concentration varied from convex through nearly linear to concave as the pump energy was varied. The use of a flame again complicates the estimate of minimum extinction.

Atkinson *et al.* in 1977 reported NO_2 absorption in a flash-lamp-pumped dye laser using a multichannel analyzer coupled to a 0.75-m spectrograph. Not only is the detector reliably linear, but spectra with and without the absorber can be conveniently subtracted, as shown in Fig. 6. For quantitative measurement, the area of the difference spectra of selected spectral features was plotted versus NO_2 partial pressure. The difference spectrum and area plots are shown in Fig. 7. The plots were linear up to a pressure which causes significant shifts in the dye-laser spectrum. This is a clear demonstration that linear results are possible and reliable for the proper experimental procedures. The spectral congestion present is so high that line-shape distortion is not observed. Substantially different results would obtain for isolated lines, such as an atomic system.

Morgan *et al.* in 1978 reported on a flash-lamp-pumped system using intracavity I_2 as the absorber and an external I_2 fluorescence cell as the detector. This is similar to the scheme devised by Hansch *et al.* (1972)

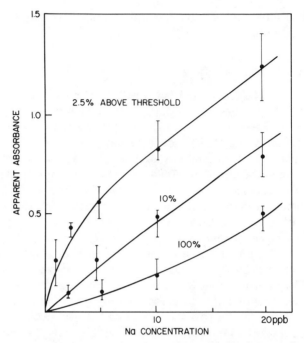

Fig. 5. Plots of quenching vs sodium concentration for varying pumping levels in a flash-lamp pumped dye laser with a seeded flame as the sample. [From Maede *et al.* (1975). Courtesy of North-Holland Publishing Company.]

described in the historical review. If saturation and quenching are avoided the spectral resolution of the detection system and the absorber should be ideally matched. Data were reported as enhancement versus single-pass extinction. If enhancement is constant a linear plot of apparent absorbance versus I_2 concentration should result. A very wide range of extinctions was measured and the enhancement changed by a factor of two over a change in extinction of three orders of magnitude. From the data shown it is not possible to determine if a linear response was present over an extinction range comparable to those in other reports.

c. Recommendations

It is apparent from these later efforts that all the reported nonlinear behavior was not attributable to photographic detection. Clearly under some conditions a linear response reproducible from day to day can be obtained. Just as clearly this property cannot be assumed. No set of controllable conditions has yet been proposed which will ensure linearity. The necessity of checking the response of each individual system would

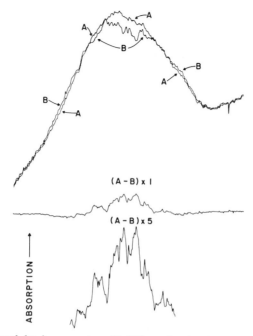

Fig. 6. Dispersed dye-laser spectra with NO_2 as the absorber. Spectra were recorded with an OMA and subtracted to yield the absorber spectrum. Spectrum A is without the absorber and spectrum B is with the absorber.

seem to be unavoidable. The array detectors provide a useful method to eliminate the uncertainty associated with photographic detection. They do, however, represent a considerable additional expense.

The lack of stability in pulsed lasers, which has limited detection, arises because of the transient nature of the gain, parasitic losses, and saturation. Continuous lasers provide a method to avoid these transient problems but they require much more effort and expense to tune across the visible. The stability has led to much higher sensitivity and should also produce a system design which can be reproduced with greater ease.

2. Continuous Lasers

The first report of continuous laser quenching for detection of narrow-band absorbers was by Hansch *et al.* in 1972. They reported an enhancement of 10^5. With 1% laser noise an extinction of 10^{-7} would be measurable. This limit was not stated explicitly nor confirmed experimentally. No statement on the functionality of quenching was made. Keller *et al.* in 1973 reported on a system similar to that of Hansch *et al.* using iodine but

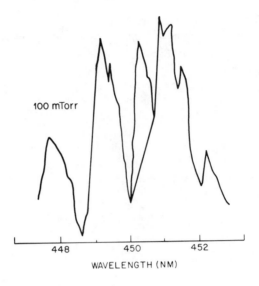

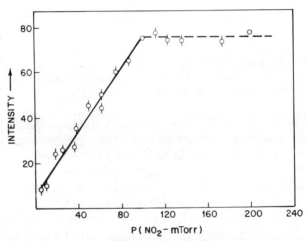

Fig. 7. Example of a single NO_2 spectral feature obtained from a difference spectrum. Integrated area plotted vs NO_2 pressure yields a straight line (Atkinson *et al.*, 1977).

with interferometric detection. An enhancement of 1000 was obtained, compared to the figure of 100 for their pulsed system. No functionality or minimum extinction was reported.

Bray *et al.* in 1977 reported on a system designed primarily to measure line positions of weak absorbers. Detection was by an OMA preceded by a 1.5-m monochromator. An enhancement of 2×10^4 was reported, but no explicit indication of minimum extinction was given. A statement that quenching was nearly linear with concentration was made but not elaborated. This device incorporated most of the features for proper design. A block diagram is shown in Fig. 8 for reference.

Maeda *et al.* in 1977 reported on a continuous dye laser which was channeled, in the same manner as the pulsed system discussed earlier. In place of the flame, an oven cell with sodium vapor was used. With this configuration, the absorption could be clearly resolved with a 1.7-m monochromator and photoelectric detection. An enhancement of 6000 was obtained and a log–log plot of I/I_0 versus concentration was shown. The plot was linear up to near 60% quenching. No value of minimum extinction was indicated.

Adamushko *et al.* (1978) reported on a cw laser channeled with an intracavity interferometer and using photoelectric recording of the dispersed emission, similar to the design of Maeda. The laser was tuned across the absorption spectrum of highly excited helium by changing the interferometer spacing. The helium states were generated by an intracavity discharge cell. A linear plot of $\Delta I/I_0$ versus discharge current was obtained. The concentration of the probed species was stated to be proportional to discharge current. The linearity extended from a quenching of 10 to 70%. A similar dependence was reported by Spiker and Shirk (1974). A minimum extinction of 1×10^{-5} was given.

In a 1979 report on the pump-power dependence of the enhancement, Harris used the original Hansch configuration of I_2 cells internal and external to a cw dye laser. The natural logarithm of fluorescence intensity

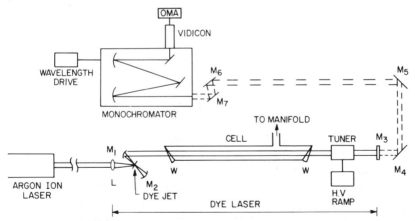

Fig. 8. Block diagram of a properly designed intracavity spectrometer for narrow-band samples. Important features are a cw laser, wedged cell windows, and OMA detection. [After Bray *et al.* (1977). Courtesy of North-Holland Publishing Company.]

was linearly proportional to intracavity I_2 partial pressure. This is equivalent to apparent absorbance linear to concentration. The range of linearity was up to 70% quenching. This plot was similarly reproduced by Harris and Weiner in 1981b. The linearity existed over a wide range of operating conditions. The enhancement, which is the slope of this plot, was shown to depend on laser linewidth and on pump power. If these factors were not held constant an apparent nonlinear concentration-dependence could be found.

Hill *et al.* in 1980 reported on a continuous dye laser designed to operate at low pressure. This design permitted operation with any convenient atmosphere, and the partial pressure of the absorber could be varied. A minimum extinction of 1×10^{-8} for O_2 bands was found. No attempt was made to quantitatively correlate quenching to absorber concentration. The importance of avoiding atmospheric absorption and etalon effects which can introduce spurious channeling was stressed.

In a recent paper, Stoeckel *et al.* (1982) investigated the time dependence of the intracavity effect. This time dependence has dominated much of the Soviet intracavity literature. Earlier efforts at investigating this time dependence were complicated by photographic detection. The latest report used a gate on the pump laser to initiate lasing, and a second gate on the OMA detector so that the dye-laser emission could be examined at any time after the beginning of pumping. The results were that the quenching from atmospheric water vapor was linearly proportional to absorbance and that the equivalent length of the cell, the enhancement, was proportional to the number of transits of the cell. In support of this statement, plots of apparent absorbance versus time were linear over a wide range of times and absorption coefficient. A minimum extinction of 3×10^{-8} with a measurement time of 1 msec was reported.

3. Summary

Again it is evident that several unresolved differences remain as to the true nature of the quenching–extinction relationship. It is clear that under certain conditions this relationship is linear. It is not clear what these necessary conditions are. The several factors that must be considered are discussed below.

Detection. Certainly the spectral resolution of the detection system could affect the measured relationship. In order for reliable quantitative measurements to be made the absorption band must be resolved. The repeated linearity of the I_2-fluorescence-detected quenching would argue toward this view. As was noted by Maede *et al.* (1977), saturation of the absorption is a very real possibility, even for low-average-power continuous lasers. A much more general solution to the problem of high-resolution detection for atomic species has been reported (Zalewski *et al.*, 1981). The change in voltage across a sodium hollow-cathode lamp when illuminated with cw dye-laser D-line radiation was used as a high-resolution detector. Hollow-cathode lamps are available for most of the elements in the periodic table. The plot of relative voltage change versus sodium concentration aspirated into an intracavity flame was distinctly nonlinear. However, similar extracavity flame data were also nonlinear, so no comment can be made as to the fundamental shape of the response.

Line Shape. The repeatedly observed anomalous line shapes further complicate quantitative studies. Le Floch *et al.* (1980) have advanced a plausible explanation for the asymmetry. The suggestion of monitoring diffracted light to compensate for the effect is reasonable but further complicates an already complex experiment. The presence of severe asymmetry did not preclude a statement (Bray *et al.*, 1977) of near-linear dependence of quenching to absorber extinction.

Dynamic Range. The maximum degree of quenching that gives reliable results is also of importance. An overall view of the applicable reports would recommend a maximum intensity reduction of 50% at the wavelength of peak absorbance. This would apply to both continuous and pulsed systems. Several authors have reported relative noise in excess of 10%. Such high noise severely limits dynamic range if the 50% loss constraint is followed.

Time Dependence. The large body of literature which places emphasis on the time development of the effect presents a dilemma. While the observations cannot be denied, the lack of inclusion of the well-documented threshold and linewidth dependence of the enhancement leaves an incomplete picture in all quarters. In addition, transfer of the significance of the time development to the condition of continuous operation has not yet been done experimentally or theoretically. The often-invoked interruption of lasing on a millisecond time scale is not present in all systems. It does, however, give a more satisfactory explanation for the existence of many modes in a homogeneously broadened laser. The alternative cause of spatial inhomogeneity for a laser with a 100-μm long gain does not fare well under quantitative scrutiny (S. J. Harris, private communication, 1981). Further experimental and theoretical work will be necessary to clarify the situation.

C. Transient Species

The time dependence of the intracavity effect also bears on the measurement of transients, for which the high sensitivity has proven useful

(Harris and Weiner, 1981b). In practice, however, there is no difference in the procedures used to detect steady-state populations of stable or unstable species. There have been several reports on detection of transient populations, all with pulsed lasers. The appropriate time-response data on the generally more sensitive continuous lasers has not been published. The time response determined to date is the rate at which quenching develops if lasing is started and the absorber population is constant. In order to detect a transient population, the laser should be at steady state with the fast generation of absorbers. Attainable sensitivity as a function of the time generation of absorber must be determined. Since the reported enhancements for cw systems are 10^3 greater than for pulsed systems, this characteristic may be very useful.

The first report of transient detection was by Atkinson *et al.* in 1973. A flash photolysis cell was mounted inside a flash-lamp-pumped dye laser. Photographic detection of the laser emission was employed. Both NH_2 and HCO radicals were studied. The lifetime of the radicals studied was much longer than the duration of laser emission. Consequently, the only provision required for detection was the proper timing of the laser trigger to the photolysis flash. No attempt was made to quantify the concentration of the transients.

Raspopov *et al.* in 1977 reported on the intracavity spectroscopy of a nitrogen discharge using a pulsed Nd–YAG laser. The duration of the discharge was 10 μsec and no detectable absorption was present 200 μsec after the discharge. The laser emission duration was 1 msec. In order to vary the time window probed by the laser, rotating slotted disks were placed both inside the dye laser and in front of the spectrograph slit. In this way 20–50 μsec segments of the time evolution of the system could be directed into the spectrograph and a variable time delay between the discharge and the onset of lasing was introduced. No quantitative data were gathered. The important observation was that quenching generated in the first 200 μsec of the laser pulse persisted until the end of laser emission. The time constant of the disappearance of the quenching was stated to be 0.1–1.0 sec although this figure was not proven experimentally. The conclusion is that the duration of the laser pulse need not be shorter than or even close to the lifetime of the absorber if one is forced to use integrated detection. It seems unlikely that quantification would be reliable under this condition.

In 1978 Reilly *et al.* published a thorough study of HCO production from formaldehyde photolysis. A flash-lamp-pumped dye laser with an intracavity gas cell was coupled to high-resolution photographic detection. The output of a second dye laser was frequency doubled to provide the photolysis energy. The photolysis laser had a baseline pulse duration of 2 μsec, while the proble laser had duration of 1 μsec. A quantitative measure of species concentration was desired and a calibration was used to verify linearity. The calibration was based on the presumption that the amount of HCO present immediately after photolysis was proportional to UV laser pulse energy. The optical density of the photograph plates at selected energies was measured as the pulse energy was varied using neutral-density filters. The plate density at the HCO bandhead was non-linear but the average density of six well-resolved rotational lines was found to be linear over a 20-fold change in photolysis energy. The plate optical density for these lines was always less than 0.54. Time resolution of the studies was limited by the width of the photolysis pulse. The success of this study certainly shows the advantages of the intracavity technique when carefully applied.

III. CONCLUSIONS

Intracavity-enhanced spectroscopy has been employed with considerable success. It has also proved to be troublesome in some cases. While certainty as to the best configuration will have to wait, some guidelines can be assembled. For narrow-band absorbers, fluorescence detection is ideal. It seems curious that with the number of sodium atom detection schemes, no report of a sodium fluorescence method has been made. Optical multichannel detection is greatly preferred over photographic detection. Continuous lasers will be 100–1000 times more sensitive than pulsed lasers and with care extinctions of 1×10^{-8} cm^{-1} are detectable. All optical elements should be wedged, preferably at several degrees. The number of intracavity elements should be the minimum possible. The problem of unwanted frequency-selective losses becomes more severe as detection becomes more sensitive. In this regard a moderate degree of paranoia is conducive to good results. Cell windows should be specified with the best available surface finish and kept scrupulously clean. Any loss on any optical element will result in spectral channeling and reduced sensitivity. The importance of avoiding parasitic losses was recently demonstrated graphically with a three-mirror cw dye laser (Antonov et al., 1982). Dramatic changes in the spectrum of the laser were induced by changing the position of the output mirror. Smooth spectra were available only when the output mirror was placed such that the total cavity length was an integral multiple of the distance between the two other mirrors. Placement of even very high-quality optical elements at incorrect positions resulted in similar effect. This channeling would severely limit sensitive detection. The work of Reilly et al. (1978) is a good example of transient detection and is an excellent model system. Higher time resolution is probably possible but further instrumental work will be necessary to determine what sensitivity will be attainable. Finally, further theoretical and experimental studies will improve the understanding of the effect and provide additional guidance.

ACKNOWLEDGMENTS

The author is grateful for many lengthy discussions of this general subject with S. J. Harris and J. S. Shirk and to G. H. Atkinson for providing a figure.

REFERENCES

Adamushko, A. V., Belokon, M. V., and Rubinov, A. N. (1978). *J. Appl. Spectrosc. (Engl. Transl.)* **28**, 287–290.

Antonov, E. N., Antsyferov, P. S., Kochanov, A. A., and Koloshnikov, V. G. (1982). *Opt. Commun.* **41**, 131–134.

Atkinson, G. H., Laufer, A. H., and Kurylo, M. J. (1973). *J. Chem. Phys.* **59**, 350–354.

Atkinson, G. H., Heimlich, T. N., and Schuyler, M. W. (1977). *J. Chem. Phys.* **66**, 5005–5012.

Batishche, S. A., Mostovnikov, V. A., and Rubinov. A. N. (1976). *Sov. J. Quantum Electron (Engl. Transl.)* **6**, 1386–1388.

Belikova, T. P., Sviridenkov, E. A., Suchkov, A. F., Titova, L. V., and Churilov, S. S. (1972). *So. Phy.—JETP (Engl. Transl.)* **35**, 1076–1079.

Bray, R. G., Henke, W., Liu, S. I., Reddy, K. V., and Berry, M. J. (1977). *Chem. Phys. Lett.* **47**, 213–218.

Chackerian, C., Jr., and Weisbach, M. F. (1973). *J. Opt. Soc. Am.* **63**, 342–345.

Crescenzi, F., and Shirk, J. S. (1979). *Opt. Commun.* **29**, 311–316.

Davies, P. B., and Evenson, K. M. (1975). *In* "Laser Spectroscopy" (S. Haroche, J. C. Pebay-Peyroula, T. W. Hansch, and S. E. Harris, eds.), pp. 132–143. Springer-Verlag, Berlin and New York.

Djeu, N. (1974). *J. Chem. Phys.* **60**, 4109–4115.

Djeu, N., Pilloff, H. S., and Searles, S. K. (1971). *Appl. Phys. Lett.* **18**, 538–540.

Green, R. B., and Latz, H. W. (1974). *Spectrosc. Lett.* **7**, 419–430.

Hansch, T. W., Schawlow, A. L., and Toschek, P. E. (1972). *IEEE J. Quantum Electron.* **8**, 802–804.

Harris, S. J. (1979). *J. Chem. Phys.* **71**, 4001–4004.

Harris, S. J., and Weiner, A. M. (1981a). *J. Chem. Phys.* **74**, 3673–3679.

Harris, S. J., and Weiner, A. M. (1981b). *Opt. Lett.* **6**, 434–436.

Harris, T. D., and Mitchell, J. W. (1980). *Anal. Chem.* **52**, 1706–1708.

Harris, T. D., and Williams, A. M. (1981). *Anal. Chem.* **53**, 1727–1728.

Hill, W. T., Abrea, R. A., Hansch, T. W., and Schawlow, A. L. (1980). *Opt. Commun.* **32**, 96–100.

Horlick, G., and Codding, E. G. (1974). *Anal. Chem.* **46**, 133–136.

Keller, R. A., Zalewski, E. F., and Peterson, N. C. (1972). *J. Opt. Soc. Am.* **62**, 319–326.

Keller, R. A., Simmons, J. D., and Jennings, D. A. (1973). *J. Opt. Soc. Am.* **63**, 1552–1555.

Konjevic, N., and Kokovic, M. (1974). *Spectrosc. Lett.* **7**, 615–620.

Konjevic, N., Orlov, N., and Trtica, M. (1977). *Spectrosc. Lett.* **10**, 311–317.

Le Floch, A., Le Naour, R., Lenormand, J. M., and Bache, J. P. (1980). *Phys. Rev. Lett.* **45**, 544–547.

Maeda, M., Ishitsuka, F., and Miyazoe, Y. (1975). *Opt. Commun.* **13**, 314–317.

Maeda, M., Ishitsuka, F., and Miyazoe, Y. (1977). *Appl. Opt.* **16**, 403–406.

Morgan, F. J., Dugan, C. H., and Lee, A. G. (1978). *Opt. Commun.* **27**, 451–454.

Pakhomycheva, L. A., Sviridenkov, E. A., Suchkov, A. F., Titova, L. V., and Churilov, S. S. (1970). *JETP Lett. (Engl. Transl.)* **12**, 43–45.

Peterson, N. C., Kurylo, M. J., Braun, W., Bass, A. M., and Keller, R. A. (1971). *J. Opt. Soc. Am.* **61**, 746–750.

Ramsey, J. M., and Whitten, W. B. (1980). *Anal. Chem.* **52**, 2192–2195.

Raspopov, N. A., Savchenko, A. N., and Sviridenkov, E. A. (1977). *Sov. J. Quantum Electron. (Engl. Transl.)* **7**, 409–411.

Reilly, J. P., Clark, J. H., Moore, C. B., and Pimentel, G. C. (1978). *J. Chem. Phys.* **15**, 4381–4395.

Schroder, H., Neusser, H. J., and Schlag, E. W. (1975). *Opt. Commun.* **14**, 395–400.

Shirk, J. S., Mitchell, J. W., and Harris, T. D. (1980). *Anal. Chem.* **52**, 1701–1705.

Siegman, A. E. (1971). "Introduction to Lasers and Masers." McGraw-Hill, New York.

Spiker, R. C., Jr., and Shirk, J. S. (1974). *Anal. Chem.* **46**, 572–574.

Stoeckel, R., Melieres, M. A., and Chenevier, M. (1982). *J. Chem. Phys.* **76**, 2191–2196.

Thrash, R. J., von Weyssenhoff, H., and Shirk, J.S. (1971). *J. Chem. Phys.* **55**, 4659–4660.

von Weyssenhoff, H., and Rehling, U. (1974). *Z. Naturforsch., A* **29A**, 256–260.

Zalewski, E. F., Keller, R. A., and Apel, C. T. (1981). *Appl. Opt.* **20**, 1584–1587.

7 ANALYTICAL APPLICATIONS OF LASER ABSORPTION AND EMISSION SPECTROSCOPY

T. D. Harris

Bell Laboratories
Murray Hill, New Jersey

F. E. Lytle

Department of Chemistry
Purdue University
West Lafayette, Indiana

I. INTRODUCTION

This chapter reviews the analytical applications of lasers to absorption and emission spectroscopy. The coverage for absorption has been restricted to condensed-phase studies in the ultraviolet–visible portion of the spectrum. Much of the discussion also applies to condensed-phase infrared studies, but gas-phase systems present different problems and approaches. The fluorescence coverage was expanded from this restriction to include gas-phase atomic and solid solution molecular methodologies. This threefold longer treatment is also indicative of the relative activity in the fields. The total number of analytically oriented high-sensitivity absorption publications is easily a factor of 10 less than similarly oriented fluorescence papers. Finally, analytical spectroscopic techniques are almost always striving for low detection limits of compounds occurring in complicated matrices. As such, sensitivity is often sacrificed for greater selectivity. This is in contrast to other areas in which the study of weak optical interactions in pure materials is of fundamental interest. As a result, many of the methods discussed in this chapter also appear in other chapters, but with a different emphasis.

II. ABSORPTION SPECTROSCOPY

A. Fundamental Considerations

Quantitative analysis by solution spectrophotometry has been in continuous practice for more than a century (Szabadvary, 1966). It has proved versatile, accurate, and sensitive. Instrumental design has remained conceptually stable since the early part of this century. However, the advent of reliable lasers has spawned a class of new techniques for measuring absorbance which provides a quantum leap in spectrophotometric sensitivity. Numerous methods have been developed for gases, liquids, and solids. This high sensitivity has not yet been given enough attention in the analytical community. The large majority of newly published spectrophotometric procedures for trace analysis continues to assume a minimum measurable absorbance of 1×10^{-2} (Kuwada *et al.*, 1978).

It is the purpose of this section to examine a representative set of these new, high-sensitivity methods with the purpose of making meaningful comparisons of their relative strengths and weaknesses and the circumstances under which each works best. The discussion will be restricted to those methods useful for absorbances of 1×10^{-4} or less in fluid solution. It will also be limited to the UV–visible portion of the spectrum. These restrictions should not be construed as bearing on the importance or utility of those areas not covered.

Absorbance methods can be easily separated into two groups, those methods that measure the transmission of the sample and those that measure either directly or indirectly the power absorbed by the sample. The latter will be labeled calorimetric techniques, while the former include conventional spectrophotometry. The transmission techniques to be discussed here are conventional spectrophotometry with and without multipass cells, laser intracavity methods, and wavelength modulation methods. The power-absorbed methods include conventional or thermocouple calorimetry, thermal lens calorimetry, and photoacoustic or acousto-optic calorimetry.

One other characteristic important in differentiating absorbance-measuring techniques is the distinction between peak-sensing methods and those sensitive to continuum absorption. Wavelength modulation spectrometry is the best example of a purely peak-sensing method. A zero response results if the transmission of the sample is unchanged with wavelength no matter how large the constant absorbance may be. In the absorbance range of concern here, all transmission techniques are in practice peak-detection methods. The reason is that transmission measurement is by nature a difference method. One is looking at very small differences between large signals. Spectrally broad or flat differences in transmission of 0.2% or less can easily result from a number of artifacts and are thus not reliable.

As is usually the case, there is no definitive best technique. The choice of the most appropriate method will be governed by the sample matrix and handling restrictions. Table I lists eight criteria by which any proposed absorbance measurement can be evaluated. These properties are sample and method dependent and will be used to determine which methods are most applicable to varying sample-imposed measurement restrictions.

A brief explanation will clarify the eight items in Table I. Tunability is a primary property for any optical method. It serves two functions. First, tunability allows measurement at the optimum wavelength for a large variety of analytes. Second, since it can be safely assumed that all samples will have measurable absorbance, the ability to tune greatly simplifies the separation of the analyte absorbance from interferences, especially dissolved impurities that vary widely from sample to sample. Many laser-

TABLE I

**Absorbance
Evaluation Criteria**

a. Tuning range
b. Pathlength
c. Reflection effects
d. Flow effects
e. Solvent effects
f. Solvent absorbance
g. Molecular scatter
h. Particulate scatter

based methods encounter great difficulty and expense when tunability is required. However, the usefulness of single-wavelength measurements cannot be ignored, as is evidenced by the popularity of single-wavelength chromatographic detectors. The pathlength dependence of the signal varies widely with measurement method. Some techniques show considerably increased signal with increased pathlength, while others show no change at all. This is an important consideration when small sample volume or experimental complexity, such as a chromatography detection, favors a small pathlength. The importance of surface reflections also varies considerably. It becomes important under those conditions for which it is difficult to keep surface reflection precisely constant throughout a measurement sequence. The importance of the response to flow is obvious in chromatography detection and flow injection applications. The increased use of flow injection as a sample handling method ensures a much wider applicability for those measurements tolerant of flow (Mottola, 1981).

The last four properties deal directly with sample characteristics. The dependence of the signal on the matrix, the solvent in our case, is most important for the calorimetric methods. The heat capacity, thermal expansion coefficient, thermal conductivity, and temperature dependence of the refractive index are important in one or more of the thermal methods. The signal intensity can differ by a factor of 50 among commonly available solvents. At the sensitivity of concern here, with rare exception, the absorbance of solvents is measurable. The ability to distinguish the analyte absorbance from the solvent absorbance will be the deciding factor in many cases. The solvent concentration normally will be at least 100 times greater than the most concentrated solutes, so the absorbance of the solvent is generally a large but constant background. Finally, the response of the methods to scatter should be considered. Transmission methods respond to all types of throughput loss and therefore to scatter. However, as with solvent absorbance, molecular scattering will likely be constant from reference to sample. Most calorimetric methods do not respond to scattering loss, but Raman scatter will yield an equivalent absorption response of 2×10^{-8}. Many methods are approaching this detection capability (Dovichi and Harris, 1981). In addition, molecular scattering processes are strongly wavelength dependent and will give steeply sloping baselines. Particulate scatter is analogous to impurity absorbance in its variation between samples. Many calorimetric methods claim immunity from particle scattering but even the most transparent solids absorb a few percent, giving a large response in some cases. In addition, if a focused laser beam is used, even small particulates can block a significant part of the beam and disrupt the measurement. With these considerations in mind, each of the listed techniques will be discussed, beginning with the transmission methods.

B. Transmission Measurements

1. Introduction

As stated, transmission methods are inherently difference techniques. The measurement of small absorbances requires detection of a very small difference between two large signals. The fundamental limit is then the uncertainty in the signal with the greatest noise. From a theoretical standpoint the minimum uncertainty is the statistical noise or shot noise of the input beam. With a few specialized exceptions, this shot noise is not the experimental limiting uncertainty. The fundamental limitations to transmission requirements have been thoroughly discussed elsewhere (Reule, 1976). In addition, there has been recent work on overcoming sample-dependent anomalies such as beam deflection and defocusing (Kaye, 1981). Two approaches to the small-difference problem can be made. One must either increase the difference or one must reduce the noise. Increasing the difference can be accomplished by increasing the pathlength or by placing the sample inside a laser cavity. Reducing the noise has a statistical limit but can be enhanced by limiting the bandwidth of detected noise, as in wavelength modulation spectrometry.

2. Increasing Sample Pathlength

First and foremost, at low absorbance, the magnitude of the transmission loss scales linearly with pathlength. Therefore, if increased pathlength is a viable option to yield a measurable signal then it is probably the most straightforward instrumental option. The analyst should be prepared for a large increase in optical artifacts. Any interference from molecular scatter or particulate scatter will also increase proportionally.

Multipass cells offer a very large pathlength and have gained considerable popularity for gas-phase infrared analysis. The enormous volume of these cells will likely prevent a similar device from being used for liquids. However, the spatial coherence of lasers provides an opportunity for simple long-path cells of reasonable volume (Kyle and Schuster, 1978). A device for this purpose has been constructed and tested for gases but the author is not aware of one for liquids. A 70-pass equivalent length was achieved in gases with only moderate effort. A similar liquid cell of modest 20 cm length would yield a 35-m equivalent path. An intensity difference of 0.5% would result from an absorbance of 1.5×10^{-6} cm^{-1}, less than most solvents.

A second example of extremely long-path cells is that of hollow glass and quartz fibers. Stone (1972a,b, 1978) has published several reports of spectra taken with liquid-filled hollow-core fibers. Pathlengths from 4.5 to 130 m were generated without difficulty. The high quality of data obtainable with these cells is shown in Fig. 1. It is obvious that exceedingly small absorptions can be detected in small volumes, 100 μL for a 30-m cell. Problems would be encountered in filling, rinsing, and refilling the same fiber and in extracting impurities from the fiber itself. However, disposable cells several meters long are quite attractive and deserve serious consideration. It should be noted that accurate estimates of absolute absorbance in long-pathlength cells are difficult but peak detection is straightforward.

Surface reflections can be a serious problem for all transmission methods. One first must consider the mismatch between sample and reference cells if the instrument calls for both. The authors are not aware of any claim of cells sufficiently matched for the accuracy of concern here. Thus, all methods require both the reference solution and the sample solution to be measured in the same cell. A slight difference in refractive index between the solutions will also result in a substantial error in absorbance. Since refractive indexes change slowly with wavelength, the error will be manifested as an offset. If the analyte has a well-known spectrum, as is usually the case in analysis, this offset probably will be identified as such, provided the measurement method is sufficiently tunable. However, if the

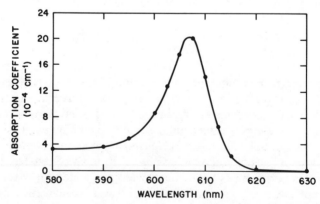

Fig. 1. Transmission spectrum of the 607-nm line of benzene. The spectrum was recorded with a 4-m fiber.

absolute transmission spectrum of a transparent material is sought, this reflection change could result in serious error. The calorimetric methods offer a more reliable measure of neat-solvent absorption.

3. Intracavity Enhancement

The last method for making the signal larger is the laser intracavity approach. Since this method is not yet widely used and is conceptually different from the others, a brief introduction is provided. A more extensive discussion is given by Harris in Chapter 6 on intracavity enhancement.

Laser intracavity-absorption enhancement has been known since at least 1967, with the invention of the dye laser. It is based on three properties of lasers, of which only two apply for most analytical situations. One instrumental configuration useful for many analytical measurements is shown in Fig. 2. Lasers are a type of oscillator, so called because of the multiple-round-trip path of the radiation, which naturally provides for multiple passes of the light through a sample placed in the optical train. In addition, the laser exhibits a threshold. This means that if more than a given loss, say 10%, is incurred in one round trip, the device ceases to operate. The threshold arises from the finite gain of the amplification medium, in our case the same 10%. The net result is that a precipitous drop in output power results when the accumulated loss is increased from 9.9 to 10.1%. The reader immediately sees that the incremental loss of 0.2% can generate a near 100% change in intensity, from a great deal to almost none. The third property of lasers is the competition effect, which

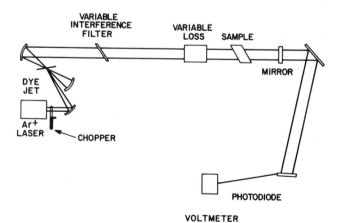

Fig. 2. Schematic diagram of a laser intracavity spectrometer designed for quantitative measurement of broadband absorbers.

takes place when the absorber linewidth is narrower than the laser linewidth. Even the broadest laser exhibits a bandwidth <5 nm. An absorption significantly narrower than this in solution is rare but there are examples, such as inner-shell transitions in rare earths (Spiker and Shirk, 1974). In order to make the discussion general, we will consider only the first two effects.

Laser intracavity enhancement is identical to other transmission techniques in its response to most of the items listed in Table I. The only difference is in the magnitude of the intensity change for a given sample transmission. In order to operate stably, the curvature and position of the laser mirrors must fall within certain ranges. These requirements are called stability criteria. If the laser design is well within the stable range, defocusing effects and beam deflection effects make very little change in the operation of the laser. This property gives intracavity sample placement a certain immunity to these problems, which can create large artifacts in conventional transmission measurements. The only report of a clearly useful design for fluid solutions was by Shirk *et al.* (1980). This spectrophotometer addressed the main disadvantage of the intracavity method, the lack of any consistent intensity–concentration dependence. The magnitude of the intensity change for a given transmission loss can be made arbitrarily large. To overcome this problem, a reference loss was used to ensure accurate results. The main advantage of the system proved to be the remarkable lack of sensitivity to very large changes in sample viscosity and refractive index. Salt solutions of several molar concentration gave identical measured absorbance to the water reference. This cannot be said for any calorimetric system. The disadvantage of the system was its experimental complexity. An extremely high degree of proficiency was required for even rudimentary sample throughput. Despite the complexity this instrument had an absorption detection limit of 2×10^{-5}, which is 1% of the water solvent. The initial application produced a blank-limited iron detection limit of 50 parts per trillion, the lowest by any method ever reported for iron (Harris and Mitchell, 1980). Since these results were obtained with a 1-cm cell, improvement should be available from a longer cell, although sample-handling procedures would have to improve proportionally.

4. Instrumental Noise Reduction

The alternative approach to the transmission signal-to-noise problem is to reduce the noise. Commercial instrumentation has made great strides in this area over the past five years. The reduction of noise and drift has been so great that double-beam designs are being abandoned in favor of more simple, higher-light-throughput single-beam systems (Kay, 1981).

An excellent example of the progress of source and electronic stability is the near shot-noise limit of commercial near-infrared reflectance instruments. In all likelihood, sample artifacts will limit the measurement long before these low noise limits. One other method of reducing the noise is to limit the bandwidth of detection and detect at frequencies with low noise. Limitation of bandwidth is a general measurement tool. The advantage of moving measurements to noise-free regions is well understood and documented (Hieftje, 1972). The noise which dominates the spectrophotometric instrumentation is low-frequency flicker or $1/f$ noise. Stated simply, there is just less noise to detect at higher frequencies, and thus smaller signals can be seen. This fact coupled with the narrow bandwidth provided by phase-sensitive detection yields a large sensitivity improvement if the signal is modulated. Unfortunately, there are only a few specialized examples for which the sample absorbance can be modulated (Hunziker, 1971; Barnes and Lytle, 1979; Levine and Bethea, 1980). However, if the absorption to be detected is peaked, the signal can be made to appear at a higher frequency by wavelength modulation (Cardona, 1969).

The added sensitivity of wavelength modulation spectrometry arises from the nonuniform distribution of source, sample, and electronic noise. The movement away from steady-state measurement along with synchronous detection provide a fundamentally quieter measurement. As stated, the measured property is now the change in absorbance with wavelength, not the absorbance itself. As a direct consequence, analytes with sharp spectra are more sensitively detected than analytes with broad spectra. In addition, all other changes of transmission with wavelength are equally detected. The major experimental difficulty is construction of an achromatic optical train. Mirrors, lenses, gratings, prisms, detectors, and cell windows all impart a signal dependent on the variation of their transmission with wavelength. This, together with the extreme chromaticity of solvent Rayleigh scatter have evidently led to lack of activity in the field. The authors are not aware of a published liquid-phase analytical scheme in which extreme sensitivity was achieved with wavelength modulation. However, it has the potential for excellent sensitivity if properly applied.

C. Power-Absorbed Measurements

The alternative solution of the general problem of high transmission is to avoid altogether measuring the small difference between large signals. Instead we can measure directly the light absorbed rather than the light transmitted. This enables measurement against a zero background, in much the same manner as fluorescence. The fate of the energy absorbed by a molecule is either to be reemitted or to be thermalized with subsequent deposition of energy in the surrounding solvent. If we wish to include all analytes we must in some manner or other measure this thermalized energy. If the molecule has a significant emission quantum yield a correction must be applied or a nonabsorbing quencher added. Because all the power-absorbed methods measure the thermalized energy, they have been labeled as calorimetric.

The historical impediment to the use of calorimetry to measure absorbance was the lack of sufficient photon flux to give a useful temperature change. The advent of lasers has removed this impediment. Even with this increased light flux, very few examples of conventional or thermocouple calorimetry for measuring solution absorbances have been published. This method has gained wide acceptance for measuring the absorbance of transparent solids (Hordvik, 1977). The major problem is that long integration times are needed to get a measurable temperature change. As a result, window absorption and wall absorption of scattered radiation cannot be distinguished from solution absorption. In addition, most very high-average-power lasers are not sufficiently tunable to give useful spectra. These problems will likely prevent any widespread use of thermocouple calorimetry for solution spectrometry.

Two other methods to detect solution absorption have been given considerable attention. The first employs modulated or pulsed irradiation, with acoustic detection of the thermally induced liquid expansion. It has been called both photoacoustic and acousto-optic spectrometry. Here it will be referred to as photoacoustic calorimetry. The second method uses the thermally induced change in refractive index and has many variations for detection. These include thermal lenses (Kliger, 1980), photodeflection (Jackson *et al.*, 1981), interferometry (Stone, 1973), optical bistability (Cremers and Keller, 1981), and induced gratings (Leach *et al.*, 1981). No doubt new variants will be published regularly for some time to come. All these techniques have similar but not identical response to the considerations in Table I. In the interest of brevity and clarity, thermal lens calorimetry will be discussed here as representative of this group, since it is the oldest and most widely studied of the list. We will begin with the acoustical methods.

1. Photoacoustic Calorimetry

The principle of acoustic detection of absorbed energy is old and widely reviewed (Rosencwaig, 1980). A thorough discussion appears in a separate chapter on this subject. The vast majority of work has concerned the absorbance of solids in closed gas cells or of the gas itself. This method is insensitive for liquids. A more recently introduced variation is the use of piezoelectric microphones in direct contact with the liquid (Lahmann *et al.*, 1977). Although the earliest use of this method was reported in 1977, the simplicity and economy of the detector has resulted in rapid popularity.

In contrast to gas-microphone photoacoustic spectrometry, only laser sources provide sufficient power to give useful signals for transparent liquids. Two different source designs have been pursued in photoacoustic calorimetry. These are a modulated continuous beam and a pulsed source. They differ in some important properties which are subsequently discussed.

For pulsed-source systems there is no change in signal with increased pathlength. The measured acoustic signal is generated by the portion of the beam closest to the microphone (Patel and Tam, 1979). In this regard, pulsed-source acoustic calorimetry offers an advantage unique among all other high-sensitivity spectrophotometry methods, the ability to time discriminate against window absorption. The discrimination obtains because the acoustic signal from the sample arrives at the detector before the acoustic signal from the window, allowing gated detection to reject the window signal. The change of signal intensity with pathlength of chopped-source systems is not clear. The original designs used piezoelectric ceramic tubes as both the sample cell and the microphone (Oda *et al.*, 1978). Later work has universally abandoned this configuration in favor of the small cylindrical microphone used in pulsed-source systems. A design which has proved to be versatile, economical, and sensitive is shown in Figs. 19–23 in the photoacoustic chapter (Chapter 1 by Tam). There seems to be no loss in sensitivity; thus no advantage accrues from the longer path length of the initial designs.

Reflection losses should cause no significant error in either design, because the beam intensity is not significantly changed between sample and reference. However, if a stray beam strikes an absorbing surface near the microphone, enormous spurious signals result. Flow can cause large problems for both pulsed and chopped systems because of flow-generated background acoustic noise. This noise is low in frequency so the higher speed of the pulsed systems allows effective filtering. Oda and Sawada (1981) have solved this problem by using high-frequency acousto-optic modulation. Judging by their results, the flow problem can be circumvented. For a fixed absorbed energy the magnitude of the acoustic signal depends on two solvent matrix properties, the heat capacity and the thermal expansion coefficient. Both vary with the nature and concentration of the solvent, requiring calibration for each solvent used.

The tolerance to solvent absorbance for all calorimetric methods depends entirely on the precision of the measurement. This is limited in most cases by source intensity fluctuations and for photoacoustic calorimetry by stray acoustical energy. Unfortunately, most lasers are by nature considerably noisier than incandescent sources. Very few instruments provide uncertainty better than 0.5%. If we wish to measure solute absorbance in water at 650 nm, reliably detectable solute absorbance must exceed 5×10^{-5}. This value is considerably above most instrumental detection limits. Transmission methods can effectively subtract the absorbance of the solvent because of their greater precision and the relative ease of maintaining the solvent absorbance at a constant level.

Photoacoustic calorimetry is in principle resistant to interference from molecular and particulate scatter because neither will significantly diminish the intensity of the absorbed radiation. However, even highly reflective surfaces absorb several percent of the energy which strikes them. If the scattered energy strikes a surface acoustically coupled to the microphone, signals several orders of magnitude larger than the analyte signal will result. In practice all solutions must be carefully filtered and the detector made as reflective as possible. In addition even small concentrations of suspended solid will absorb radiation and generate interfering acoustic signals (Oda *et al.*, 1980).

At the present time pulsed photoacoustic calorimetry provides the best combination of sensitivity and tunability. Measurement of absorbances down to 10^{-7} cm^{-1} with broad tunability has been demonstrated. The method also lends itself to easy calibration, which is of crucial importance to analytical accuracy. The only drawback is the precision, which is a few percent relative to the signal. In a highly absorbing solvent such as water this creates a considerable restriction on usable sensitivity.

2. Thermal Lens Calorimetry

A second method to probe the energy absorbed by a solution takes advantage of the change in refractive index which results from the temperature change. The thermal lens method of interest here probes the refractive index gradient produced by the natural intensity distribution of a laser beam. A detailed treatment is given in Chapter 3, by Fang and Swofford, and only a brief description of the method will be given.

Absorbance measurement by thermal lens formation relies on two factors. The first is that a laser can be configured to produce a beam with a well-defined Gaussian spatial distribution of energy, with the maximum intensity at the beam center. The second is that the amount of radiation absorbed is proportional to the input intensity. Therefore, a solution illuminated with a Gaussian beam will absorb more energy from the center of the beam than from the edges. If the proper optical configuration is used, a temperature gradient, and consequently a refractive index gradient, sufficiently large to defocus the beam is generated. This defocusing is used as a probe of the solution absorbance. The physical arrangement used to accomplish the measurement is shown in Fig. 3.

In practice this method requires a focused beam and the majority of the effect is generated near the focus. As a result no significant improvement is obtained with increased pathlength. Changes in reflection between sample and reference are not generally large enough to cause significant error and stray reflections generally cause no interference. Since the technique is based on the generation of a smooth temperature gradient, flow causes enormous changes in the signal. This problem was investigated thoroughly and reported by Harris and Dovichi (1981). The conclusions were that slow flow, 1 mL/min, could be dealt with, but at some sacrifice in sensitivity.

A major consideration in thermal lens calorimetry and all related optical-probe methods is the change in the signal with solvent matrix. These techniques share with photoacoustic calorimetry dependence on heat capacity and thermal expansion coefficient, although the latter is now cast as the change in refractive index with temperature as opposed to the change in volume with temperature. In addition, the thermal conductivity, which is the mechanism for removal of the temperature gradient on which the techniques are based, is also important. Each of the three solvent properties reduces the response for water compared to organic solvents. Solvent absorbance affects these optical probe methods in the same adverse manner as for acoustically probed methods. A well-defined Gaussian beam must be available for thermal lens measurements. Beams of this stability are available only from continuous lasers, which have relatively narrow tuning ranges of 20–60 nm per dye. As a result the flexibility of the method is limited, especially if UV measurements are needed. Molecular scatter is likely not to ever be a problem. Particulate scatter can be a serious impediment because of the small radius of the focused beam. Even a single dust particle or a bubble will cause a large interference. The net result is a necessity for filtering all samples.

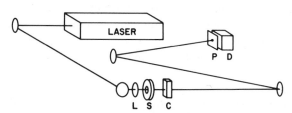

Fig. 3. Instrumental arrangement for performing thermal lens calorimetry: (L) lens; (S) shutter; (C) sample cell; (P) pinhole aperture; (D) detector.

Most analytical applications of thermal lens calorimetry have used line-tunable gas lasers. These lasers are relatively simple to operate and usually have good transverse mode structure. They do not offer the flexibility of continuous tunability and large spectral gaps are present with any single device. Under optimum conditions a usable thermal lens measurement can be made on even the most transparent solvents. The practical limit to sensitivity is precision, as is the case for photoacoustic calorimetry. The absorbance detection limit in nonpolar solvents will be in the range of 1×10^{-7}. The problem of solvent absorbance was addressed in a clever two-cell design (Dovichi and Harris, 1980). Its purpose was to subtract solvent absorbance by taking advantage of the cell position-dependent response. Because of the onset of nonlinear behavior with a strong lens this configuration has been abandoned in favor of a transient signal detection scheme which gives greater dynamic range (Dovichi and Harris, 1981). The transient detection configuration should be the most useful for analytical measurements. Of the laser-based methods discussed here, thermal lens calorimetry is probably the least expensive way to achieve high sensitivity.

D. Recommendations

As is always the case, there are trade-offs in choosing the most appropriate technique. Conventional spectrophotometric instruments are cost effective, broadly tunable, and increasingly sensitive. Thermal lens methods are simple and relatively inexpensive for line-tunable visible-wavelength applications. Photoacoustic methods are somewhat more expensive but offer wide tunability and sensitivity together. Both photoacoustic and thermal lens methods suffer considerable sensitivity loss in aqueous systems and require recalibration for any significant change in solvent properties. They do offer the most straightforward method for measuring neat-solvent absorption spectra. Laser intracavity absorption works well for aqueous systems and is relatively immune to changes in the solvent. It is complex and will be used best during the method-development stage where changes in the analyte matrix can be made without recalibration. Once conditions are established, then use of thermal lens or photoacoustic calorimetry will achieve higher sample throughput. Long-path cells have been ignored and should be attractive for inexpensive highly sensitive measurement in which the analytical peak is riding on a relatively flat background. The cost factors for all these methods will continue to change unpredictably, especially the lasers themselves. It is hoped that with these brief guidelines the analyst will consider the extraordinary sensitivity now available with spectrophotometry and be able to make intelligent choice of the most appropriate method.

III. FLUORESCENCE SPECTROSCOPY

A. Direct Excitation of Atoms

1. Introduction

Atomic fluorescence is a specific form of photoluminescence where the sample consists of gas-phase atoms. As such, the general instrument is composed of an excitation source, a device for producing and/or maintaining gas-phase atoms, and an emission detector. Monochromating devices may be used to restrict the bandwidth of the source and/or the emission reaching the detector. The transfer optics are usually designed with throughput as the primary design criterion, although high f numbers may be used to reduce background interferences.

In all nonsaturating, linear photoluminescence techniques, the instrumental sensitivity is proportional to the source photon flux density irradiating the sample. Thus, in principle, the laser should provide a dramatic improvement in such methodologies since its bandwidth can be made equal to or less than the atomic linewidth without a serious reduction in photons. In addition, the spatial coherence of the beam makes the transfer and focusing of the source radiation very efficient. The major historical impediment to the rapid introduction of lasers into instrumental systems was the difficulty of matching the source frequency to the narrow atomic absorption lines. This problem was obviated with the development of the broadband dye laser by Sorokin and Lankard (1966), or perhaps more realistically, the grating tunable dye laser by Soffer and McFarland (1967).

The year 1971 marked the first use of lasers in analytical atomic fluorescence. Denton and Malmstadt (1971) employed a commercial, frequency-doubled, Q-switched ruby laser to pump a home-built dye laser; Fraser and Winefordner (1971) employed a commercial N_2-driven dye laser; and Kuhl and Marowsky (1971) employed a home-built flash-lamp-pumped dye laser (Kuhl *et al.*, 1972). When compared to nonlaser approaches, the first two instruments (employing a flame reservoir) showed little or no improvement in the detection limits of Al, Ga, Cr, Fe, In, Mn, Sr (Fraser and Winefordner, 1971), and Ba (Denton and Malmstadt, 1971). The third instrument (employing a vapor-cell reservoir) showed a 60× improvement in the detection limit of Na. The following year, Jennings and Keller (1972) demonstrated a 16 fg cm^{-3} detection limit for Na using a cw dye laser [about 500× improvement over the results reported by Kuhl and Marowsky (1971)]. Piepmeier (1972) proposed the technique of saturated excitation and, in the first problem-oriented application, Gibson and Sandford (1972) used an atomic fluorescence LIDAR technique to detect sodium in the atmosphere. The next three years produced about one notable experimental advance apiece. In 1973, Kuhl and Spitschan utilized a frequency-doubled dye laser to excite Pb (283.3 nm), Mg (285.2 nm), and Ni (305.1 nm). In 1974, Neumann and Kriese used a graphite furnace atomizer to quantitate Pb at a level of 0.2 pg total. The technique combined frequency doubling and near-saturated excitation in a low-scattering environment to achieve these results. Finally in 1975, Fairbank *et al.* combined wavelength-modulated cw excitation with sophisticated signal processing to detect 10^2 sodium atoms per cubic centimeter ($\sim$0.004 ag cm^{-3}). This paper, perhaps more than any other, stimulated a general frenzy of experimental design effort aimed at increasing sensitivity while at the same time reducing interferences. Officially the search for a single-atom detection scheme had begun.

2. Instrumentation

a. Source

Both pulsed and cw dye lasers have been successfully employed as excitation sources. The choice between these two general classes is rather straightforward. If the instrument is to be used as a general-purpose analytical tool, a pulsed system should be considered because of its broad spectral range. If the instrument can be operated in a narrow tuning range and if a greater signal processing flexibility is desired, a cw system should be considered.

Within the pulsed class of lasers, virtually every form of excitation has been employed. Already mentioned was an instrument utilizing the output of frequency-doubled ruby. Other pump lasers have included doubled Nd–YAG and nitrogen. Flash-lamp-based systems have included both commercial and homemade versions. No general compelling reason can be given for choosing one system over any other.

A typical nitrogen-based system is that discussed by Weeks *et al.* (1978). The authors reported using thirteen dyes to cover the wavelength range from 357 to 687 nm with a hole from 430 to 440 nm. Although not much was said about the doubling range of the instrument it can be assumed to approximate 217–340 nm, leaving a second spectral hole from 340 to 357 nm. The lowest-wavelength excitation reported was for Cd at 228.8 nm, and the highest wavelength produced by doubling was for Cu at 324.7 nm. The bandwidth in the UV was predicted to be less than 0.01 nm.

A typical flash-lamp-based system is that discussed by Hosch and Piepmeier (1978). Although no dye but rhodamine 6G seems to be used by these and other authors, the manufacturer lists a wavelength range from 265 nm to 2.6 μm. On the other hand, this laser has been frequency doubled to excite Ni at 300.2 nm and Sn at 300.9 nm (Epstein *et al.*, 1980a).

The only Nd–YAG-based system is that discussed by Bolshov *et al.* (1976). The pump laser was a commercial device (assumed to be of Russian origin) producing 370 kW of 532-nm radiation in 13-nsec pulses and at a 12.5-Hz repetition rate. The apparently home-built rhodamine 6G dye laser produced 80-kW, 8-nsec pulses with a bandwidth of 0.03 nm. KDP was used to double the visible output into the range of 280–310 nm. It is unclear in this and most of the other papers describing the use of frequency doubling exactly what average power irradiated the sample.

Within the cw class of lasers, only an argon ion output has been used to pump the two dyes, rhodamine 110 and rhodamine 6G. Although no one has reported frequency doubling, almost all have used some form of frequency narrowing. A typical system is that discussed by Green *et al.* (1976). A commercial 5-W all-lines argon ion laser was used to pump a commercial dye laser. The wavelength range was 535–620 nm with rhodamine 110, and 570–650 nm with rhodamine 6G. The bandpass was reduced to 0.003 nm by using a 0.5-mm etalon and tuned over a ~0.05-nm interval by tilting the etalon with a stepping motor. 100 mW of power was achieved at the 553.5-nm barium wavelength, while 300 mW was achieved at the 589-nm sodium wavelength. Again because most of the cited authors entered the field quite early on, there apparently has been no use of the newer, higher-power, narrow-frequency ring lasers.

b. Sample Configuration

Virtually every standard form of creating an atom reservoir has been combined with laser excitation. The choice is fairly straightforward. If the sample needs to be chemically destroyed to produce a reasonable level of free atoms, then devices such as flames or plasmas should be considered. The basic trade-off with this approach is the high blank level produced by the resultant matrix. If extreme sensitivity is required and if the sample simply needs to be converted to the vapor phase, a vacuum cell should be considered. For intermediate types of problems a furnace technique could be tried as a suitable compromise.

A typical flame-based system is that described by Weeks *et al.* (1978) in which commercial nebulizer/burner assemblies were used with home-built capillary heads. Both air/acetylene and nitrous oxide/acetylene were used to produce a flame, and an inert (argon) gas sheath was used to separate the flame from the laboratory atmosphere. The gas flow rates were ~0.04

L sec⁻¹ of acetylene and ~0.2 L sec⁻¹ of air for the air/acetylene flame
and ~0.1 L sec⁻¹ of acetylene and 0.18 L sec⁻¹ of nitrous oxide for the
nitrous oxide/acetylene flame. For Al, Bi, and Cr it was noted that a fuel-
rich flame gave better results, whereas for Ni, Ag, and Cu operation was
fuel lean. The fluorescence was measured ~25 mm above the 19-mm-diam
burner head. The optical alignment with respect to the flame was not
critical because of the 5-mm-diam laser beam used.

A typical vapor cell is that described by Brod and Yeung (1976). In
their work, a pure metal sample was placed in a 40-cm-long by 5-cm-diam
cell containing scatter-reducing light stops. The cell was then evacuated
and set to the desired temperature by monitoring Chromel–Alumel ther-
mocouples and controlling the current supplied to an external heating
tape. The resultant fluorescence was observed at 90° to the laser beam.
The actual sample concentration was computed from established vapor-
pressure curves.

A typical furnace-based system (shown in Fig. 4) is that described in
detail by Hohimer and Hargis (1978). The atomizer uses a vitreous carbon, slotted evaporation boat which has a sample capacity of 60 μL. This

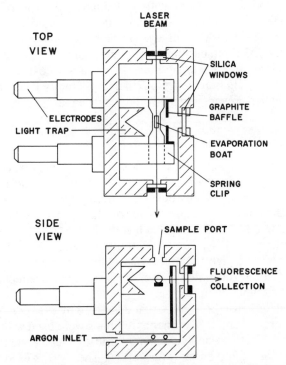

Fig. 4. A vitreous carbon atomizer useful for laser excited atomic fluorescence. [After
Hohimer and Hargis (1978). Courtesy of Elsevier Scientific Publishing Company.]

boat is heated by a programmable power supply which automatically cycles through the drying (120 sec at $\geq 100°C$), ashing (90 sec at $\sim 400°C$), and atomizing (7–15 sec at $\sim 1400°C$) stages. The total time required to heat and cool a sample and a blank was ~ 10 min. Apertures at either end of the furnace maintained alignment of the excitation beam at 2 mm above the surface of the boat. The fluorescence was observed at right angles to the beam direction.

It should be noted that although both pulsed (Epstein *et al.*, 1980b) and cw (Pollard *et al.*, 1979) lasers have been combined with an inductively coupled plasma, the technique has shown little or no advantage over other methods of excitation.

c. Signal Processing

Although oscilloscopes have been used as signal processors (Bolshov *et al.*, 1976; Brod and Yeung, 1976) the vast majority of the instruments combined a boxcar integrator with a pulsed laser, and a lock-in amplifier with a cw laser. The basic advantage accured by boxcar integration is that discrimination against the continuous flame background via a low duty cycle. Although such a device could potentially discriminate against scatter interference, the time resolution of the instruments must not have been sufficient to show any net improvement. The basic advantage accrued by lock-in amplification is discrimination against the flame background. This device can also discriminate against scatter interference when it is combined with wavelength modulation or used in a second-harmonic mode.

Of those authors reporting the use of a boxcar integrator, the majority make no mention of ratioing the output against any kind of a reference signal. In contrast to this, Hosch and Piepmeier (1978) used a photodiode-based laser energy detection system to monitor the power irradiating the sample. A PDP-11/20 was used to perform the ratioing and format the results. The most elegant correction method is probably that reported by Hohimer and Hargis (1977, 1978). In their instrument a reference vapor cell containing pure analyte is used in a second channel of the boxcar. This approach allows both wavelength and amplitude stability to be monitored. Although no specific statement is made concerning ratioing, this obviously is a front-panel option on most commercial devices.

Of those authors reporting the use of a lock-in amplifier, the majority use a simple chopper arrangement. Three variations of this approach have been reported. Green *et al.* (1976) ratioed the output of the lock-in against a photodiode reference signal. This compensated for amplitude drift in the dye laser. Brod and Yeung (1976) wavelength-modulated the dye laser to produce a lock-in output which discriminated against the scatter background. And finally, Frueholz and Gelbwachs (1980) used the second harmonic of the chopping frequency to improve the signal-to-background ratio for a sample irradiated under conditions of saturation.

d. Optics

Most of the instruments reported in the literature utilize a liberal number of beam restricting apertures (baffles) or field-of-view restricting spatial filters. The apertures are mainly used to help align the optics and to reduce the amount of dye spontaneous emission reaching the sample. The spatial filters are used to reduce the amount of flame background reaching the detector. In addition to these elements, it seems prudent to add light traps to terminate the laser and as an emission view stop on the back side of the sample. Good examples are given by Hohimer and Hargis (1978) and Epstein *et al.* (1980a).

No consensus exists as to whether the laser beam should simply traverse the atom reservoir, be focused into it, or be expanded and collimated through it. However, a multipass cell comprised of a toroidal and flat mirror pair has been shown by Epstein *et al.* (1980c) to increase the instrument sensitivity by a factor of three over a single-pass arrangement. Nothing was specifically mentioned about its effect on the level of the recorded flame background.

A single lens or lens pair is usually used to transfer the emission to a monochromator or detector. Only Hosch and Piepmeier (1978) reported using a mirror pair, and this was done more for high spatial resolution than a reduction in chromatic aberrations.

The emission of interest has primarily been separated from the background radiation by the use of grating monochromators. Since the requisite bandwidth is provided by the laser, high-throughput devices were preferred over those with good spectral resolution. An excellent example of the relaxed requirements on the emission half of the instrument is given by Weeks *et al.* (1978), who quantified Mn (403.08 nm) in the presence of a thousandfold excess of Ga (403.30 nm) using a monochromator with a 1.6-nm bandpass.

The only report specifically mentioning the use of polarization bias to reduce the scatter background is that of Gelbwachs *et al.* (1977).

3. Example Measurements

a. *Capitalizing on Power per Unit Bandwidth*

In 1977, Gelbwachs *et al.* introduced a technique that they called Saturated Optical NonResonant Emission Spectroscopy (SONRES). The basic concept (shown in Fig. 5) is to find a set of transitions for one element sufficiently close in energy that collisions will thermalize the population of the nearly degenerate excited states. If this separation is simultaneously large enough to allow irradiation at one line and observation at the other line, the scatter blank will be minimized while essentially maintaining sensitive resonance excitation. Such a scheme describes the well-known process of the thermally assisted anti-Stokes fluorescence (Sychra *et al.*, 1975). The clever variation in the technique is the realization that the laser, saturating one transition, by default saturates the second transition and optimizes the number of photons emitted by the sample.

The first experiment utilizing SONRES (Gelbwachs *et al.*, 1977) reported a detection limit of 10^4 sodium atoms/cm^3 in a flame. Unfortunately, this signal was due to a sodium contamination in the air and no value was given for the working lower limit of detection with an aspirated solution. When the atom reservoir was changed to a vapor cell containing a sodium–argon mixture, the detection limit was reduced to a value near 10 atoms/cm^3. In a follow-up paper (Gelbwachs *et al.*, 1978) these same authors discussed the technique in more detail and demonstrated single-atom detection. Finally, they predicted that commercially available lasers could be used to apply SONRES to more than 45 elements.

Another example of saturated excitation is that given by Frueholz and Gelbwachs (1980), who used the process of saturation to change the Fourier components of the detected signal. Experimentally, a cw dye laser was nearly sinusoidally chopped and focused into a flame atomizer. Scattered light will have a frequency spectrum with the primary contribution arising from the chopping rate. On the other hand, saturated fluores-

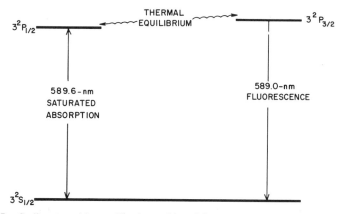

Fig. 5.   Sodium transitions utilized to achieved Saturated Optical NonResonant Emission Spectroscopy (SONRES).

cence will have significant contributions from harmonics due to the non-linear nature of the process. For an incident laser power of 6.8 kW cm^{-2} and second-harmonic detection with a lock-in amplifier, the signal-to-background ratio improved by more than a factor of 40 over detection of the fundamental frequency. Although this specific technique may not revolutionize the field of atomic fluorescence, the concept is a very important one which can be applied to a variety of analytical problems.

b. Capitalizing on Longitudinal Mode Control

Goff and Yeung (1978) reported an instrument based on an electronically tuned cw dye laser. A block diagram is shown in Fig. 6. The commercially available electro-optic device was driven by a 1-kHz square wave which produced lasing at two wavelengths separated by 10 Å. The basic concept is to have a constant flame background, a constant scatter blank, and a modulated fluorescence signal. Thus, a lock-in amplifier can, in principle, eliminate the two major contributions to the blank and enhance the working lower limit of detection. In studies using this instrument, the lower limit of detection was improved by a factor of 30 for the determination of barium using a flame atomizer. Although the work of Fairbank *et al.* (1975) also used wavelength modulation, it was at a 1.5-Hz rate, which is far too low for flame studies. As a matter of fact, Goff and Yeung suggest that the optimum modulation frequency might be in the megahertz range in order to reduce the contribution from flame noise. A sure-fire approach would be the use of an rf spectrum analyzer on the photomultiplier output to locate the best frequency. Caution has to be

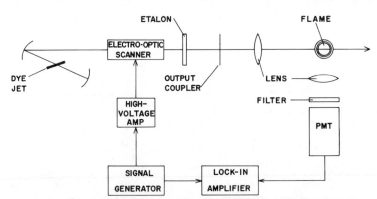

Fig. 6. An atomic fluorescence spectrometer based on a wavelength-modulated dye laser. [After Goff and Yeung (1978). Reprinted with permission from *Anal. Chem.* **50,** 625. Copyright 1968, American Chemical Society.]

used in the megahertz domain since laser mode beating can sometimes reach down to values well under that predicted by the cavity length.

c. Capitalizing on Small Focal Volume

Although laser excitation combined with an inductively coupled plasma has not yielded any significant improvements in practical detection limits, Omenetto et al. (1980) have shown that the ability to tightly focus and position the excitation beam can yield useful plasma diagnostics. In their work, a pulsed dye laser was used to profile the density of barium ions and atoms in a commercially available plasma torch. The key point of the paper is that such profiling can be done without the complication of an Abel inversion.

In a totally different use of the laser's focusing abilities, Measures and Kwong (1979) and Kwong and Measures (1980) have combined two lasers to construct a Trace-element Analyzer Based on Laser Ablations and Selectively Excited Radiation (TABLASER). This technique is used for examining ~ 100-μm spots on solid samples. The ablation is achieved with a 10–20 mJ Q-switched ruby laser, while fluorescence excitation was achieved with a pulsed dye laser. Because the sample is held in a vacuum and because the atomized sample is ablated ~ 6 mm into the cell before it reaches the probe beam, the technique is relatively free of chemical matrix effects.

d. Capitalizing on a Highly Collimated Beam

Although, as already mentioned, the first practical use of laser-excited atomic fluorescence involved remote sensing, the only recent paper based on such an approach is that of Mayo et al. (1976), who developed an instrument suitable for semiconductor processing ovens. In this work a cw dye laser was collimated and sent through an open contaminated quartz tube heated to 1000°C. The wavelength was adjusted to a sodium resonance and the resulting emission was collected with a telescope set about 1 m away from the oven. The minimum detectable level of sodium was estimated to be $\sim 5 \times 10^5$ atoms cm^{-3}.

4. Representative Analytical Results

Table II shows the best lower limit of detection for the results presented in the references listed in Sections II.A.2 and II.A.3. There are three general conclusions. First, laser excitation does not always produce the lowest detection limit. For all of the elements where conventional sources prevail, the excitation was in the ultraviolet and most often filters

TABLE II

Representative Analytical Results

Element	Atomizer[a]	Laser[b]	LLD[c] (ng mL⁻¹)	LDR[d]	Nonlaser Flame[e] (ng mL⁻¹)	Reference
Ag	F	P	4	4.2	*0.08	Weeks et al. (1978)
Al	F	P	*0.6	5.7	30	Weeks et al. (1978)
Ba	F	P	8	5	200	Weeks et al. (1978)
		C	*2	>5		Green et al. (1976)
Bi	F	P	*3	5.2	100	Weeks et al. (1978)
Ca	F	P	*0.08	4.9	0.3	Weeks et al. (1978)
Cd	F	P	8	3.5	*0.04	Weeks et al. (1978)
Co	F	P	1000	>3.3	*1.5	Weeks et al. (1978)
Cr	F	P	1	5.5	*0.3	Weeks et al. (1978)
Cs	G	P	*0.02	3	—	Hohimer and Hargis (1977)
Cu	F	P	1	5	*0.3	Weeks et al. (1978)
Fe	F	P	0.2	~6	0.6	Epstein et al. (1980c)
	G	P	*0.025	~6		Bolshov et al. (1976)
Ga	F	P	*0.9	5.4	7.2	Weeks et al. (1978)
In	F	P	*0.2	6.2	15	Weeks et al. (1978)
Li	F	P	*0.5	>4.3	—	Weeks et al. (1978)
Mg	F	P	0.2	5	*0.05	Weeks et al. (1978)
Mn	F	P	*0.4	5.4	0.5	Weeks et al. (1978)
Mo	F	P	*12	>4.9	200	Weeks et al. (1978)
Na	F	P	<0.1	>5.7	—	Weeks et al. (1978)
		C	2	>3	—	Green et al. (1976)
Ni	F	P	*0.5	~6	1	Epstein et al. (1980a)
Pb	F	P	13	>4.9	10	Weeks et al. (1978)
	G	P	*0.0025	~7		Bolshov et al. (1976)
Sn	F	P	*3	~6	200	Epstein et al. (1980a)
Sr	F	P	*0.3	5	0.8	Weeks et al. (1978)
Ti	F	P	*2	5.2	4000	Weeks et al. (1978)
Tl	F	P	4	4.9		Weeks et al. (1978)
	G	P	*0.0005	6	5	Hohimer and Hargis (1978)
V	F	P	*30	>4.5	250	Weeks et al. (1978)

* Best detection limit for a given element.
[a] F = flame, G = graphite furnace.
[b] P = pulsed laser, C = continuous wave.
[c] LLD = lower limit of detection. Usually at S/N = 2 or 3.
[d] LDR = logarithm of the linear dynamic range.
[e] All data taken from Chap. 5 of Sychra et al. (1975).

were used for isolation of the emission. Thus, the use of a laser could possibly provide greater selectivity. Second, the graphite furnace yields the best detection limits for all elements where flame data were available for comparison. Because of its lack of utility in solving general analytical problems, vapor-cell data have not been included in the table even though it always wins. Finally, a cw laser should be used only when barium is the element of interest.

5. Recommended Configurations

The only specifications made for the laser are pulsed excitation and a bandwidth comparable to an atomic linewidth, i.e., 0.001–0.003 nm at 500 nm. The user should not necessarily purchase a laser system just because it has appeared in a published article. As mentioned before, many of the commercial lasers found in the literature are older engineering versions, and virtually every company has newer generations that are more targeted toward turnkey operation.

For an atomizer, the graphite furnace seems to yield the best results. Therefore, the sample optics of Hohimer and Hargis (1978) is recommended (refer to Fig. 4). If for some reason a flame atomizer must be used, the sample optics system of Epstein et al. (1980a) is recommended. It is shown in Fig. 7. For general work, the emission should be isolated with a high-throughput device such as an ISA DH-10 double monochromator (f/3.5) and the boxcar amplifier ratioed with the system mentioned by Hosch and Piepmeier (1978). For fixed-element analyses of a routine nature, the emission isolation and ratioing system of Hohimer and Hargis

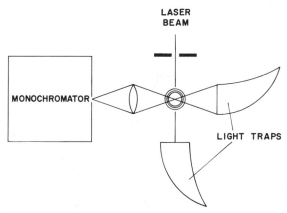

Fig. 7. Example optical system suitable for laser-excited atomic fluorescence in flames. [After Epstein et al. (1980a).]

(1978) should provide excellent results. Two final suggestions: First, adjust the polarization of the laser to minimize scattered light, and second, if a flat retroreflector is deemed necessary, be very careful since narrow-band lasers can set up a standing wave in the sample and negate any anticipated increase in the signal.

If a cw dye laser must be used, the basic system of Green *et al.* (1976) is sufficient. For increased versatility, a krypton ion pump laser should be considered, and for reliable single-mode excitation a ring dye laser would be desirable. Again, a high-throughput monochromator such as the DH-10 should be considered. If amplitude modulation does not produce sufficient background suppression, the electro-optic tuning of Goff and Yeung (1978) could be utilized.

B. Direct Excitation of Molecules in Solution

1. Introduction

The instrumental considerations for molecular fluorescence in solution are similar to those described in Section III.A.1. Also as in atomic fluorescence, the instrumental sensitivity is proportional to the source photon flux density irradiating the sample. However, the broad overlapping absorption bands reduce the utility of the laser's monochromaticity. For similar reasons, plus facile dark relaxation pathways, the wavelength of excitation does not ordinarily have to match any particular transition. Thus, hardware availability should not have played a major role in the development of methodology.

The year 1971 again marked the first use of lasers in analytical molecular fluorescence. Jankow *et al.* (1971) employed a commercial He–Cd device as the source, and a modified Cary model 14 spectrophotometer as the monochromator/detector combination. Example results included microliter sample volumes, polarization measurements, emission spectra, and a quinine bisulfate detection limit of $2 \times 10^{-13} M$. In addition, the authors noted that the laser should not impinge upon the cell wall or a strong silica fluorescence would interfere with the measurement. This same year, Measures and Bristow (1971) used a frequency-doubled ruby laser to remotely obtain the fluorescence spectrum of water, oil film, and rock samples. Although these papers delineated many of the actual and potential applications in analytical chemistry, they did not signal the beginning of a large effort in areas other than remote sensing. The primary reason for this lag was the well-known blank-limited nature of the measurement. That is, even with gas-discharge-lamp excitation, most real analyses are characterized by a sample matrix which generates an optical signal (scattered radiation plus impurity emission) equal to analytes having a concentration quantum yield near 10^{-12} (Matthews and Lytle, 1979). Thus, a more intense source simply increases the magnitude of the interfering radiation at the same rate as the signal. Almost all of the laser-based schemes demonstrated as general analytical tools, have either involved remote sensing or achieved their success by providing additional degrees of selectivity. Examples have been the combination of liquid chromatography with fluorescence detection (Section III.B.3), line narrowing techniques (Section III.C), and the temporal separation of the signal from the blank (Section III.D).

2. Instrument

a. Source

Both pulsed and cw dye lasers have been successfully employed as excitation sources. The choice between these two classes is not as clear-cut as in atomic spectroscopy. First, if the instrument is to be used to obtain excitation spectra, a pulsed system should be considered because of its broad spectral range. Second, if small focal volumes are required a laser with good spatial coherence might be chosen. Finally, for quantitative work either type can be used as long as the output wavelength matches a strong absorption.

Within the pulsed class, the vast bulk of studies have been performed with nitrogen lasers (Bradley and Zare, 1976; Van Geel and Winefordner, 1976), nitrogen-driven dye lasers (Van Geel and Winefordner, 1976; Harrington and Malmstadt, 1975; Perry *et al.*, 1977; Richardson and George, 1978; Ishibashi *et al.*, 1979; Miyaishi *et al.*, 1981), and frequency-doubled, nitrogen-driven dye lasers (Richardson and Ando, 1977; Strojny and de-Silva, 1980). As in atomic fluorescence, only one system has employed a Nd–YAG laser, and that study simply used it in a fixed, frequency-quadrupled mode (Sato *et al.*, 1978). A typical system capable of generating tunable ultraviolet radiation is that described by Richardson and Ando (1977). Doubling was primarily achieved through the use of angle-tuned KDP crystals. However, 254.0 nm was achieved by the use of lithium formate monohydrate. Typical peak powers of the doubled dye output were 2–5 kW in ~8-nsec pulses. Some examples of systems studied with their excitation wavelengths and dyes were: anthracene, 254.0 and 258.7 nm (Coumarin 500); benzene, 259.95 nm (Coumarin 500); pyrene and naphthalene, 273 nm (Coumarin 495); and fluoroanthene, 287.0 nm (rhodamine 6G).

An atypical system capable of recording excitation spectra is that described by Perry *et al.* (1977). Both the nitrogen and dye lasers were homemade and placed under control of a microprocessor. The system was capable of automatic tuning from 360 to 650 nm via adjustment of the diffraction grating angle and a mechanical interchange of sixteen dye cells. The spectrum could be scanned as rapidly as 2 nm sec^{-1}, except for cell changes, which required 5 sec. The instrument was fully automatic with the microprocessor being used to control the laser locally or as a slave for a larger measurement system. This paper certainly presaged modern commercial turnkey dye-laser systems capable of similar performance capabilities.

Within the cw class, non-time-resolved studies have been split between helium–cadmium lasers with principal lines at 325 and 442 nm (Jankow *et al.*, 1971; Diebold and Zare, 1977; Imasaka and Zare, 1979; Diebold *et al.*, 1979a) and argon ion lasers with principal lines from 457 to 515 nm (Richardson and George, 1978; Hirschfeld, 1976; Luciano and Kingston, 1978; Hershberger *et al.*, 1979; Lidofsky *et al.*, 1979) and frequency doubled 515 nm (Crepeau *et al.*, 1976). All systems appear to be operated in the form "as delivered" and merit no further discussion.

b. Sample Configuration

The bulk of all reported studies have employed a standard 1-cm-square fluorimeter cell. Next most common would be the usage of a capillary cell to reduce the sample volume. Most such cells are very simply constructed from quartz or glass tubing, although a 1-mm-square version has been reported (Jankow *et al.*, 1971).

When attention has been devoted specifically to cell design, one of the major goals has been the reduction of both scattered radiation from interfaces and the silica luminescence that reaches the detector. One simple approach is to use a rather large cell. To this end, Ishibashi *et al.* (1979) describe the use of a 4-cm-diam by 6-cm-high cylinder. Naturally, the trade-off is an increased sample size. Three sets of authors have reported using excitation through the bottom or top of the cell. Jankow *et al.* (1971) used vertical excitation for both standard and capillary cells, Strojny and deSilva (1980) used standard cells, while Matthews and Lytle (1979) constructed a cell from a 1-cm-diam quartz tube epoxied to a 1.2-cm^2 plate cut from a Corning 7-54 filter. Although Corning filters are often notorious emitters, no luminescence was detected from the 7-54 material.

The same considerations have been given to microcell design. As an example, that described by Hershberger *et al.* (1979) uses a sheath flow principle to both control the sample volume and remove the window material from the region of observation. Another approach is that described by Sepaniak and Yeung (1980) in which fiber-optic observation was coupled to a normal capillary cell. This arrangement made the signal insensitive to the exact positioning of the laser beam while rejecting scatter and luminescence from the capillary wall. Perhaps the ultimate of such endeavors is the flowing liquid droplet technique developed by Zare and co-workers (Diebold and Zare, 1977; Diebold *et al.*, 1979a,b; Lidofsky *et al.*, 1979) for examining the eluate from liquid chromatographs. Although this would seem to be the method of choice Yeung and Sepaniak (1980) list several problems including the effect on droplet size of bubbles, gradient elution, and thermal lensing.

c. Signal Processing.

The most common signal processing arrangements are a boxcar integrator or sampling oscilloscope combined with a pulsed laser and a picoammeter, or photon counter combined with a cw laser. More unusual combinations have been gated photon counting combined with low-repetition-rate pulsed lasers (Ishibashi *et al.*, 1979; Miyaishi *et al.*, 1981; Yamada *et al.*, 1981), lock-in amplification combined with cw lasers (Diebold and Zare, 1977; Diebold *et al.*, 1979a), an A/D converter system for multiplexing centrifuge samples (Crepeau *et al.*, 1976), and computerized diode array for remote detection (O'Neil *et al.*, 1980).

Of those authors reporting the use of a boxcar integrator, only two specifically mention ratioing the output against any kind of a reference signal. Both Richardson and Ando (1977) and Strojny and deSilva (1980) used a photomultiplier to act as trigger and a reference for a dual-channel boxcar. Again, in both systems the intensity reference is derived from a beam splitter positioned between the doubling crystal and the sample. While this scheme is usually not recommended for normal fluorimetric instruments, the nearly fixed polarization of the laser output yields accurate results. Care would have to be exercised in remembering that the ultraviolet beam is polarized at 90° to the fundamental for situations where intercomparisons were made.

The one instrument employing a lock-in amplifier is unique in that the He–Cd laser was intracavity modulated by acousto-optics (Diebold and Zare, 1977; Diebold *et al.*, 1979a). Electronically, a master 50-kHz oscillator is fed to a differencing amplifier of the modulator drive and to the reference input of the lock-in amplifier. The photomultiplier anode was connected directly to the signal input, and hard copy was obtained via a strip-chart recorder. A block diagram is shown in Fig. 8.

d. Optics

As in atomic fluorescence, most of the instruments utilized baffles or field-of-view restricting spatial filters. The laser is usually terminated in

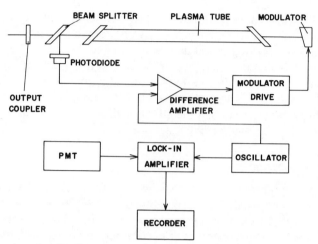

Fig. 8. A molecular fluorescence spectrometer based on an amplitude-modulated He–Cd laser. [After Diebold *et al.* (1979a).]

some sort of a low-scatter beam stop. Instead of a view stop, many setups used spherical mirrors in an attempt to double the fluorescence collection efficiency. Because of the blank-limited nature of the measurement, very few attempts have been made to multiple-pass the laser through the sample.

Both filters and monochromators have been used to isolate the fluorescence from the exciting radiation. Richardson and Ando (1977) specifically mention the selection of low-fluorescing filters, while Imasaka and Zare (1979) have used liquid chemical filters to avoid this difficulty. Since most molecular solution fluorescence is characterized by broad bands, the reported monochromators had bandpasses in the 4–8 nm range. Miyaishi *et al.* (1981) and Perry *et al.* (1981) used double monochromators. This was most likely done to reduce the stray laser scatter reaching the detector.

Very few authors indicated that an attempt was made to monochromate the output of the laser prior to sample irradiation. This is strange since it is a well-known need in Raman spectroscopy for both dye-laser and plasma-tube excitation sources. It should be noted that graphite tubes are notorious in this respect since they are intense blackbody radiators and the ~1 m by ~2 mm bore produces a highly collimated beam.

The use of lenses to gather the fluorescence was ubiquitous whereas their usage with the excitation beam was split about evenly. The only system sufficiently novel to mention was that of Hershberger *et al.* (1979) utilizing a Leitz microscope. One note of caution: since two-photon-induced fluorescence is proportional to the power of the laser squared, a tight focusing lens should not be used with the excitation beam.

3. Example Measurements

a. Capitalizing on Small Focal Volume

Crepeau *et al.* (1976) combined ultraviolet laser excitation with an analytical ultracentrifuge to permit longitudinal fluorescence monitoring in macromolecular sedimentation studies. The source was composed of an argon ion laser whose 515-nm output was frequency doubled to 257 nm. The fundamental and second harmonic were separated by the use of a constant-deviation prism and iris diaphragm. The resultant ultraviolet radiation was focused with a lens to produce a 50-μm beam waist radially propagating across a 1.2-cm-diam centrifuge cell. Scanning the length of the cell was achieved by mounting the laser focusing assembly on a motorized translation stage. The resultant fluorescence was collected by an end-on photomultiplier. Excitation radiation was blocked by a combination of spatial and spectral filters. Since the centrifuge spins at a rather high rate the signal processing circuitry was quite complex, requiring synchronizing circuitry and digital data processing. Although the paper dealt primarily with a detailed description of the instrument, bovine-serum albumin-sedimentation velocity experiments were performed utilizing only 20 μg mL^{-1} of reagent.

b. Capitalizing on a Highly Collimated Beam

Remote sensing via fluorescence spectroscopy has become a mature analytical technique. Much of this work has been done by various individuals at the Canada Centre for Remote Sensing and has been concerned primarily with oil spills (O'Neil *et al.*, 1980; Bristow, 1978). Other representative studies have involved oil film thickness (Visser, 1979) and coal-processing effluents (Capelle and Franks, 1979).

The instrument of Sato *et al.* (1978) will be discussed in detail since it is mostly constructed from commercially available components. The source was comprised of both the second (530 nm) and fourth (265 nm) harmonics of a Nd–YAG laser. The ultraviolet line was used to excite fluorescence, while the green line was used to produce Raman bands. The backscatter and fluorescence were received by a 50-cm-aperture Cassegranian telescope and focused onto the slit of a 0.3-m polychromator. A silicon-intensified and gated OMA was used to detect the dispersed emission. The entire system was computer controlled. Gating the OMA allowed the instrument to be used during the daylight hours without the background radiation overwhelming the desired information. Typical spectra from oils are shown in Fig. 9.

A related instrument described by O'Neil *et al.* (1980) used a nitrogen laser and a custom-built intensified diode array capable of monitoring sixteen wavelength intervals. This more complete instrument provided altitude data as well as the potential for two-channel fluorescence decay measurements. This particular fluorosensor was tested by actual flights over oil and dye spills. Cross-correlation techniques were shown to be capable of identifying oil types.

c. Capitalizing on Chromatographic Specificity

Zare and co-workers have pioneered the combination of chromatography with laser-excited fluorimetry. The first such work utilized thin-layer chromatography to identify and quantitate subnanogram levels of aflatoxins (Berman and Zare, 1975), while later work utilized liquid chromatography to isolate aflatoxins (Diebold and Zare, 1977; Diebold *et al.*, 1979a), the mycotoxin zearalenone (Diebold *et al.*, 1979b), and insulin (Lidofsky *et al.*, 1979). The determination of aflatoxin B_1 in corn (Diebold *et al.*, 1979a) is worth further description because it presents results from an analysis of real samples. The outline of the procedure is as follows. Aflatoxin B_1 is extracted from 50 g of corn using water–methanol and cleaned

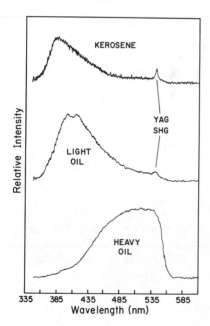

Fig. 9. Fluorescence spectra of 20-μm-thick oils obtained in the daylight and from a distance of 33 m. [After Sato *et al.* (1978).]

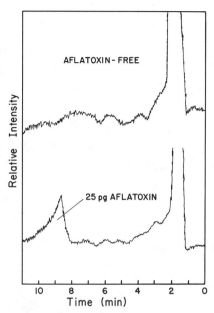

Fig. 10. Chromatograms of aflatoxin-free corn and corn spiked with 25 pg of aflatoxic B_1. [After Diebold *et al.* (1979a).]

up by thin-layer chromatography. The plate position having an R_f corresponding to B_1 is removed and extracted with chloroform. After removal of the solvent by drying, the residue is converted to the more fluorescent B_{2A} derivative by treatment with $1N$ HCl. After the HCl is removed by drying, the residue is dissolved in a water–ethanol solvent and injected onto a reverse-phase HPLC column. At the end of the column, a flowing droplet of eluant is irradiated by a modulated He–Cd laser and the fluorescence detected by a lock-in amplifier. Chromatograms are shown in Fig. 10 for aflatoxin-free corn, and corn spiked with 25 pg of B_1. The lower limit of detection is estimated to be about 10 pg.

4. Representative Analytical Results

Table III shows the best lower limit of detection for the results presented in the references cited in Sections III.B.1–III.B.3. A comparison was not made to data obtained by conventional instrumentation because of the difficulty in searching the literature. In several instances, the paper cited in the table presents such an overview of alternative methodologies for the specific compound.

TABLE III

Representative Analytical Results—Molecular Fluorescence

Compound	Solvent	Cell configuration[a]	Laser[b]	LLD (ng mL^{-1})[c]	LDR[d]	Reference
acridine	dioxane or benzene	M	He–Cd	179	---	Jankow et al. (1971)
aflatoxin B$_1$	water–ethanol	FD	He–Cd	0.1	>3	Diebold et al. (1979a)
amphetamine (fluorescamine deriv.)	fluorescamine reaction mixture	S	N$_2$/dye	2500	~1	Strojny and deSilva (1980)
anthracene	water	S	N$_2$/dye/SHG	<0.004	3	Richardson and Ando (1977)
benzene	water	S	N$_2$/dye/SHG	19	3	Richardson and Ando (1977)
carprofen	acetic acid–ethanol	S	N$_2$/dye/SHG	2.5	>1	Strojny and deSilva (1980)
europium (III)	ethanol	S	N$_2$	0.002	---	Yamada et al. (1981)
fluoranthene	water	S	N$_2$/dye/SHG	0.001	>2	Richardson and Ando (1977)
fluorescein	water	C	N$_2$/dye	2×10^{-5}	---	Ishibashi et al. (1979)
flurazepan (9-acridine deriv.)	acetic acid–ethanol	S	N$_2$/dye	0.5	>1	Strojny and deSilva (1980)
γ-globulin	aqueous	MS	Ar$^+$	1 molecule[e]	---	Hirschfeld (1976)
glucose-6-phosphate	aqueous	S	He–Cd	0.52	>1	Imasaka and Zare (1979)
insulin	phosphate buffer	FD	Ar$^+$	0.4	>1	Lidofsky et al. (1979)

402

α-ketoglutaric acid	aqueous	S	He–Cd	0.584	--	Imasaka and Zare (1979)
NADP	aqueous	S	He–Cd	0.007	>1	Imasaka and Zare (1979)
naphthalene	water	S	N_2/dye/SHG	0.0013	>3	Richardson and Ando (1977)
pyrene	water	S	H_2/dye/SHG	5×10^{-4}	>3	Richardson and Ando (1977)
pyrido[2,1-b]quinazoline	methanol	S	N_2/dye/SHG	0.5	>1	Strojny and deSilva (1980)
quinazoline	0.1 NaOH	S	N_2/dye	<10	~2	Strojny and deSilva (1980)
quinine sulfate	0.1 N H_2SO_4	S	N_2/dye	1.0	>1	Strojny and deSilva (1980)
riboflavin	water	C	N_2dye	6×10^{-4}	--	Ishibashi et al. (1979)
rhodamine B	water	S	Ar^+	5×10^{-4}	~3	Richardson and George (1978)
rhodamine 6G	ethanol	S	N_2	3.9×10^{-4}	>1	Bradley and Zare (1975)
salicylic acid	0.1 NaOH	S	N_2/dye/SHG	1	>1	Strojny and deSilva (1980)
uranyl ion[f]	water/CaF_2	P	Ar^+	1×10^{-5}	--	Perry et al. (1981)
zearalenone	ether–hexane	FD	He–Cd	5 ppb in corn	>4	Diebold et al. (1979b)

[a] M = capillary; FD = flowing drop; S = 1-cm-square cell; C = 4-cm-diam × 6-cm cylinder; MS = microscope slide; P = CaF_2 pellet. He–Cd = helium–cadmium; N_2 = nitrogen; N_2/dye = nitrogen-pumped dye; SHG = second-harmonic generation (frequency doubling); Ar^+ = argon ion.

[b] LLD = lower limit of detection, usually at S/N = 2 or 3.

[c] LDR = logarithm of the linear dynamic range.

[d] This does not represent ordinary methodology. See text Section III.B.4 for a detailed discussion.

[e] Although this analysis is based on irradiation of a pelleted solid, it is included to demonstrate several ancillary points.

The best detection limit shown in Table III (one molecule of γ-globulin) is the result of a very clever, albeit restricted, chemical approach to enhancing sensitivity. In this study Hirschfeld (1976) bound eighty to one hundred molecules of fluorescein isothiocyanate to one molecule of MW 20,000 polyethyleneimine. This highly fluorescent tag was then bound to one molecule of γ-globulin and irradiated with a light flux sufficiently intense to photobleach the system during observation. Such an approach produced the maximum number of fluorescence photons per molecule (about one million) as the sample was mechanically swept through the laser beam.

The best detection limits for normal methodology (10^{-3}–10^{-5} ppb) are for single compounds prepared by dilution in prepurified solvents. In general, these same performances would not be expected with real samples. The only exception to the above observation is that for the uranyl ion (Perry et al., 1981), where actual water samples containing uranium at the 10^{-2} ppb level were run. This determination had both specificity and preconcentration built in. First, the uranium was coprecipitated with CaF_2, and second, the resultant solid was fired for 3 h at 800°C to destroy most of the organic interferences.

The best detection limits which could be expected for most real samples (1–10^{-2} ppb) resulted from the combination with some additional forms of specificity, or chemical amplification. For example, some approaches use enzymes for glucose, NADP, and ketoglutanic acid (Imasaka and Zare, 1979), immuno assay for insulin (Lidofsky et al., 1979), and liquid chromatography for aflatoxins (Diebold et al., 1979a). The combination with liquid chromatography should be particularly fruitful. This field has recently been reviewed by Yeung and Sepaniak (1980).

The compounds resulting in reported detection limits greater than 1 ppb have all been involved in the analysis of messy samples. As examples, carpofen, pyrido-[2,1-b]quinazoline, and demoxepam were doped into drug-free plasma and subsequently recovered by standard techniques including thin-layer chromatography and solvent extraction (Strojny and deSilva, 1980). The most complicated example is that for zearalenone in corn, which involved many separations and one derivatization (Diebold et al., 1979a).

After examining the data shown in Table III, there are two observations concerning practicality that are worth noting. First, unless additional forms of selectivity (chemical or instrumental) are combined with those normally associated with steady-state fluorimetry, it is doubtful that chemically complicated samples will benefit much from laser excitation. Second, Hirschfeld's (1976) demonstration of photobleaching sheds some doubt as to the ability to construct a methodology capable of detecting a single molecule by spontaneous fluorescence. A simple calculation will demonstrate this point. Suppose that the efficiency of emitted photons to instrumental counts is an excellent value of 10^{-3}. This would restrict any photochemical side reaction to $<10^{-3}$ if a signal-to-noise ratio greater than unity is to be obtained by "cycling" the molecule through its excited state one thousand or more times. As in many other analytical trace techniques, small side reactions may ultimately control the lower limit of detection.

5. Recommended Configurations

For purely quantitative applications where the laser wavelength itself does not provide a significant percentage of the selectivity, no compelling reason exists to choose a pulsed laser over a cw device. However, if one of the experimental goals is to obtain excitation spectra or optimize the analysis via excitation wavelength then the choice would lean heavily toward a pulsed dye-laser system. As in atomic fluorescence, the user should not necessarily purchase a laser system just because it has appeared in a published article. New turnkey systems are available that have the ability to scan wavelengths under microprocessor control, thus facilitating the recording of excitation spectra.

For a general instrument design, that of Richardson and Ando (1977), shown in Fig. 11, is a good model. The optical components are modular,

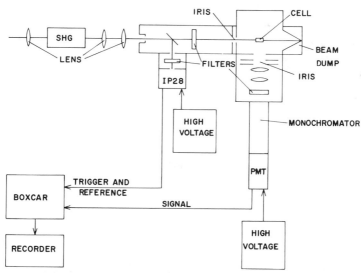

Fig. 11. Example instrumental configuration useful for laser-excited molecular fluorescence in solution. [After Richardson and Ando (1977). Reprinted with permission from *Anal. Chem.* **49**, 955. Copyright 1977, American Chemical Society.]

allowing for experimental flexibility, and care has been taken with the lensing, baffling, etc. The boxcar integrator was used in a reference mode which helps partially normalize for laser fluctuations. The addition of a computer to control the laser and data gathering would be well worth the investment.

Further sophistication that may be necessary depends upon a complete analysis of the blank. Major components of this reading will be rf interference, cell-induced scatter and cell luminescence, solvent or reagent luminescence, or interferences in the sample matrix. Rf interference can be reduced or eliminated by the extensive use of caging. The entire nitrogen laser and power supply can be placed inside a solid metallic box. All interconnections to the outside world should be made either through optical couplers or 60-Hz notch isolation filters. If at all possible, deflect the laser beam at 90° so that it travels through two perpendicular holes in the shielding.

Cell scatter and luminescence can be reduced by the use of a design with windows far removed from the observation volume, or vertical excitation through the cell bottom. If at all possible, solutions should be centrifuged or filtered to remove fine particulate.

Reagent luminescence from many organic solvents can be most easily handled by proper choice of vendor (Matthews and Lytle, 1979). For ethanol the usual suggested approach is a fractional distillation, either neat (Bradley and Zare, 1976) or with 1 g L^{-1} cf added KOH (Diebold *et al.*, 1979a). For water, start with deionization, then carbon filtration, and finally go to distillation in glass (Bradley and Zare, 1976). One unusual approach was for $6M$ NaOH where the solvent was placed in strong sunlight for one day (Imasaka and Zare, 1979). This reduced the background luminescence by a factor of five! For glassware, one school of thought recommends washing with concentrated nitric acid, followed by rinsing with hydrofluoric acid (Diebold *et al.*, 1979a). The alternative proposal is to wash with soap, sodium hydroxide, chromic acid in an ultrasonic cleaner, and then rinse with water (Ishibashi *et al.*, 1979).

When the sample matrix itself proves to be the major source of interference, some form of separation should be attempted, or perhaps another technique altogether.

C. Line Narrowing Techniques

1. Introduction

Molecular fluorescence spectroscopy suffers from highly congested spectra in the gas phase, and broad and often featureless bands in fluid solution. As a result, the potential selectivity of monochromatic laser excitation is difficult to realize in a working analysis. Line narrowing techniques address this problem by removing thermal effects and/or site inhomogeneities. The various approaches covered in this review are spectral simplification by rotational cooling in supersonic jets, creating site uniformity by the use of matrix isolation or Shpol'skii matrices, and the photoselection of a subset of sites in crystal or glass matrices. With each technique it is possible to obtain fluorescence data comprised of fewer spectral features, and, in addition, have these features occur with an instrument-limited linewidth. This combination of attributes allows the direct quantification of very complex multicomponent mixtures without the need to resort to separation methodology.

Since these techniques differ primarily in the manner in which the sample is handled or prepared, very little comment needs to be made concerning the requisite laser equipment. Suffice it to say that monochromaticity is uniquely important. Most instruments employed either intrinsically narrow gaseous laser transitions or dye lasers with intracavity etalons. The ability to continuously tune the excitation wavelength is also of importance, but many analytically useful studies have been done with fixed-line sources. Signal processing will, in most cases, resemble that described in Section III.B.2, while the optics will be slightly complicated by the need to get in and out of high-vacuum chambers or cryostats.

2. Rotational Cooling

In a supersonic molecular beam, cooling is provided by expanding a helium-diluted vapor-phase sample molecule from a high-pressure region through an orifice into a vacuum. This free-jet expansion has been shown to produce translational temperatures near 0.006 K and rotational temperatures near 0.5 K (Levy *et al.*, 1977). The narrow velocity distribution and the fact that the laser excitation is perpendicular to the direction in which the beam is traveling produce spectra containing very little Doppler broadening. In addition, the rotational temperature simplifies the excitation spectrum by keeping the sample in very low quantum states. As an example, NO_2 seeded into a helium beam has been shown to have less than 2% of its population rotationally or vibrationally excited (Levy *et al.*, 1977).

An excellent example of a compound demonstrating the dramatic spectral simplification possible with rotational cooling is free-base phthalocyanine. Levy, Fitch, and co-workers have published a series of articles dealing with its spectral properties. One of these (Fitch *et al.*, 1980) shows the fluorescence excitation spectrum superimposed on the static gas

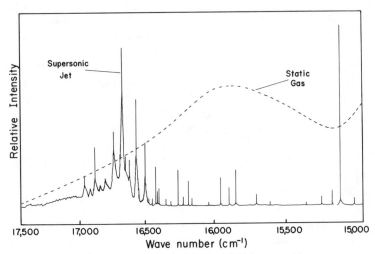

Fig. 12. Fluorescence excitation spectrum of free-base phthalocyanine cooled in a supersonic jet. [After Fitch *et al.* (1980).]

absorption data (see Fig. 12). Under proper expansion conditions the bandwidths can be as narrow as a fraction per centimeter, providing an increase of three to four orders of magnitude in spectral resolution. This resolution and simplification produces a concomitant increase in specificity.

Although very little research has been done with a specific analytical goal, Small and co-workers (Brown *et al.*, 1981) have proposed using the technique as a detector coupled to a gas chromatograph. These authors point out that the most successful variant of the technique would be one where excitation would be to states higher than S_1^*. This would allow a fixed-frequency laser to excite a variety of eluted compounds, as long as fluorescence emission was the observed quantity. To demonstrate that excitation into higher excited vibronic levels of S_1^* would not unduly broaden the spectrum, they examined naphthalene fluorescence excited at 299.0 and 294.8 nm and obtained laser-limited bandwidths of 3 cm^{-1}.

3. Site Selection Spectroscopy in Inorganic Crystals

The lanthanides and actinides have intrinsically narrow and simple f → f* spectra. In fluid solution at room temperature it is no surprise that these bands are broadened to ~300 cm^{-1}, reducing the potential selectivity in a multicomponent analysis. However, even in a solid CaF$_2$ matrix at 13 K where thermal effects are minimized, the multiplicity of crystal sites produces a complicated pattern of narrow lines. In a very beautiful series of papers Wright and co-workers have outlined the sample preparation techniques necessary to produce simple spectra composed of very sharp lines. As an example, Fig. 13 shows the excitation spectrum of Er^{3+} in unignited CaF_2, and in CaF_2 which has been ignited in a furnace at 500°C for 3 h and then cooled very slowly (Gustafson and Wright, 1977). The above line narrowing experiment has been extended to include the determination of Pr^{3+}, Nd^{3+}, Sm^{3+}, Eu^{3+}, Tb^{3+}, Dy^{3+}, Ho^{3+}, Er^{3+}, and Tm^{3+}. Of the remaining lanthanides, La^{3+} and Lu^{3+} were not determined because they have no $f \rightarrow f^*$ transitions; Ce^{3+}, Gd^{3+}, and Yb^{3+} have transitions out of the range of the instrument; and Pm^{3+} is radioactive (Gustafson and Wright, 1979).

Line narrowing can be applied to the determination of a nonfluorescent species as long as its incorporation into the crystal lattice results in a spectroscopically unique site for the probe ion. Thus, associative clustering produces sites which fluoresce at their own wavelength, and displays a signal strength dependent upon the concentration of analyte. This approach has been used for both the determination of an anion, PO_4^{3-} (Wright, 1977), and several cations, La^{3+}, Ce^{3+}, Gd^{3+}, Lu^{3+}, Y^{3+}, Th^{4+} (Johnston and Wright, 1979), Sc^{3+}, and Zr^{4+} (Johnston and Wright, 1981).

4. Site Selection Spectroscopy in Vitreous Solvents

A typical linewidth for an organic compound in a vitreous solvent at 2 K is 300 cm^{-1}. Owing primarily to the work of Personov and co-workers

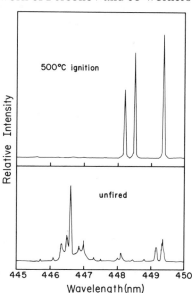

Fig. 13. Fluorescence excitation spectrum of Er^{3+} in unignited CaF_2 and in CaF_2 ignited at 500°C for 3 h. [After Gustafson and Wright (1977). Reprinted with permission from *Anal. Chem.* **49**, 1680. Copyright 1977, American Chemical Society.]

(1972, 1974; Personov and Kharlamov, 1973), it has become established that this width is due almost entirely to inhomogeneities in the solvent cages (sites). The statistical distribution of these sites throughout the sample produces the spectrum as measured by a broadband excitation source. If a laser (or other suitably narrow-band source) is used to excite only certain sites in the ensemble, the resultant fluorescence will be composed of narrow-width lines. These lines in turn are composed of two parts, a very sharp phonon-free line, and a much broader phonon wing appearing on the low-energy side. For an excellent discussion of the photophysics of this effect see McCoglin *et al.* (1978).

Four general observations can be made. First, as the temperature is increased the phonon wing grows in intensity compared to the phonon-free line. At approximately 40 K the narrow-line emission is gone. This is in contrast to the inorganic crystal approach where 77 K still produced analytically useful spectra. Second, as the excitation energy increases above that of the 0–0 transition, the lines broaden. About 1500 cm^{-1} of extra energy are sufficient to eliminate the characteristic narrow features. Originally thought to be thermal in nature, this effect is now considered to be due to the excitation of several vibronic transitions in a molecule, with subsequent relaxation to a multiplicity of sites (Brown *et al.*, 1978; Bykovskaya *et al.*, 1981). Third, because of nonphotochemical hole burning (Hayes and Small, 1978) one might not want to use an extremely narrow laser bandwidth (Bykovskaya *et al.*, 1981). In general, hole burning should not be a problem in quantification since excitation widths up to 0.5 nm have produced usable spectra (Cunningham *et al.*, 1975), and calibration data obtained with excitation by ultraviolet laser lines was unaffected (Brown *et al.*, 1978). Finally, narrow-line spectra are obtained only when there are no large physical changes in the host solvent cage during the excitation and emission steps. As a result, phenyl and methyl groups (Eberly *et al.*, 1974; Flatsher *et al.*, 1976), hydrogen bonding (Eberly *et al.*, 1974), and charged species (McCoglin *et al.*, 1978) all yield broadened spectra. An example of a spectrum for pyrene is shown in Fig. 14.

The analytical potential of optical site selection spectroscopy in vitreous solvents was first discussed by Fünfschilling and Williams (1977), who mentioned that the spectra were highly reproducible and had an intensity proportional to concentration. The first detailed analytical paper was that by Brown *et al.* (1978), who developed a solvent system suitable for aqueous PAH studies and produced calibration curves for binary mixtures. In a later paper, these same authors refined the method and presented data for both a synthetic fourteen-component mixture and a sample of solvent-refined coal (Brown *et al.*, 1980). Most recently, Personov and co-workers have turned to analytical studies in which qualitative

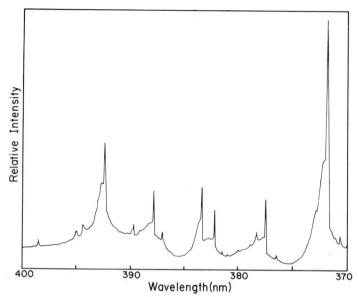

Fig. 14. Line-narrowed fluorescence emission spectrum of pyrene in a glycerol: ethanol glass at 4.2 K. [After Brown *et al.* (1978). Reprinted with permission from *Anal. Chem.* **50,** 1394. Copyright 1978, American Chemical Society.]

identification of gasoline components and quantification via standard addition were reported (Bykovskaya *et al.*, 1981).

An outline of the experiment is as follows (Brown *et al.*, 1980). The sample is dissolved in a $\sim 1:1:1$ glycerol : water : dimethylsulfoxide glass-forming solvent. Polystyrene culture tubes are used to contain the glass. An inexpensive double-nested 3-L-capacity glass liquid-helium Dewar with quartz optical windows is utilized. The sample cooldown procedure involves lowering the samples from a point 8 in. above the liquid helium to just above the helium in ~ 15 min. At this time the glass is formed and the tubes can be lowered rapidly to the bottom of the Dewar. The sample is then excited with a dye laser driven by a nitrogen laser. The emission is collected and dispersed by a 0.3-m monochromator. The detector is a photomultiplier. The output and reference currents are measured with gated integrators. The necessary ratioing and control logic is provided by an LSI-11 microcomputer. The instrument also allows for some degree of time resolution.

An alternative instrument that might be considered is that described by Fünfschilling and Williams (1976). In their device the cw dye-laser excitation source is wavelength modulated to produce a derivative spectrum as the narrow emission is swept back and forth across the monochromator output slit. This approach offers the obvious advantage of eliminating the broad emission from compounds excited from above the 0–0 transition and reducing the contribution due to the phonon wing. Unfortunately, this idea will be difficult to implement for compounds requiring frequency-doubled radiation. The wavelength dependence of the doubling efficiency will probably produce a residual amplitude modulation that will be picked up by the lock-in amplifier.

5. Site Selection in Shpol'skii Solvents

When a polycyclic aromatic hydrocarbon (PAH) is dissolved in an *n*-alkane solvent and then frozen to liquid-helium temperatures, the absorption and fluorescence spectra change from broad bands to sharp (~10 cm^{-1}) richly structured features. This phenomenon is called the Shpol'skii effect, and is due to a reduction (compared to vitreous solvents) in the total number of solvent cage orientations (sites) that are possible. In addition the 1–5 major site orientations have very narrow energy spreads, producing reasonably resolved bands. The technique has been used to analytical advantage for many years, with recent examples employing mercury lamp (Colmsjö and Sternberg, 1979) and x-ray (Woo *et al.*, 1980) excitation.

Although the advantages of narrow-band laser excitation of PAH compounds in Shpol'skii matrices have been known for some time (Vo-Dinh and Wild, 1973; Abram *et al.*, 1974), it is only within the past two years that Fassel and co-workers have demonstrated this combination analytically. Their major thrust has been in the application of the technique to quantify the components in complicated mixtures. Examples include pyrene, 4-methyl pyrene, benzo(*a*)pyrene, and benzo(*k*)fluoranthane in a solvent-refined coal liquid (Yang *et al.*, 1980), and a twelve-component synthetic mixture containing benz(*a*)anthracene and eleven of its alkylated derivatives (Yang *et al.*, 1981a). Figure 15 demonstrates the spectral resolution possible with a ternary mixture composed of benzo(*a*)pyrene, benzo(*k*)fluoranthene, and benzo(*g*,*h*,*i*)pyrene. The calibration curve for benzo(*a*)pyrene in this mixture was linear from 0.1 to 100 mg mL^{-1}.

Although the Shpol'skii technique seems to engender a large number of stated shortcomings [see Yang *et al.* (1981)], two serious objections are historical problems with nonreproducibility of relative site populations and inner filter effects, and intramolecular interaction. The first of these seems to have been minimized by the use of dilute (<10^{-6} *M*) solutions and a controlled cooling rate (Yang *et al.*, 1981a), while the second has been eliminated through the use of deuterated internal standard reference compounds (Yang *et al.*, 1981b). Comparison of the fluorescence data

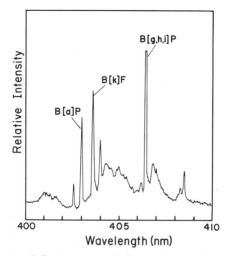

Fig. 15. Line-narrowed fluorescence emission spectrum of a ternary mixture composed of benzo[*k*]fluoranthene, and benzo[*g,h,i*] pyrene in an *n*-octane Shpol'skii solvent. [After Yang *et al.* (1981a). Reprinted with permission from *Anal. Chem.* **53**, 894. Copyright 1981, American Chemical Society.]

obtained in one hour for benzo(*a*)pyrene and perylene in various liquid fuels with results obtained by separation methodology requiring three man days yields excellent agreement (Yang *et al.*, 1981b).

6. Matrix Isolation Spectroscopy

In matrix isolation the sample is vaporized and then mixed with a very large excess of inert diluent gas. The resulting mixture is then deposited at temperatures of 11–15 K on a sapphire substrate. The spectroscopic study is then performed with the sample as a frozen solid. The major advantage of the technique is the elimination of solute–solute interactions and the ability to examine unstable molecules. As a side benefit, because the inert gas (N_2, Ar) is a poor solvent, the site energy distribution is typically 100 cm^{-1}. This represents a line narrowing of about a factor of 3–5 over vitreous solvents.

The first analytical use of laser-excited fluorescence with matrix isolation was by Shirk and Bass (1969). This work did not attract much attention, more than likely because its published scope was restricted to inorganic molecules. In addition, although matrix isolation fluorimetry has been used for the analysis of complicated mixtures (Wehry and Mamantov, 1979), Wehry and co-workers (1981) state that for most work involving solvents such as N_2 or Ar, classical excitation yields perfectly acceptable resolution.

The revival of lasers in matrix isolation was due to three variations of the basic experiment. The first, chronologically, involved pulsed excitation to temporally aid in the separation of components (Dickinson and Wehry, 1979). This type of experiment will be discussed in Section III.D. The second was the use of n-alkane solvents coupled with site-selective narrow-band excitation (Maple et $al.$, 1980). As expected the linewidth of the resulting fluorescence dropped dramatically when shifting to the Shpol'skii solvent. As an example, the principal bands in perylene were 135 cm^{-1} wide in N_2 and 1.3 cm^{-1} wide in n-heptane. Another interesting and useful feature of the matrix isolation approach is the fact that the relative band heights change with the solvent. Thus a qualitative identification feature is built into the technique.

The third variation of the normal matrix isolation experiment is the use of argon or fluorocarbon solvents to permit bandwidth-reducing site selection spectroscopy of polar compounds (Maple and Wehry, 1981). The normal difficulty of doing such an experiment with vitreous solvents is twofold; first, polar compounds tend to form aggregates; and second, polar compounds require strong solvent–solute interactions to form non-aggregated solutions at low temperatures. Obviously matrix isolation eliminates both of these difficulties. Figure 16 shows the matrix isolation, site-selected spectrum of 2,7-dihydroxynaphthalene in perfluoro-n-hexane. The bands are typically 15–35 cm^{-1} wide, permitting the quantification of species in a nine-component mixture of hydroxy-substituted naphthalenes.

7. General Considerations

Nothing much needs to be said about inorganic systems. With the elements for which the crystalline host works, employ the technique. It is both sensitive and selective, albeit a little time consuming. On the other hand, nothing much can be said about rotational cooling because of the dearth of analytical applications. It does, however, offer tremendous potential. For molecular line narrowing in low-temperature solvents three possible approaches exist. A suggested set of differentiating criteria might be the following: (i) For a first attempt start with the vitreous solvents. A large comparative data set exists in the literature and the technique experimentally resembles low-temperature phosphorescence so far as quantification, etc., are concerned. These solvents have been demonstrated to handle a wide range of structural types and the resultant line narrowing has been successfully employed in mixture analyses. (ii) If the sample contains nearly equal-sized polycyclic aromatic hydrocarbons and if vitreous solvents do not provide sufficient spectral resolution, then try to

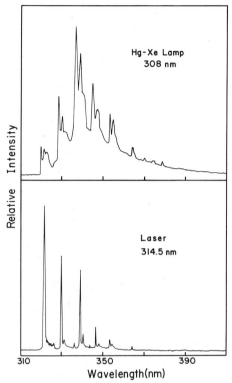

Fig. 16. Line-narrowed, matrix isolation fluorescence emission spectrum of 2,7-dihydroxynaphthalene in perfluoro-*n*-hexane. Lamp excitation results are shown to emphasize the advantage of laser excitation. [After Maple and Wehry (1981). Reprinted with permission from *Anal. Chem.* **53**, 266. Copyright 1981, American Chemical Society.]

use a Shpol'skii solvent. The main aggravation will be learning the experimental nuances necessary to control the site distribution and utilize internal standards. (iii) If polar compounds are to be studied or if a broad range of problems are to be tackled, use matrix isolation. This suggestion is made because the instrument is extremely flexible and can be used to research extensively the analytical problem before more routine methodology is developed. (iv) If phosphorescence spectra are to be utilized (nonfluorescing compounds), then a Shpol'skii solvent should be considered. Dinse and Winscom (1979) have reported such line narrowing where triplet population was achieved by pumping the $S_0 \rightarrow S_1^*$ transition. This is in direct contrast with vitreous solvents, where direct $S_0 \rightarrow T_1$ excitation is required [by Bykovskaya *et al.* (1981)].

Two notes of caution have to be made concerning these last three line-narrowing techniques. First, there is a real sensitivity/selectivity trade.

This is because the homogeneous population is much, much lower than the inhomogeneous one. Thus, as the laser line is made narrower and narrower, the potentially fluorescing population decreases. Second, the selectivity of the techniques has to be evaluated very carefully. One cannot simply count up resolution elements and square the result (excitation/emission intensity matrix). This is because the entire fluorescence spectrum will slide with the excitation frequency over the inhomogeneous $0 \rightarrow 0$ transition linewidth. Also, excitation spectra will begin to wash out about 1500 cm^{-1} above this level. Here both matrix isolation and Shpol'skii solvents have a slight edge over vitreous solvents.

D. Time-Resolved Fluorimetry

1. Introduction

The initial use of laser excitation in time-resolved fluorimetry occurred in the late sixties (Mack, 1968; Merkelo *et al.,* 1969). However, these experiments had little or no analytical perspective. On the other hand, it is somewhat amusing that the early (1972–1974) analytical applications of the technique involved the rejection of fluorescence from Raman spectra (Reed *et al.,* 1972; Yaney, 1972; Hirschfeld, 1973; Van Duyne *et al.,* 1974). The last year of this period saw the first such experiments described where fluorescence was the signal. Sacchi *et al.* (1974) quite accurately described the potential benefits of time resolution, while Lytle and Kelsey (1974) and Measures *et al.* (1974) independently demonstrated these points with experimental results.

As with line-narrowing techniques, the major goal of time-resolved fluorimetry is an increase in specificity. This can be achieved in two ways. First, the temporal behavior of the analyte can be utilized to distinguish its emission from other components of the sample. This is normally achieved by observing the luminescence decay subsequent to impulse excitation. Second, the signal can purposely be caused to appear in one region of time while the major interferences will hopefully appear either in some other region of time or spread evenly over all time. This is normally achieved by recording a fluorescence spectrum within a narrow-temporal-resolution element. Because resolution is involved, a concomitant decrease in sensitivity is generally observed. Thus, the large number of photons available from the laser can serve simply to bring the signal back to the level obtained with steady-state discharge-lamp excitation.

Unlike line narrowing, the samples are almost always held in room-temperature cuvettes and require little discussion. However, the measurement is very hardware oriented with three basic excitation schemes being employed—impulse excitation with the measurement in the time domain, sinusoidal excitation with the measurement in the phase-angle domain, and correlation techniques with the measurement in the phase-delay domain. Within each of these classes there exists a nearly infinite variety of experimental approaches which involve changes in the laser, the photodetector, and the amplifier/data handling unit. Several of the more common variants will be outlined in Section III.D.5.

2. Fluorescence Decays

Several of the simplest applications of time resolution are perhaps the most generally useful. An example would include creating additional confidence about the chemical nature of dilute solutions by comparing both the fluorescence decay and spectrum with corresponding data from more concentrated solutions. A second example concerns the ability to detect the presence of more than one emitter. This is achieved by recording the decay as a function of fluorescence wavelength or detecting nonexponential behavior (Lytle and Kelsey, 1974; Jones, 1975) in the decay. A final, very powerful use of the measured lifetime is to correct fluorescence intensities for samples that are in the presence of an unknown amount of quencher (Hieftje and Haugen, 1981a). In such systems the quantum yield is a function of quencher concentration, thus destroying the ability to construct calibration curves. Since the lifetime and quantum yield are arithmetically linked, the change in value from the pure compound to the quenched solution can be used to normalize the intensities and permit reliable quantitation.

The fluorescence lifetime can be used as a characteristic parameter in differentiating among several closely related compounds. Rayner and Szabo (1978) used both decays and spectra to classify oil spills into the three broad categories of light refined oil (>10 nsec, $\lambda_{max} = 400$ nm), crude oil (<10 nsec, $\lambda_{max} = 475–500$ nm), and heavy oil (<1 nsec, $\lambda_{max} = 530$ nm). Richardson and co-workers (1979) used lifetimes to distinguish among amino acids. They noted that the dipeptide Gly-Gly has a value which is intermediate between those for glycine (3.8 nsec) and tripeptide Gly-Gly-Gly (6.2 nsec). In addition, it was found that a noticeable change in lifetime occurred between free fluorescein (4.8 nsec), fluorescein-labeled antibody (3.0 nsec), and the subsequent antibody–antigen complex (3.9 nsec).

In chromatography, time resolution can be used to restrict the signal being measured to certain lifetime domains (Richardson *et al.*, 1980).

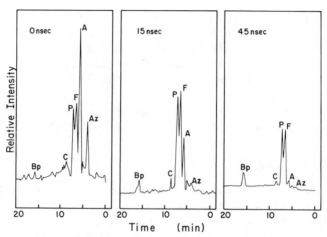

Fig. 17. Liquid chromatograms of six PAH standards illustrating the use of time resolution: A (anthracene), AZ (azulene), Bp (1,12-benzoperylene), C (chrysene), F (fluoranthene), and P (pyrene). [After Richardson *et al.* (1980), courtesy of Elsevier Scientific Publishing Company.]

Figure 17 shows the HPLC chromatograms resulting from the elution of six polycyclic aromatic compounds. Increasing the temporal delay to 15 nsec improves the signal-to-noise ratio by decreasing the scatter background, while a further increase to 45 nsec results in the emphasis of long-lived emitters. An alternative approach would involve measuring both the wavelength range of emission and the lifetime, as the components are eluted from the column. These two parameters plus the retention time should permit the rapid identification of closely related compounds.

A final use of the fluorescence lifetime is in a simultaneous kinetic analysis based on a multiple exponential decay. This is particularly useful for families of compounds where the spectral differences are minimal. First proposed by Lytle and co-workers (1973) for ligand luminescence chelates, the technique was experimentally developed and popularized by Morishige and co-workers (Hiraki *et al.*, 1978; Onoue *et al.*, 1979). The general concept behind the approach is the fact that although the metal ion only slightly perturbs the spectral distribution of the fluorescence, the heavy-atom effect enhances intersystem crossing, systematically varying the lifetime. The application of such methodology to organic systems tends to be more haphazard because of a lack of a general theory to correlate lifetimes to structure. Even so, Cline-Love and Upton (1980) have achieved good two-component analyses of atabrine and its derivatives.

3. Time-Resolved Raman Spectra

The lower limit of detection in Raman spectroscopy is often determined by residual solution fluorescence. This unwanted signal is, in general, difficult to eliminate because of the huge difference in the cross sections for the scattering and emission processes. One approach to minimizing this artifact is based on the difference in the temporal behavior between scattered and emitted photons. That is, if the sample were to be irradiated by a short-duration light pulse, the Raman signal would follow the intensity–time profile of the excitation whereas the fluorescence would be extended to longer times.

As mentioned earlier, the first analytically oriented publications based on time-resolved fluorimetry were aimed toward eliminating this emission. Reed *et al.* (1972) demonstrated a system employing a cavity-dumped argon ion laser and unspecified gated detection electronics. Their system suffered from a 12–15 nsec excitation pulse but did provide a sixfold increase in the Raman to fluorescence ratio for a sample with a 7.5-nsec fluorescence lifetime. Yaney (1972) described a system based on a 100-nsec Q-switched Nd–YAG laser and fast photon counting for the duration of the pulse. This system operated at 450 Hz and worked best with long lifetimes. A 63-fold increase in the Raman to fluorescence ratio was reported for a sample with a 125-nsec fluorescence lifetime. The published abstract of a paper by Hirschfeld at the 1973 annual meeting of the Optical Society of America (Hirschfeld, 1973) indicated an improvement in the ratio, but no details were given. And finally, Van Duyne *et al.* (1974) developed a system based on a mode-locked argon ion laser as the source and a time-to-amplitude converter connected to a level discriminator for gated detection of the signal. Their system was limited primarily to short lifetimes because of the high and fixed repetition rate of the source. The improvement in the Raman to fluorescence ratio, while not stated, appears to have been between twofold and threefold for a sample with a 4.5-nsec lifetime.

The next several years saw improvements in each of these instrumental approaches. Yaney (1976) published an extensive update on his instrumental system. Examples of spectra for nylon, acrylic, benzene with acridine orange, doped SrF_2 crystals, and various forms of aluminum oxide and hydroxide were given. This paper also demonstrated a decrease of about 17% in the background from a short-lived fluorophore. Burgess and Shephard (1977) reduced the timing jitter associated with the time-to-amplitude converter and used a mode-locked argon ion laser to obtain sixfold signal-to-background improvements for both acridine orange and rhodamine 6G (3.9 nsec). And Mulac *et al.* (1978) employed a cavity-dumped argon ion laser to obtain N_2 Q-branch Raman data in a luminous flame background. The improvement in the Raman to background ratio due to time resolution was by a factor of 30.

Two alternative methodologies were published in 1976. Nemanich *et al.* (1976) used an air turbine dentist's drill with an 80-tooth, 12.7-mm-diam clock gear to chop a focused laser beam from 100 Hz to 400 kHz. Using phase-sensitive lock-in detection, this approach was capable of almost complete suppression of luminescence with lifetimes greater than 1 μsec. The use of acousto-optics to modulate the laser intensity (Lytle *et al.*, 1979) and the new breed of commercially available 50-MHz lock-in amplifiers should make this approach usable with nanosecond emitters. However, the inability to photon count may restrict the range of applicable systems. Harris *et al.* (1976a) used a mode-locked, cavity-dumped argon ion laser in combination with sampled photon counting to achieve very high rejection efficiencies. Raman spectra were obtained for a test solution of benzene doped with the fluorophores acridine orange (4.5 nsec) and rubrene (15.6 nsec). The magnitude of the background rejection was determined by comparing the level of fluorescence in a time-resolved experiment to that for continuous excitation using the 992-cm^{-1} scattering peak of the solvent as an internal standard. The resulting Raman to fluorescence enhancements were 34 and 115 for the fast and slow emitters, respectively. Figure 18 shows the dramatic background reduction achieved for rubrene in benzene. For this solution the fluorescence was visually very bright. The major drawback to the approach is the slow cycle time of the sampling oscilloscope, ~40 kHz. Thus the average power irradiating the sample was ~1 mW, drastically reducing the sensitivity of the measurement. As a practical consequence, samples were limited to very high concentrations or neat liquids. In a recent variation of this experiment, Gustafson and Lytle (1982) have combined a frequency-doubled synchronously pumped cw dye laser with megahertz-gated photon counting to permit the gathering of Raman data for millimolar nucleotide solutions. The basic trade-off was a poorer time resolution, resulting in a Raman-to-background improvement of only 35 for a test solution of rubrene in benzene.

4. Time-Resolved Fluorescence Spectra

The lower limit of detection in fluorescence spectroscopy is often determined by Rayleigh and Raman scatter from the solvent in which the sample is dissolved. This unwanted signal is quite often difficult to eliminate spectrally without resorting to double monochromators on both the excitation and emission branches of the fluorimeter. The first work in the

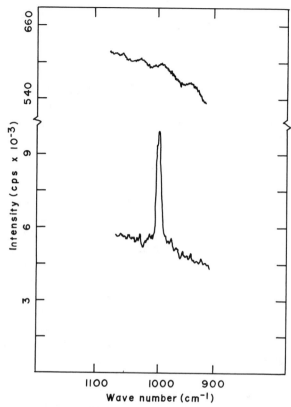

Fig. 18. Temporal rejection of fluorescence from a Raman spectrum. All data are for a 1×10^{-4} M solution of rubrene in benzene excited by 514.5-nm radiation. The top curve is the result of a cw experiment while the bottom curve is for the time-resolved case. [From Harris *et al.* (1976a). Reprinted with permission from *Anal. Chem.* **48,** 1937. Copyright 1976, American Chemical Society.]

area (Lytle and Kelsey, 1974) was concerned primarily with this problem. Unlike the time-domain experiments described in Section III.D.3, scatter can be totally separated from fluorescence once the excitation pulse has terminated. The only compound which these authors studied that appears in Table III is rhodamine B. The lower limit of detection for obtaining a 4-nm-bandpass time-resolved spectrum of this species with a signal-to-noise ratio of 2 was 10^{-2} ng mL^{-1}. This is about a factor of 20 higher than the value in the table and can be accounted for by the use of a monochromator versus a filter.

The reduction of solvent-impurity emission can be achieved as long as the sample has a lifetime ≥ 5 nsec. Matthews and Lytle (1979) have shown that in most cases the best commercially available solvents have a spectrum consisting of spectrally sharp Raman bands and temporally fast (<2 nsec) fluorescence. Of all of the solvents surveyed, carbon tetrachloride had the lowest emission, while spectral-grade benzene had the largest background. Although the bulk of solution luminescence is in the nanosecond domain requiring fast electronics, most systems dealing with analyte decays >100 nsec employ some form of temporal rejection from scatter or matrix impurities. As an example, the inorganic crystal site selection experiments described in Section III.C.3 held off the amplifier until 30 μsec after the sample was irradiated (Miller *et al.*, 1977).

Time resolution has been combined with both matrix isolation and line narrowing in vitreous solvents. Dickinson and Wehry (1979) used a mode-locked, cavity-dumped argon ion laser in combination with a sampling oscilloscope to provide additional selectivity in the matrix isolation of complicated polycyclic aromatic hydrocarbon samples. As an example, Fig. 19 compares the steady-state and time-resolved fluorescence for a mixture of benzo[*a*]pyrene (78 nsec) and benzo[*k*]fluoranthene (13 nsec). Brown *et al.* (1981) used a pulsed dye laser and gated integration to separate the site-selected fluorescence of anthracene (~10 nsec) and pyrene (~200 nsec). Such combination techniques may yield instruments with extremely high selectivities, albeit optically and electronically complicated. In an extension of this type of measurement, Knorr and Harris (1981) have demonstrated how gathering the fluorescence spectra at different, discrete points in time and processing the data by a matrix least-squares routine could aid in the quantitation of components in a mixture.

For binary mixtures, sinusoidal excitation followed by phase-sensitive detection can completely separate the emission of the two components. Lakowicz and Cherek (1981a,b) have demonstrated this ability by separating the signals from indole (9.0 nsec) and dimethylindole (2.5 nsec). Mitchell and Spencer (1981) have also examined this possibility, and separated anthracene (4.5 nsec) from perylene (5.1 nsec) using a 30-MHz modulation. There is no intrinsic reason why scatter cannot be eliminated from fluorescence, particularly since the reverse experiment has already been demonstrated (Nemanich *et al.*, 1976).

Using a completely different approach from those described above, Haugen and Lytle (1981) have developed the technique of time filtering to obtain the concentration of dilute fluorophore solutions. The general idea is to construct a photon acceptance window having detector transit-time-spread-limited edges. With such an instrument they were capable of dramatically reducing the scatter contribution to the blank and consequently achieved a lower limit of detection for rubrene of 1.8×10^{-13} M (1×10^{-4} ng mL^{-1}). One of the fascinating conclusions made in this study was the

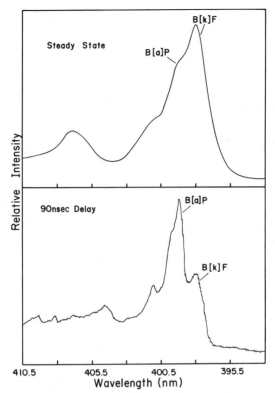

Fig. 19. Steady-state and time-resolved matrix isolation fluorescence spectra of a two-component mixture: B[*a*]P (benzo[*a*]pyrene) and B[*k*]F (benzo[*k*]fluoroanthene). [After Dickinson and Wehry (1979). Reprinted with permission from *Anal. Chem.* **51,** 778. Copyright 1979, American Chemical Society.]

fact that sub–part per billion detection limits could be obtained without any form of spectral selectivity at all.

5. Instrumental Considerations

The temporal behavior of fluorescence can be measured in either the time domain or the frequency domain. The final results should be independent of the approach, differing only in how the data are obtained, the relative ease of performing certain kinds of experiments, and the type of determinate errors to which the methodology is susceptible. In the frequency domain two basic choices exist. The sample can be excited with sinusoidally modulated radiation and the phase lag of the resultant sinusoidal fluorescence determined. Alternatively, periodic arbitrarily shaped excitation can be employed and the autocorrelation or cross-correlation computed over one period of delay.

In the time domain, either the photodetector current or the photon flux can be measured as a function of time. This obviously requires very fast electronics to observe nanosecond signals. Once the province of only those well versed in the associated instrumental art, present commercially available technology permits an individual of average experimental prowess to utilize any one of the many possible approaches. The measurement of anode current with a boxcar integrator, gated integrator, or sampling oscilloscope (all functionally identical devices) represents the vast bulk of the papers referenced in this review. It is a simple, effective method of measuring either decays or time-resolved spectra. The major drawback to measuring current is the very poor time response of virtually all photomultipliers, and the fact that many tubes are badly wired. As a result the output often exhibits a large amount of ringing (Harris *et al.*, 1976b). The location of photons in time accrues the advantage that the measurement is now limited by the transit-time spread of the detector and not the current response. The two parameters can differ by as much as a factor of five–ten, e.g., the RCA 8850 has values of 5 nsec and ~800 psec at the FWHM. The effect that this has on the measured fluorescence is shown in Fig. 20. It is an incredible misconception that in time-resolved spectroscopy, photon counting is performed for sensitivity. This can be seen to be nonsense when it is realized that the RCA 8850 has an average single-photon pulse height of several hundred millivolts. Many more than ten or twenty photons can overload the input of a typical sampling oscilloscope.

Three basic schemes for locating photons in time are time-to-amplitude conversion (TAC), sampled photon counting, and gated photon counting. The TAC approach must account for over 90% of the published instrumentation designs. Its primary advantage is the ability to locate the photon independent of where within the acceptance window it appears. The primary disadvantage is the inability to handle more than one photon at a time. Representative publications are those by Spears *et al.* (1978) and Koester and Dowben (1978). Both of these designs utilized picosecond, mode-locked, cavity-dumped dye lasers resulting in measurements which were detector transit-time limited. A secondary advantage of this class of laser is the ability to operate the TAC at its maximum data rate near 130–140 kHz (Haugen *et al.*, 1979). The use of the inverted TAC configuration (Van Duyne *et al.*, 1974; Burgess and Shephard, 1977) to increase the data throughput should be attempted only with extreme caution since the resultant signal is nonlinear in intensity (Haugen *et al.*, 1979). Although at least one instrument has used a N_2-driven dye laser in combination with a TAC (Meltzer and Wood, 1977), it must take aeons to gather a usable number of counts.

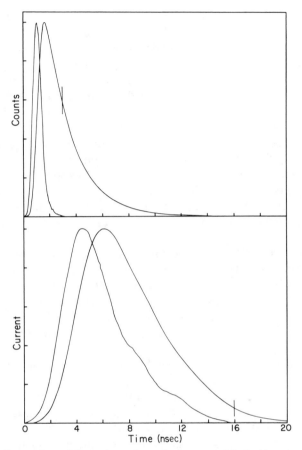

Fig. 20. Comparison of the analog and transit-time-spread-limited impulse responses of an RCA 8850 photomultiplier. In each case a 2-nsec fluorescence decay is shown to demonstrate how this change increases the quality of the data. The vertical line represents the point in time from which a log plot would yield a straight line.

In sampled photon counting the anode current is examined and its value compared to a discriminator reference in order to make a photon/nonphoton decision. Although this does not produce transit-time-spread-limited data, it does reduce the FWHM a bit and makes the impulse response much more symmetric. In addition, by using multilevel discrimination and photomultipliers with well-defined pulse height spectra, this mode of operation permits data rates of ∼150% the excitation rate without pulse pileup (Gustafson *et al.*, 1978). Finally, with the use of two sampling channels it is possible to obtain transit-time-spread-limited results (Harris and Lytle, 1977). An example of an instrument combining a frequency-doubled, synchronously pumped dye laser with this detection scheme has been described by Harris *et al.* (1977). The primary advantage of the sampling approach is its ability to behave as a boxcar integrator, a single-channel photon counter, or a dual-channel device. The primary disadvantage is the low trigger cycle rate of ~40 kHz, and the ability to locate only a photon centered in the acceptance window.

Gated photon counting was developed for two reasons, both involving data throughput. The first of these has to do with compensating for the low repetition rate of most pulsed lasers by allowing the data to be collected over the entire span of time during which the sample is emitting. To this end, the technique resembles a time-filtered experiment much more than a time-resolved one. Chief proponents of this approach are Ishibashi and co-workers (Imasaka *et al.*, 1979; Miyaishi *et al.*, 1981), who used custom-built electronics to perform the gating. They achieved a ~4-nsec FWHM impulse response. A second reason for using gated photon counting is the ability to trigger the circuit rapidly, capitalizing on the high repetition rate of mode-locked, cavity-dumped (pulse-picked) cw lasers. Gustafson and Lytle (1980) utilized a system composed totally of commercially available components to allow excitation rates as high as 2.7 MHz. The impulse response was ~4.8 nsec. The chief advantage of this approach is the high data rate which is possible. The chief disadvantage is a temporal resolution limited by the current response of the photomultiplier.

The techniques utilizing sinusoidal excitation vary primarily with the manner in which the fluorescence phase lag is measured. Govindjee *et al.* (1972) utilized a cw mode-locked argon ion laser, rf filters, and a vector voltmeter. Haar and Hauser (1978) used electronic mixing to locate the phase null, while Menzel and Popovic (1978) used photodetector mixing to achieve the same goal. In perhaps the most quoted configuration, Spencer and Weber (1969) heterodyned the fluorescence modulation down to ~40 Hz and determined the phase lag by a zero crossing measurement. In a very interesting variant Schlag and co-workers (1974) combined photon counting with digital correlation hardware to determine the phase lag. The primary advantage of sinusoidal excitation is the ability to determine precisely subnanosecond lifetimes, while a secondary advantage is the ability to null out completely the response of one component using phase-sensitive techniques. The primary historical disadvantage has been the restriction to one or two modulation frequencies. Fortunately, the combination of lasers with electro-optics (Haar and Hauser, 1978; Heidt, 1976) or acousto-optics (Lytle *et al.*, 1979) has produced sources capable of arbitrarily generating any frequency from 0.01 Hz to 1.2 GHz, and the availability of commercial 50-MHz lock-in amplifiers has simplified the necessary electronics over a very usable portion of this range.

The techniques involving correlation can be classified into electronic mixing, optical mixing, and excited-state population mixing. Both Hieftje and co-workers (1977; Ramsey *et al.*, 1979; Hieftje and Haugen, 1981b) and Harris and co-workers (Dorsey *et al.*, 1979) have employed electronic mixing techniques. Two of these experiments combined the pseudo-random-mode noise of a cw laser with a spectrum analyzer (Hieftje *et al.*, 1977) and a microwave mixer (Dorsey *et al.*, 1979). This last instrument utilized a 12-mW He–Cd laser and a Beck (1976) wired RCA 931A to achieve a scatter autocorrelation response of 0.88 nsec. In the third variation, a mode-locked argon ion laser was used to both excite the sample and provide, via a photodiode, a subnanosecond gating pulse to a microwave mixer (Ramsey *et al.*, 1979). The scatter cross-correlation response was 1.2 nsec.

An example of optical mixing is given by Frigo *et al.* (1977). In their instrument, spontaneous fluorescence is combined with the picosecond output of a synchronously pumped cw dye laser and subsequently mixed in an ADP nonlinear summing crystal. The resultant ultraviolet radiation is isolated with a monochromator and quantified with a photomultiplier. This instrument is laser-pulse-width limited and time resolution is achieved via an optical delay line. The experiment is conceptually similar to the electronic version described by Ramsey *et al.* (1979). An example of excited-state population mixing is given by Barnes and Lytle (1979). In their instrument, one laser is used to pump the sample into its excited state and the second is used to probe the extent to which the population is redistributed. Thus, the sample itself transfers the amplitude modulation from the pump beam to the probe beam.

The primary advantage of several of the variants of correlation techniques is their superior time resolution. Indeed, the last two are Heisenberg limited! The primary disadvantage is the lack of easy control over the free temporal range; i.e., long-lived emitters are tough to monitor. Other variants have specific problems. Examples are detector overloading and measurements that are source shot-noise limited instead of fluorescence shot-noise limited.

6. Recommended Configurations

We feel strongly biased toward the combination of acousto-optics with cw lasers. A system composed of one ion laser, one dye laser, a standing-wave mode locker, and a traveling-wave Bragg cell can produce a tunable wavelength output with pulse widths ranging from cw to picoseconds and repetition rates from single shot to ~100 MHz. Table IV lists combinations that have been successfully assembled and their associated output

TABLE IV

Common Ion Laser/Dye Laser Characteristics

| Ion laser | | | Dye laser | | | Typical characteristics | | |
ML[a]	CD	PP	ML	CD	PP	Pulse width	Repetition rate	Energy per pulse
(cw pump)			---	yes	---	10 nsec–cw[b]	10 MHz–SS	50 nJ
---	yes	---				10 nsec–cw[b]	10 MHz–SS	150 nJ
yes	---	---	(not used)			250 psec	100 MHz	10 nJ
yes	yes					1 nsec	10 MHz–SS	25 nJ
yes	---	yes				250 psec	10 MHz–SS	0.3 nJ
			yes[c]	---	---	10 psec	100 MHz	3 nJ
yes	---	---	yes[c]	yes	---	25 psec	10 MHz–SS	10 nJ
			yes[c]	---	yes	10 psec	10 MHz–SS	0.1 nJ

[a] Abbreviations: CD = cavity dumped; cw = continuous wave; ML = mode locked; PP = pulse picked; SS = single shot.

[b] Used to obtain high-frequency modulation by heterodyning techniques.

[c] Mode locked by synchronous pumping.

characteristics. The resultant instrument is expensive but flexible. The major trade-off is a smaller range of wavelength tuning compared to pulsed dye lasers.

We also feel strongly biased toward the use of a sampling oscilloscope as the amplifier–signal averager. This device can be used as a boxcar integrator, a single-channel photon counter, or a dual-channel transit-time-spread-limited photon counter. It can also be used to make the initial timing adjustments on all of the electronics. As mentioned before, its primary drawback is the 40-kHz trigger cycle rate.

One alternative technology should be monitored very closely. It may soon be possible to combine a microprocessor-controlled pulsed dye laser with a subnanosecond digital transient recorder capable of detecting photons. This instrument would then circumvent the primary disadvantage of such lasers, their extremely low repetition rate. In addition, it may not require an optics afficionado to keep the equipment operable.

REFERENCES

Abram, I., Auerbach, R. A., Birge, R. R., Kobler, B. E., and Stevenson, J. M. (1974). *J. Chem. Phys.* **61**, 3857–3858.

Barnes, W. T., and Lytle, F. E. (1979). *Appl. Phys. Lett.* **34**, 509–511.

Beck, G. (1976). *Rev. Sci. Instrum.* **47**, 537–541.

Berman, M. R., and Zare, R. N. (1975). *Anal. Chem.* **47**, 1200–1201.

Bolshov, M. A., Zybin, A. V., Zybina, L. A., Koloshnikov, V. G., and Majorov, I. A. (1976). *Spectrochim. Acta, Part B* **31B**, 493–500.

Bradley, A. B., and Zare, R. N. (1976). *J. Am. Chem. Soc.* **98**, 620–621.

Bristow, M. P. F. (1978). *Remote Sens. Environ.* **7**, 105–127.

Brod, H. L., and Yeung, E. S. (1976). *Anal. Chem.* **48**, 344–348.

Brown, J. C., Edelson, M. C., and Small, G. J. (1978). *Anal. Chem.* **50**, 1394–1397.

Brown, J. C., Duncanson, J. A., Jr., and Small, G. J. (1980). *Anal. Chem.* **52**, 1711–1715.

Brown, J. C., Hayes, J. M., Warren, J. A., and Small, G. J. (1981). *In* "Lasers in Chemical Analysis" (G. M. Hieftje, J. C. Travis, and F. E. Lytle, eds.), Chapter 12. Humana Press, Clifton, New Jersey.

Burgess, S., and Shephard, I. W. (1977). *J. Phys. E* **10**, 617–620.

Bykovskaya, L. A., Personov, R. I., and Romanovskii, Yu. V. (1981). *Anal. Chim. Acta* **125**, 1–11.

Capelle, G. A., and Franks, L. A. (1979). *Appl. Opt.* **18**, 3579–3586.

Cardona, M. (1969). "Modulation Spectroscopy." Academic Press, New York.

Cline-Love, L. J., and Upton, L. M. (1980). *Anal. Chem.* **52**, 496–499.

Colmsjö, A., and Stenberg, U. (1979). *Anal. Chem.* **51**, 145–150.

Cremers, D. A., and Keller, R. A. (1981). *Tech. Dig. Conf. Lasers Electroopt. 1981*. Optical Society of America, Abstract WJ6.

Crepeau, R. H., Conrad, R. H., and Edelstein, S. J. (1976). *Biophys. Chem.* **5**, 27–39.

Cunningham, K., Morris, J. M., Fünfschilling, J., and Williams, D. F. (1975). *Chem. Phys. Lett.* **32**, 581–585.

Denton, M. B., and Malmstadt, H. V. (1971). *Appl. Phys. Lett.* **18**, 485–487.

Dickinson, R. B., and Wehry, E. L. (1979). *Anal. Chem.* **51**, 778–780.

Diebold, G. J., and Zare, R. N. (1977). *Science* **196**, 1439–1441.

Diebold, G. J., Karny, N., Zare, R. N., and Seitz, L. M. (1979a). *J. Assoc. Off. Anal. Chem.* **62**, 564–569.

Diebold, G. J., Karny, N., and Zare, R. N. (1979b). *Anal. Chem.* **51**, 67–69.

Dinse, K. P., and Winscom, C. J. (1979). *J. Lumin.* **18/19**, 500–504.

Dorsey, C. C., Pelletier, M. J., and Harris, J. M. (1979). *Rev. Sci. Instrum.* **50**, 333–336.

Dovichi, N. J., and Harris, J. M. (1980). *Anal. Chem.* **52**, 2338–2342.

Dovichi, N. J., and Harris, J. M. (1981). *Anal. Chem.* **53**, 106–109.

Eberly, J. H., McCoglin, W. C., Kawaoka, K., and Marchetti, A. P. (1974). *Nature (London)* **251**, 215–217.

Epstein, M. S., Bradshaw, J., Bayer, S., Bawer, J., Boightman, E., and Winefordner, J. D. (1980a). *Appl. Spectrosc.* **34**, 372–376.

Epstein, M. S., Nikdel, S., Bradshaw, J. D., Kosinski, M. A., Bower, J. N., and Winefordner, J. D. (1980b). *Anal. Chim. Acta* **113**, 221–226.

Epstein, M. S., Bayer, S., Bradshaw, J., Voigtman, E., and Winefordner, J. D. (1980c). *Spectrochim. Acta, Part B* **35B**, 233–237.

Fairbank, W. M., Jr., Hänsch, T. W., and Schawlow, A. L. (1975). *J. Opt. Soc. Am.* **65**, 199–204.

Fitch, P. S. H., Haynam, C. A., and Levy, D. H. (1980). *J. Chem. Phys.* **73**, 1064–1072.

Flatscher, G., Fritz, K., and Freidrich, J. (1976). *Z. Naturforsch., A* **31A**, 1220–1227.

Fraser, L. M., and Winefordner, J. D. (1971). *Anal. Chem.* **43**, 1693–1696.

Frigo, N. J., Daly, T., and Mahr, H. (1977). *IEEE J. Quantum Electron.* **QE-13**, 101–109.

Frueholz, R. P., and Gelbwachs, J. A. (1980). *Appl. Opt.* **19**, 2735–2741.

Fünfschilling, J., and Williams, D. F. (1976). *Appl. Spectrosc.* **30**, 443–446.

Fünfschilling, J., and Williams, D. F. (1977). *Photochem. Photobiol.* **26**, 109–113.

Gelbwachs, J. A., Klein, C. F., and Wessel, J. E. (1977). *Appl. Phys. Lett.* **30**, 489–491.

Gelbwachs, J. A., Klein, C. F., and Wessel, J. E. (1978). *IEEE J. Quantum Electron.* **QE-14**, 121–25.

Gibson, A. J., and Sandford, M. C. W. (1972). *Nature (London)* **239**, 509–511.

Goff, D. A., and Yeung, E. S. (1978). *Anal. Chem.* **50**, 625–627.

Govindjee, Hammond, J. H., and Merkelo, H. (1972). *Biophys. J.* **12**, 809–814.

Green, R. B., Travis, J. C., and Keller, R. A. (1976). *Anal. Chem.* **48**, 1954–1957.

Gustafson, F. J., and Wright, J. C. (1977). *Anal. Chem.* **49**, 1680–1689.

Gustafson, F. J., and Wright, J. C. (1979). *Anal. Chem.* **51**, 1762–1774.

Gustafson, T. L., and Lytle, F. E. (1980). *Appl. Spectrosc.* **34**, 185–189.

Gustafson, T. L., and Lytle, F. E. (1982). *Anal. Chem.* **54**, 634–637.

Gustafson, T. L., Lytle, F. E., and Tobias, R. S. (1978). *Rev. Sci. Instrum.* **49**, 1549–1550.

Haar, H.-P. and Hauser, M. (1978). *Rev. Sci. Instrum.* **49**, 632–633.

Harrington, D. C., and Malmstadt, H. V. (1975). *Anal. Chem.* **47**, 271–276.

Harris, J. M., and Dovichi, N. J. (1981). *Anal. Chem.* **53**, 689–692.

Harris, J. M., and Lytle, F. E. (1977). *Rev. Sci. Instrum.* **48**, 1469–1476.

Harris, J. M., Chrisman, R. W., Lytle, F. E., and Tobias, R. S. (1976a). *Anal. Chem.* **48**, 1937–1943.

Harris, J. M., Lytle, F. E., and McCain, T. C. (1976b). *Anal. Chem.* **48**, 2095–2098.

Harris, J. M., Gray, L. M., Pelletier, M. J., and Lytle, F. E. (1977). *Mol. Photochem.* **8**, 161–174.

Harris, T. D., and Mitchell, J. W. (1980). *Anal. Chem.* **52**, 1706–1708.

Haugen, G. R., and Lytle, F. E. (1981). *Anal. Chem.* **53**, 1554–1559.

Haugen, G. R., Wallin, B. W., and Lytle, F. E. (1979). *Rev. Sci. Instrum.* **50**, 64–72.

Hayes, J. M., and Small, G. J. (1978). *Chem. Phys.* **27**, 151–157.

Heidt, G. K. (1976). *Appl. Spectrosc.* **30**, 553–554.

Hershberger, L. W., Callis, J. B., and Christian, G. D. (1979). *Anal. Chem.* **51**, 1444–1446.

Hieftje, G. M. (1972). *Anal. Chem.* **44**(6), 81A–88A.

Hieftje, G. M., and Haugen, G. R. (1981a). *Anal. Chim. Acta* **123**, 255–261.

Hieftje, G. M., and Haugen, G. R. (1981b). *Anal. Chem.* **53**(6), 755A–765A.

Hieftje, G. M., Haugen, G. R., and Ramsey, J. M. (1977). *Appl. Phys. Lett.* **30**, 463–466.

Hiraki, K., Morishige, K., and Nishikawa, Y. (1978). *Anal. Chim. Acta* **97**, 121–128.

Hirschfeld, T. (1973). *J. Opt. Soc. Am.* **63**, 1309.

Hirschfeld, T. (1976). *Appl. Opt.* **15**, 2965–2966.

Hohimer, J. P., and Hargis, P. J., Jr. (1977). *Appl. Phys. Lett.* **30**, 344–346.

Hohimer, J. P., and Hargis, P. J., Jr. (1978). *Anal. Chim. Acta* **97**, 43–49.

Hordvik, A. (1977). *Appl. Opt.* **16**, 2827–2833.

Hosch, J. W., and Piepmeier, E. H. (1978). *Appl. Spectrosc.* **32**, 444–446.

Hunziker, H. E. (1971). *IBM J. Res. Dev.* **15**, 10–26.

Imasaka, T., and Zare, R. N. (1979). *Anal. Chem.* **51**, 2082–2085.

Imasaka, T., Ogawa, T., and Ishibashi, N. (1979). *Anal. Chem.* **51**, 502–504.

Ishibashi, N., Teiichiro, C., Imasaka, T., and Kunitake, M. (1979). *Anal. Chem.* **51**, 2096–2099.

Jackson, W. B., Amer, N. M., Boccara, A. C., and Fournier, D. (1981). *Appl. Opt.* **20**, 1333–1344.

Jankow, R., Kilham, O., Renken, W., and Bender, R. (1971). *Anal. Biochem.* **43**, 300–324.

Jennings, D. A., and Keller, R. A. (1972). *J. Am. Chem. Soc.* **94**, 9249–9250.

Johnston, M. V., and Wright, J. C. (1979). *Anal. Chem.* **51**, 1774–1780.

Johnston, M. V., and Wright, J. C. (1981). *Anal. Chem.* **53**, 1054–1060.

Jones, P. F. (1975). *ACS Symp. Ser.* **13**, 183–196.

Kaye, W. (1981). *Anal. Chem.* **53**, 369–374.

Kliger, D. S. (1980). *Acc. Chem. Res.* **13**, 129–134.

Knorr, F. J., and Harris, J. M. (1981). *Anal. Chem.* **53**, 272–276.

Koester, V. J., and Dowben, R. M. (1978). *Rev. Sci. Instrum.* **49**, 1186–1191.

Kuhl, J., and Marowsky, G. (1971). *Opt. Commun.* **4**, 125–128.

Kuhl, J., and Spitschan, H. (1973). *Opt. Commun.* **7**, 256–259.

Kuhl, J., Marowsky, G., Kuntsmann, P., and Schmidt, W. (1972). *Z. Naturforsch., A* **27A**, 601–604.

Kuwada, K., Motomizur, S., and Toei, K. (1978). *Anal. Chem.* **50**, 1788–1792.

Kwong, H. S., and Measures, R. M. (1980). *Appl. Opt.* **19**, 1025–1027.

Kyle, T. G., and Shuster, B. G. (1978). *Appl. Opt.* **17**, 2659–2660.

Lahmann, W., Ludewid, H. J., and Welling, H. (1977). *Anal. Chem.* **49**, 549–551.

Lakowicz, J. R., and Cherek, H. J. (1981a). *J. Biol. Chem.* **256**, 6348–6353.

Lakowicz, J. R., and Cherek, H. J. (1981b). *J. Biochem. Biophys. Methods* **5**, 19–35.

Leach, R. A., Pelletier, M. J., and Harris, J. M. (1981). *182nd Natl. Meet. Am. Chem. Soc., 1981* ACS Abstract ANAL66.

Levine, B. F., and Bethea, C. G. (1980). *Appl. Phys. Lett.* **36**, 245–247.

Levy, D. H., Wharton, L., and Smalley, R. E. (1977). *In* "Chemical and Biochemical Applications of Lasers" (C. B. Moore, ed.), Vol. II, Chapter 1. Academic Press, New York.

Lidofsky, S. D., Imasaka, T., and Zare, R. N. (1979). *Anal. Chem.* **51**, 1602–1605.

Luciano, M. J., and Kingston, D. L. (1978). *Rev. Sci. Instrum.* **49**, 718–721.

Lytle, F. E., and Kelsey, M. S. (1974). *Anal. Chem.* **46**, 855–860.

Lytle, F. E., Storey, D. R., and Juricich, M. E. (1973). *Spectrochim. Acta, Part A* **29A**, 1357–1369.

Lytle, F. E., Pelletier, M. J., and Harris, T. D. (1979). *Appl. Spectrosc.* **33**, 28–32.

McCoglin, W. C., Marchetti, A. P., and Eberly, J. H. (1978). *J. Am. Chem. Soc.* **100**, 5622–5626.

Mack, M. E. (1968). *J. Appl. Phys.* **39**, 2483–2485.

Maple, J. R., and Wehry, E. L. (1981). *Anal. Chem.* **53**, 266–271.

Maple, J. R., Wehry, E. L., and Mamantov, G. (1980). *Anal. Chem.* **52**, 920–924.

Matthews, T. G., and Lytle, F. E. (1979). *Anal. Chem.* **51**, 583–585.

Mayo, S., Keller, R. A., Travis, J. C., and Green, R. G. (1976). *J. Appl. Phys.* **47**, 4012–4016.

Measures, R. M., and Bristow, M. (1971). *Can. Aeronaut. Space J.* December, pp. 421–422.

Measures, R. M., and Kwong, H. S. (1979). *Appl. Opt.* **18**, 281–286.

Measures, R. M., Houston, W. R., and Stephenson, D. G. (1974). *Opt. Eng.* **13**(6), 494–501.

Meltzer, R. S., and Wood, R. M. (1977). *Appl. Opt.* **16**, 1432–1434.

Menzel, E. R., and Popovic, Z. D. (1978). *Rev. Sci. Instrum.* **49**, 39–44.

Merkelo, H., Hartman, S. R., Mar, T., Singhal, G. S., and Govindjee (1969). *Science* **164**, 301–302.

Miller, M. P., Tallant, D. R., Gustafson, F. J., and Wright, J. C. (1977). *Anal. Chem.* **49**, 1474–1482.

Mitchell, G. W., and Spencer, R. D. (1981). *Pittsburgh Conf. Anal. Chem. Appl. Spectrosc. 1981* Abstract No. 521.

Miyaishi, K., Kunitake, M., Imasaka, T., and Ishibashi, N. (1981). *Anal. Chim. Acta* **125**, 161–168.

Mottola, H. A. (1981). *182nd Natl. Meet. Am. Chem. Soc., 1981* ACS Abstract ANAL 1–5, 20–25, 41–45, 58–63.

Mulac, A. J., Flower, W. L., Hill, R. A., and Aeschliman, D. P. (1978). *Appl. Opt.* **17**, 2695–2699.

Nemanich, R. J., Solin, S. A., and Doehler, J. (1976). *Rev. Sci. Instrum.* **47**, 741–744.

Neumann, S., and Kriese, M. (1974). *Spectrochim. Acta., Part B* **29B**, 127–137.

Oda, S., and Sawada, T. (1981). *Anal. Chem.* **53**, 471–474.

Oda, S., Sawada, T., and Kamada, H. (1978). *Anal. Chem.* **50**, 865–867.

Oda, S., Sawada, T., Moriguchi, T., and Kamada, H. (1980). *Anal. Chem.* **52**, 650–653.

Omenetto, N., Nikdel, S., Reeves, R. D., Bradshaw, J. B., Bower, J. N., and Winefordner, J. D. (1980). *Spectrochim. Acta, Part B* **35B**, 507–517.

O'Neil, R. A., Buja-Bijunas, L., and Rayner, D. M. (1980). *Appl. Opt.* **19**, 863–870.

Onoue, Y., Morishige, K., Hiraki, K., and Nishikawa, Y. (1979). *Anal. Chim. Acta* **106**, 67–72.

Patel, C. K. N., and Tam, A. C. (1979). *Appl. Phys. Lett.* **34**, 467–470.

Perry, D. L., Klainer, S. M., Bowman, H. R., Milanovich, F. P., Hirschfeld, T., and Miller, S. (1981). *Anal. Chem.* **53**, 1048–1050.

Perry, J. A., Bryant, M. F., and Malmstadt, H. V. (1977). *Anal. Chem.* **49**, 1702–1710.

Personov, R. I., and Kharlamov, B. M. (1973). *Opt. Commun.* **7**, 417–419.

Personov, R. I., Al'shits, E. I., and Bykovskaya, L. A. (1972). *Opt. Commun.* **6**, 169–173.

Personov, R. I., Al'shits, E. I., Bykovskaya, L. A., and Kharlamov, B. M. (1974). *Sov. Phys.—JETP. (Engl. Transl.)* **38**, 912–917.

Piepmeier, E. H. (1972). *Spectrochim. Acta, Part B* **27B**, 431–443.

Pollard, B. D., Blackburn, M. B., Nikdel, S., Massoumi, A., and Winefordner, J. D. (1979). *Appl. Spectrosc.* **33**, 5–8.

Ramsey, J. M., Hieftje, G. M., and Haugen, G. R. (1979). *Appl. Opt.* **18**, 1913–1920.

Rayner, D. M., and Szabo, A. G. (1978). *Appl. Opt.* **17**, 1624–1630.

Reed, P. R., Landon, D. O., and Moore, J. F. (1972). Spex Industries, Metuchen, New Jersey (personal communication).

Reule, A. G. (1976). *J. Res. Natl. Bur. Stand., Sect. A* **80A**, 609–624.

Richardson, J. H., and Ando, M. E. (1977). *Anal. Chem.* **49**, 955–959.

Richardson, J. H., and George, S. M. (1978). *Anal. Chem.* **50**, 616–620.

Richardson, J. H., Steinmetz, L. L., Deutscher, S. B., Bookless, W. A., and Schmelzinger, W. L. (1979). *Anal. Biochem.* **97**, 17–23.

Richardson, J. H., Larson, K. M., Haugen, G. R., and Johnson, D. C. (1980). *Anal. Chim. Acta* **116**, 407–411.

Rosencwaig, A. (1980). "Photoacoustics and Photoacoustic Spectroscopy." Wiley (Interscience), New York.

Sacchi, C. A., Svelto, O., and Prenna, G. (1974). *Histochem. J.* **6**, 251–258.

Sato, T., Suzuki, Y., Kashiwagi, H., Nanjo, M., and Kakui, Y. (1978). *Appl. Opt.* **23**, 3798–3803.

Schlag, E. W., Selzle, H. L., Schneider, S., and Larsen, J. G. (1974). *Rev. Sci. Instrum.* **45**, 364–367.

Sepaniak, M. J., and Yeung, E. S. (1980). *J. Chromatogr.* **190**, 377–383.

Shirk, J. S., and Bass, A. M. (1969). *Anal. Chem.* **41**(11), 103A–106A.

Shirk, J. S., Harris, T. D., and Mitchell, J. W. (1980). *Anal. Chem.* **52**, 1701–1705.

Soffer, B. H., and McFarland, B. B. (1967). *Appl. Phys. Lett.* **10**, 266–267.

Sorokin, P. P., and Lankard, J. R. (1966). *IBM J. Res. Dev.* **10**, 162–163.

Spears, K. G., Cramer, L. E., and Hoffland, L. D. (1978). *Rev. Sci. Instrum.* **49**, 255–262.

Spencer, R. D., and Weber, G. (1969). *Ann. N.Y. Acad. Sci.* **158**, 361–376.

Spiker, R. C., and Shirk, J. S. (1974). *Anal. Chem.* **45**, 572–574.

Stone, J. (1972a). *Appl. Phys. Lett.* **20**, 239–240.

Stone, J. (1972b). *IEEE J. Quantum Electron.* **QE-8,** 386–388.

Stone, J. (1973). *Appl. Opt.* **12,** 1828–1830.

Stone, J. (1978). *Appl. Opt.* **17,** 2876–2877.

Strojny, N., and deSilva, J. A. F. (1980). *Anal. Chem.* **52,** 1554–1559.

Sychra, V., Svoboda, V., and Rubeska, I. (1975). "Atomic Fluorescence Spectroscopy," p. 23. Van Nostrand-Reinhold, Princeton, New Jersey.

Szabadvary, F. (1966). "History of Analytical Chemistry." Pergamon, Oxford.

Van Duyne, R. P., Jeanmarie, D. L., and Shriver, D. F. (1974). *Anal. Chem.* **46,** 213–222.

Van Geel, T. F., and Winefordner, J. D. (1976). *Anal. Chem.* **48,** 335–338.

Visser, H. (1979). *Appl. Opt.* **18,** 1746–1749.

Vo-Dinh, T., and Wild, U. P. (1973). *J. Lumin.* **6,** 296–303.

Weeks, S. J., Haraguchi, H., and Winefordner, J. D. (1978). *Anal. Chem.* **50,** 360–368.

Wehry, E. I., and Mamantov, G. (1979). *Anal. Chem.* **51**(6), 643A–656A.

Wehry, E. L., Gore, R. R., and Dickinson, R. B., Jr. (1981). *In* "Lasers in Chemical Analysis" (G. M. Hieftje, J. C. Travis, and F. E. Lytle, eds.), Chapter 10. Humana Press, Clifton, New Jersey.

Woo, C. S., D'Silva, A. P., and Fassel, V. A. (1980). *Anal. Chem.* **52,** 159–164.

Wright. J. C. (1977). *Anal. Chem.* **49,** 1690–1702.

Yamada, S., Miyoshi, F., Kanc, K., and Ogawa, T. (1981). *Anal. Chim. Acta* **127,** 195–198.

Yaney, P. P. (1972). *J. Opt. Soc. Am.* **62,** 1297–1303.

Yaney, P. P. (1976). *J. Raman Spectrosc.* **5,** 219–241.

Yang, Y., D'Silva, A. P., Fassel, V. A., and Iles, M. (1980). *Anal. Chem.* **52,** 1350–1351.

Yang, Y., D'Silva, A. P., and Fassel, V. A. (1981a). *Anal. Chem.* **53,** 894–899.

Yang, Y., D'Silva, A. P., and Fassel, V. A. (1981b). *Anal. Chem.* **53,** 2107–2109.

Yeung, E. S., and Sepaniak, M. J. (1980). *Anal. Chem.* **52**(13), 1465A–1481A.

INDEX

435